THE PHYSICS OF
Neutrinos

THE PHYSICS OF Neutrinos

Vernon Barger
Danny Marfatia
Kerry Whisnant

PRINCETON UNIVERSITY PRESS • PRINCETON AND OXFORD

Published by Princeton University Press, 41 William Street, Princeton, New Jersey 08540

In the United Kingdom: Princeton University Press, 6 Oxford Street, Woodstock, Oxfordshire OX20 1TW

Library of Congress Cataloging-in-Publication Data
Barger, V. (Vernon), 1938–
The physics of neutrinos / Vernon Barger, Danny Marfatia, Kerry Whisnant.
p. cm.
Includes bibliographical references and index.
ISBN 978-0-691-12853-5 (hardcover : alk. paper) 1. Neutrinos. I. Marfatia, Danny, 1972– II. Whisnant, Kerry Lewis. III. Title.
QC793.5.N42B37 2012
539.7′215–dc23
2012007537

British Library Cataloging-in-Publication Data is available

This book has been composed in Sabon LT Std
Printed on acid-free paper. ∞

press.princeton.edu

Typeset by S R Nova Pvt Ltd, Bangalore, India
Printed in the United States of America

10 9 8 7 6 5 4 3 2 1

To our families

- Contents -

Preface xi

1 Introduction 1

2 Neutrino Basics 11
2.1 Dirac and Majorana Neutrinos 11
2.2 Neutrino Counting 12
2.3 Neutrinos from Weak Decays 14
2.4 Neutrino Cross Sections 16
2.5 Neutrino Detectors 24
2.6 Neutrino Beams 28

3 Neutrino Mixing and Oscillations 33
3.1 Vacuum Oscillations 33
3.2 Matter Effects on Oscillations 36
3.3 Solar Neutrino Oscillations 38
3.4 Long-baseline Oscillations through the Earth 41
3.5 Matter Effects for Sterile Neutrinos 42
3.6 Decoherence 43

4 Solar Neutrinos 45
4.1 Origin of Solar Neutrinos 45
4.2 Solar Neutrino Experiments 46
4.3 KamLAND 49
4.4 Solar/Reactor Neutrino Parameters 49
4.5 Flux-independent Tests 53
4.6 Future Experiments 56
4.7 Geoneutrinos 57

5 Atmospheric Neutrinos 59
5.1 Atmospheric Neutrino Experiments 59
5.2 Matter Effects for Atmospheric Neutrinos 63
5.3 Long-baseline Neutrino Experiments 64

6 Global Three-neutrino Fits 68

7 Absolute Neutrino Mass 71
7.1 Beta Decay 71
7.2 Cosmological Limits 72
7.3 Neutrinoless Double-beta Decay 73

8 Long-baseline Neutrino Oscillations 76
8.1 Conventional Neutrino Beams 77
8.2 Reactor Experiments 80
8.3 Superbeams 85
8.4 Neutrino Factories 87
8.5 Beta Beams 91
8.6 Comparing Long-baseline Experiments 92
8.7 T and CPT Symmetries 97

9 Model Building 99
9.1 The Seesaw Mechanism 99
9.2 Patterns of Neutrino Masses and Mixings 102
9.3 GUT Models 105
9.4 Non-GUT-specific Models 107
9.5 Leptogenesis 114

10 Supernova Neutrinos 116
10.1 General Description of a Supernova 116
10.2 Neutrino Fluxes from the SN Core 118
10.3 Flavor Swapping from Collective Effects 119
10.4 MSW Conversions in a Supernova 120
10.5 Detection of Supernova Neutrinos 122
10.6 Supernova Relic Neutrinos 124

11 High-energy Astrophysical Neutrinos 126
11.1 Cosmogenic Neutrinos 126
11.2 IceCube 128
11.3 Waxman–Bahcall Flux 132
11.4 Ultra High-energy Neutrino Cross Sections 133
11.5 Z-burst Mechanism 134
11.6 Astrophysical Neutrino Flavor Content 135
11.7 Neutrinos from Dark Matter Annihilation 138

12 Beyond Three Neutrinos 147
12.1 LSND Experiment 147
12.2 MiniBooNE Experiment 152
12.3 Mass-varying Neutrinos 158
12.4 Neutrino Decay 161

12.5 Neutrino Decoherence 163
12.6 Lorentz Invariance Violation 164
12.7 Non-standard Neutrino Interactions 166
12.8 Heavy Majorana Neutrinos at Colliders 169
12.9 Neutrino Magnetic Moment 170
12.10 Fourth Generation Neutrino 171

13 Summary and Outlook 172

References 177

Index 221

- Preface -

The overall thrust of the book is how a bottom-up understanding of the physics of neutrinos is being achieved. The current state of the physics of massive neutrinos is surveyed with both historical and forward-looking vantage points. The complementarity of particle physics, nuclear physics, astrophysics and cosmology in contributing to a fuller understanding of the physics of neutrinos is integrated in the developments. The survey begins with a review of neutrino production and detection methods and the basic phenomenology of neutrino oscillations in vacuum and their modifications in matter. The experimental evidence for oscillations of neutrinos from solar, reactor, atmospheric and accelerator sources is then documented. Further, the road map is laid out for future experimental determinations of the unknown neutrino parameters in low-energy experiments (beta-decay, neutrinoless double-beta decay, reactor) and in accelerator experiments (superbeams, beta beams and neutrino factories). The interplay of conventional neutrino physics with cosmology (Big Bang Nucleosynthesis, Cosmic Microwave Background radiation, leptogenesis) and astrophysics (supernovae, highest-energy cosmic rays) is discussed and the window to the universe opened by neutrino telescopes is explored. As well, a compendium of theories of neutrino mass, their underlying frameworks, and their future tests is given.

We wish to express gratitude to our research collaborators over a period of several decades. Those research interactions enriched our knowledge of the subject and formed our perspectives. We also thank Professors Paul Langacker and Muneyuki Ishida and anonymous readers for valuable comments on the manuscript, and gratefully acknowledge research support from the U.S. Department of Energy, the U.S. National Science Foundation, our Universities, the Wisconsin Alumni Research Foundation and the Vilas Trust. We thank the Aspen Center for Physics, the University of Hawaii-Manoa, and the Kavli Institute for Theoretical Physics, Santa Barbara for their hospitality.

An extensive list of references to the scientific literature on neutrinos is provided. However, it is inevitable that relevant citations will have been missed and we regret any such inadvertent omissions.

Vernon Barger
Danny Marfatia
Kerry Whisnant

THE PHYSICS OF
Neutrinos

1

Introduction

The unfolding of the physics of neutrinos has been a premier scientific achievement of the 20th century. The hallmark of this decades-long endeavor has been the intertwined contributions of experiment and theory in its advancement. This fascinating history has been the subject of many treatises. Our aim is to give an overview of the aggregate knowledge of neutrino physics today and to mark future pathways for still deeper understanding. In this enterprise we bring together, under one broad umbrella, what has been learned and what is now being pursued about neutrinos in a diversity of subareas–particle physics, nuclear physics, astrophysics, and cosmology. Neutrinos are of key importance in understanding the nature of our universe and there is a new synergy of these branches of physics in their study. A brief flashback to major milestones along the road of neutrino discovery is an appropriate beginning and the subject of this introduction.

The nuclear model of the atom circa 1930 was atomic electrons bound to a positive nucleus by the electromagnetic force. The nucleus was believed to be composed of both protons and electrons, in numbers such that the atomic number A and the nuclear charge Z were accounted for. A challenge to this description was that radioactive nuclei were observed to undergo spontaneous beta-decay $A \to A' + e$. By energy and momentum conservation, all the emitted electrons should have the same energy, but a continuous electron energy spectrum was observed. This totally unexpected phenomenon caused both Niels Bohr and Paul Dirac to consider the extreme possibility that energy was not conserved. Another apparent difficulty of the nuclear model was the "false" statistics of the ^{14}N and ^{19}Li nuclei. Because ^{14}N has 7 atomic electrons, its nucleus, supposedly consisting of 14 protons and 7 electrons, should have spin-$\frac{1}{2}$, but scattering experiments showed it to have integer spin. Wolfgang Pauli of Eidgenössische Technische Hochschule (ETH), Zurich, saw a way out of this conundrum. He proposed, in a letter to a conference that he was unable to attend, his desperate remedy: nuclei also have very light neutral constituents of spin-$\frac{1}{2}$, which he called neutrons [1]. His neutrons could solve the spin-statistics problem and explain the continuous beta spectrum, since the neutrons would be emitted in conjunction with electrons, $A \to A' + e + n$, so the energy spectrum of the emitted electrons would not be monoenergetic. To be consistent with the observed electron energy sprectrum, the mass of his neutron had to be less than one percent of the proton mass. Pauli was embarrassed by his rash proposal because he thought that his neutron could never be detected, because of the weakness of its interaction. Pauli's nuclear model was complex: the nucleus would consist of protons, electrons, and neutrons: e.g., 14 protons, 7 electrons, and 7 neutrons in the ^{14}N nucleus.

In 1932 James Chadwick, then at the Cavendish Laboratory of the University of Cambridge in England, discovered the neutron [2], but it was not the weakly interacting particle emitted in beta decays. Instead, the neutron was a strongly interacting neutral companion of the proton, and the nuclear model simplified to protons and neutrons bound by the strong force: 7 protons and 7 neutrons in the ^{14}N nucleus.

In 1934 Enrico Fermi, then at the University of Rome, reformulated Pauli's idea that a very light neutral particle was involved in radioactive decays. He renamed it the neutrino (the "little neutral one" in Italian). In his famous theory of beta decay [3], Fermi invoked antiparticles (predicted by Dirac in 1931), Pauli's emitted particle (the antineutrino), and quantum field theory (in which particles can be destroyed or created). In the weak interaction according to Fermi, neutrons decay to protons via a nonrenormalizable four-fermion interaction, $n \rightarrow p + e^- + \bar{\nu}_e$ where $\bar{\nu}_e$ is the electron-antineutrino. The electron and the antineutrino are created as a pair, rather than being emitted from the nucleus. Moreover, the process obtained by crossing initial and final lines in a Feynman diagram have the same strength. Thus, Fermi's theory predicts the inverse process $\bar{\nu}_e + p \rightarrow e^+ + n$, with an interaction of the same strength as that of neutron decay. The reality of the neutrino could thus be tested by observing this inverse reaction with an intense neutrino beta decay source from reactors.

In 1955, $\bar{\nu}_e$ scattering events were observed by Frederick Reines and Clyde Cowan, Jr., American physicists working at the Los Alamos National Laboratory, via the inverse beta decay process in an experiment at the Savannah River reactor in South Carolina [4]. The reactor provided an intense antineutrino flux of $5 \times 10^{13}/\text{cm}^2/\text{s}$. Scintillators in a tank of water were used to observe the oppositely directed gamma rays from positron annihilations and a time-delayed (by 200 μs) 2.2 MeV gamma ray from the capture of the neutron on cadmium in the water. The measured inverse beta decay cross section was later found to be consistent with the prediction, indicating that the antineutrinos had been detected.

In 1956, T. D. Lee of Columbia University and C. N. Yang, then of Brookhaven National Laboratory (BNL), interpreted the decays of two species of neutral kaons observed in experiments at BNL as a breakdown of the law of parity (P) conservation (invariance under spatial inversion) [5]. They suggested radioactive beta-decay experiments as a further test. Shortly thereafter, C. S. Wu of Columbia University carried out an experiment on the radioactive beta decays of ^{60}Co that confirmed parity violation [6].

The idea of a maximal parity violating $V - A$ chiral structure of the weak interaction (with vector and axial vector currents of equal strength) originated in 1957–1958 by George Sudarshan and Robert Marshak [7], of Harvard University and the University of Rochester, respectively, and by Richard Feynman and Murray Gell-Mann [8], of Caltech, at a time when some experiments favored a scalar-tensor interaction. According to the $V - A$ theory the neutrino is left-handed and the antineutrino is right-handed. This was confirmed in 1958 by Maurice Goldhaber,

Lee Grodzins, and Andrew Sunyar at BNL by studying the circular polarization and resonant scattering of gamma rays following orbital electron capture in a metastable state of ^{152}Eu [9].

A major experimental leap forward occurred in 1962, when a team led by Leon Lederman, Melvin Schwartz, and Jack Steinberger used charged pions produced by the Alternating Gradient Synchrotron at the BNL to establish the existence of the muon-neutrino (ν_μ) [10]. Charged pions decay dominantly to muons and an associated neutrino. The interactions of these neutrinos in a 10-ton spark chamber were found to produce muons but not electrons.

In 1964, James Cronin and Val Fitch showed that, in the decays of the particles called neutral kaons, not only was the parity symmetry violated, but also the combination CP was violated [11], where C is the charge conjugation symmetry. This CP symmetry breaking is very small but could have created an initial asymmetry between matter and antimatter at the beginning of the universe (at the level of one part in a billion), which after matter-antimatter annihilation leads to the preponderance of matter in the known universe [12]. In the last decade, the BaBar [13] and Belle [14] experiments have shown that CP is violated in the B mesons decays, and much more strongly.

The question of whether neutrinos had mass persisted for decades. A direct probe is the energy spectrum of the electron emitted in beta decay, since a finite neutrino mass would cause a truncation of the spectrum at its endpoint. Experiments on tritium beta decays placed increasingly more restrictive upper bounds and currently restrict the neutrino mass to be less than a few electron-volts [15, 16].

The prescient idea of neutrino oscillations was made by Bruno Pontecorvo in 1957 [17], who proposed the idea of transitions between neutrinos and antineutrinos as an analogy to the K^0–$\bar{K}^0$ oscillations observed in the neutral kaon system. This process later became known as oscillations into sterile states. For oscillations to occur among different neutrino types, the neutrinos must have different masses and the quantum mechanical wave functions of the observed neutrino flavors (i.e., the neutrinos associated with the electron, muon, and tau) must be linear superpositions of the neutrino mass eigenstates. In 1962, the Japanese theorists Ziro Maki, Masami Nakagawa and Shoichi Sakata represented the mixing of two neutrinos by a 2×2 mixing matrix (now called the MNS matrix after the names of the pioneer theorists) [18].

For the oscillations of two neutrinos in vacuum, the probability of a neutrino produced by the weak interaction as a flavor eigenstate ν_α being detected as the same flavor at a distance $L = ct$ from the source is

$$P(\nu_\alpha \to \nu_\alpha) = 1 - \sin^2 2\theta \sin^2\left(\frac{\delta m^2 L}{4E}\right), \tag{1.1}$$

where L is the distance from the source to the detector, E is the neutrino energy, θ is the angle that describes the mixing between the flavor eigenstates and the mass eigenstates ν_1 and ν_2, and $\delta m^2 = m_2^2 - m_1^2$ is the mass-squared difference between the mass eigenvalues. This probability, that the initial neutrino is observed as the same flavor, is known as the "survival" probability. The deviation of the survival probability from unity is sometimes called the "disappearance" probability. The "appearance" probability that a new flavor is observed is given by

$$P(\nu_\alpha \to \nu_\beta) = \sin^2 2\theta \sin^2\left(\frac{\delta m^2 L}{4E}\right), \tag{1.2}$$

where $\beta \neq \alpha$. The sum of the survival and appearance probabilities is necessarily unity. The oscillation probability has a sinusoidal dependence with an amplitude that depends on the neutrino mixing angle and a wavelength that depends on the mass-squared difference and neutrino energy. The L/E dependence of the oscillation argument is characteristic of vacuum neutrino oscillations due to neutrino masses and mixing.

Later, after the mixing of 3 generations of quarks was described by the CKM matrix [19] (named after Cabibbo, Kobayashi and Maskawa), the MNS matrix was extended to a 3×3 matrix appropriate to three generations of neutrinos. The third lepton (the tau) was discovered at the Stanford Linear Accelerator Center in the mid-seventies by Martin Perl and collaborators [20] and the tau-neutrino was discovered in 2000 at Fermilab by the DONUT (Direct Observation of the NU Tau) collaboration [21]. Measurements of the invisible width of the weak neutral Z-boson at the Large Electron Positron Collider at CERN determined in 1989 that the number of neutrinos coupled to the Z-boson was 2.984 ± 0.008, as anticipated from 3 generations of leptons.

Looking for neutrino oscillation effects was a huge experimental challenge, since the neutrino mixing angles and mass-squared differences were a priori unknown parameters. A range of dedicated accelerator searches for evidence of neutrino oscillations over four decades placed only upper bounds on the oscillation probabilities. It turned out that astrophysical sources, cosmic rays, and the sun led to the discoveries of the phenomena.

The first indications that neutrino oscillations may in fact occur was an apparent deficit in the flux of ν_e with MeV energies that originate from the nuclear fusion chain in the core of the the Sun (called solar neutrinos) and detected via the charged-current (CC) weak interaction. In 1964, an experiment was proposed by Raymond Davis, Jr. of BNL to extract and count radioactive isotopes of argon created when neutrinos interacted with chlorine atoms in a 10^5-gallon tank of perchloroethylene [22]. The first results from the experiment, located in the Homestake Mine in South Dakota, reported an upper bound [23] for solar neutrinos that was a factor of two to three times below predictions of the Standard Solar Model (SSM) developed concurrently by John Bahcall of the Institute for Advanced Study and collaborators [24], which gave predictions for the solar neutrino flux based on the fusion reactions in the solar core. The first observation of ν_e from the Sun [25] was experimental confirmation that these fusion reactions were indeed occurring, although neutrinos were not seen at the rate predicted by the SSM. This became known as the solar neutrino problem.

Vladimir Gribov and Pontecorvo suggested in 1968 that the apparent deficit of solar neutrinos could be due to neutrino oscillations [26], whereby the ν_e produced in the fusion processes in the sun oscillated to ν_μ during their propagation to earth. The Homestake experiment operated for more than thirty years; as the experiment and the theory improved over the years, both the observed and theory values went down, but the deficit persisted, although a neutrino oscillation interpretation of the flux deficit was initially met with skepticism by many particle physicists.

The evidence for solar neutrino oscillations continued to build through the 1990s as experiments with sensitivities to different MeV energy ranges all found rate deficits of 0.3 to 0.7 compared to the SSM. The low energy solar neutrinos from the primary *pp* fusion process were measured in the SAGE [27], GALLEX [28] and GNO [29] radiochemical experiments based on the neutrino capture reaction $\nu_e + {}^{71}\mathrm{Ga} \to {}^{71}\mathrm{Ge} + e^-$ with a threshold of about 0.23 MeV. The solar neutrino flux at high energies, 4 to 15 MeV, was measured in water Cherenkov detectors (Kamiokande and Super-Kamiokande [30–32] in Japan) and in heavy water in the Sudbury Neutrino Observatory (SNO) [33] in Canada.

The definitive proof that oscillations are the right interpretation of the solar flux discrepancies came from the neutrino neutral-current (NC) measurements of the SNO experiment [33, 34] that determined the combined flux of all three neutrinos, as well as the ν_e flux from the charged-current process. The survival probability of solar ν_e was thus determined from the measured charged-current to neutral-current flux ratio, independent of the solar flux calculations in the SSM.

A crucial aspect in interpreting solar neutrino oscillations is the effect of matter on neutrino propagation. As the ν_e travel through the dense solar core, they undergo coherent forward $\nu_e + e \to \nu_e + e$ scattering, as first discussed by Wolfenstein [35]. Matter effects can produce large changes in the oscillation amplitude and wavelength compared to vacuum oscillations, as first shown by Barger, Whisnant, Pakvasa and Phillips [36] who studied a medium of constant density (appropriate for neutrinos propagating through the mantle of the earth in long-baseline neutrino experiments). They found a resonant enhancement that depended on the neutrino energy. Mikheyev and Smirnov later applied the enhancement at a given neutrino energy to the propagation of solar neutrinos through the varying electron density in the sun [37]. A matter enhancement can be realized only for neutrinos or antineutrinos, but not both.

Because of the prevailing prejudice that neutrino mixing would be small, there was a strong theoretical bias in favor of a resonant solar solution, which was the original solution to the solar neutrino problem proposed by Mikheyev and Smirnov (the so-called Mikheyev-Smirnov-Wolfenstein or MSW solution). Initial studies assumed adiabatic propagation of neutrinos through the sun [38, 39], but it was subsequently realized that for small mixing angles nonadiabatic propagation was also possible [40–42]. In addition to this small mixing angle solution (known as SMA), other solutions with matter effects and a large vacuum mixing angle were later identified that could account for the solar neutrino flux suppression [43].

The other solutions were named LMA (large mixing angle), LOW (low δm^2, low probability) [44], QVO (quasi-vacuum oscillations) [45] and VO (vacuum oscillations) [46]. These solutions correspond to isolated islands in the $(\delta m^2, \tan^2\theta)$ parameter space of the solar neutrino oscillations. The flat energy spectrum relative to the SSM and the absence of a significant day/night difference caused by earth-matter effects [32, 47], favored the LMA solution with adiabatic propagation. The SNO salt phase data [48] in conjunction with other solar neutrino data selected the LMA solution uniquely at a high confidence level. The mass-squared difference indicated by the solar neutrino data is $\sim 8 \times 10^{-5}$ eV^2 and the mixing is large but not maximal, $\theta \simeq 34^\circ$. The large size of the mixing angle was surprising since all quark mixing angles were known to be small.

The averaged probability of vacuum neutrino oscillations accounts for the suppression by approximately a factor of two of the low energy neutrinos, while the suppression of the high energy neutrinos from ^{8}B decay by approximately a factor of three is caused by matter effects with an adiabatic level crossing of the transition of $\nu_e \to \nu_\mu$. The Borexino experiment [49], with a liquid scintillator detector in the Gran Sasso Laboratory in Italy, is doing real time detection of the solar neutrino flux from the ^{7}Be line at 0.86 MeV via elastic scattering of neutrinos on electrons. Their measured oscillation probability is consistent with the predicted oscillation probability in the transition region from matter effects to averaged vacuum oscillations [50–53].

An amazing orthogonal confirmation of the solar neutrino oscillations comes from the energy dependence of the flux of antineutrinos with MeV energies from reactors (called reactor neutrinos). Assuming CPT invariance the probabilities of $\nu_e \to \nu_e$ and $\bar{\nu}_e \to \bar{\nu}_e$ oscillations should be equal at the same values of L/E. In the KamLAND reactor experiment [54, 55] nuclear reactors in Japan are distributed such that a centrally placed detector can measure the L/E dependence of the antineutrino flux. At the average distance $L \sim 180$ km of the reactors from the KamLAND detector and the typical energies of a few MeV of the reactor $\bar{\nu}_e$, the experiment has very good sensitivity to the δm^2 value of the LMA solar solution. The KamLAND [56, 57] data show precisely the L/E dependence of the oscillation probability expected from the solar LMA solution, a dramatic vindication of the oscillation interpretation of the solar neutrino problem. The KamLAND determination of the δm^2 value is a factor of about 3 more precise than the value inferred from the solar neutrino data, but the solar neutrino analysis better determines the mixing angle. Thus, the two probes are very complementary.

Underground water Cherenkov detectors of many-kiloton size that were built primarily to search for proton decay (not found to a sensitivity of around 10^{34} years) turned out to be key neutrino observatories. The Kamiokande detector was constructed in Japan, and the IMB (Irvine-Michigan-Brookhaven) experiment was located in a salt mine near Lake Erie, USA. Fortuitously, both experiments observed neutrino events from a supernova explosion in the Large Magellanic Cloud, SN1987A [58, 59]. The time-energy spectrum of the neutrino events confirmed the basic tenets of the physics of supernova. Neutrino observations of a future supernova in our galaxy can yield fundamental insights about the neutrino dynamics in the explosion.

The first confirmed neutrino oscillations were of neutrinos of GeV energies that originated in the weak decays of pions, kaons, and muons produced by the interactions of cosmic rays with the earth's atmosphere (called atmospheric neutrinos). In the early studies of atmospheric neutrino events by the Kamiokande [60] and IMB [61] experiments (c.1988), the electron to muon event ratio was found to be about a factor of 2 above expectations. A deficit of ν_μ compared to flux calculations was found for neutrinos produced in the atmosphere on the other side of the earth from an underground detector (upward events, with large L), but not for events on the same side (downward events, with small L). This result was interpreted as evidence for oscillations with neutrino mass-squared difference $\delta m^2 \sim 10^{-2}$ eV^2 and near maximal neutrino mixing [62]. However, due to the prevailing theoretical prejudice at the time that neutrino mixing angles would be small like quark mixings, this interpretation of the atmospheric neutrino data did not receive widespread acceptance.

The conclusive evidence that atmospheric ν_μ oscillate, and ν_e do not, came in 1988 from the Super-Kamiokande experiment [63]. With the capability to make high-statistics measurements of the zenith angle (or, equivalently, path distance) and energy distributions of both electron and muon events, the Super-K experiment convincingly established that the observed L/E dependence was consistent with $\nu_\mu \to \nu_\tau$ oscillations due to neutrino masses and mixing, with approximately maximal mixing at a mass-squared-difference scale $\delta m^2 \sim 2.5 \times 10^{-3}$ eV^2.

It was originally thought that the energy and angular resolutions of the atmospheric neutrinos in the Super-K experiment would be too coarse to allow the first minimum in the $\nu_\mu \to \nu_\mu$ oscillation to be resolved and hence that accelerator-based long-baseline (LBL) experiments would be essential to make the important confirmation of ν_μ oscillations and rule out non-standard interpretations, such as neutrino decay [64, 65] or neutrino decoherence [66, 67]. Unexpectedly, Super-K succeeded in reconstructing the L/E distribution of atmospheric ν_μ events and strongly disfavored the non-oscillation alternatives. Other experiments that measured the atmospheric neutrino flux (the MACRO [69] and Soudan-2 detectors [68]) with different detector technologies found results in accord with Super-K.

Neutrinos produced by accelerators and detected at long baselines from the sources—the K2K experiment [70] from KEK to Super-K in Japan and the MINOS experiment [71] from Fermilab to the Soudan mine in Minnesota—have independently confirmed and improved the measurement of the atmospheric oscillation parameters, just as reactor experiments improved our knowledge of solar neutrino parameters. These long-baseline experiments verified the depletion of events at the first oscillation minimum. So far, the Super-K, K2K, and MINOS experiments have only measured the disappearance of ν_μ. The detection of $\nu_\mu \to \nu_e$ appearance oscillations remains as an important goal of MINOS and future accelerator based neutrino oscillation experiments.

The four parallel paths of experimental endeavor (atmospheric, solar, reactor, and accelerator neutrinos) have conclusively established oscillations of the three types of neutrinos (e, μ, τ). The CHOOZ [72] and Palo Verde [73] reactor neutrino experiments found no disappearance of $\bar{\nu}_e$ at an L/E similar to that in atmospheric neutrino experiments, confirming evidence that the primary oscillations of atmospheric neutrinos is $\nu_\mu \to \nu_\tau$. It is interesting that all of the solid evidence for neutrino oscillations comes from measurements of survival probabilities. We note, however, that the long-baseline OPERA experiment [74] from CERN to the Gran Sasso Laboratory in Italy has reported one tau-appearance event [75] from a ν_μ beam, which could also confirm $\nu_\mu \to \nu_\tau$ oscillations.

Since solar ν_e and reactor $\bar{\nu}_e$ oscillate with one characteristic δm^2 and atmospheric and accelerator ν_μ (but not ν_e) oscillate with a different δm^2, a full three-neutrino description of oscillations is clearly needed. The oscillations of three neutrinos are described by the 3×3 MNS mixing matrix, with three mixing angles (θ_{12}, θ_{23}, and θ_{13}) and a CP-violating phase (δ) [18, 76–79]. Two of the angles, θ_{12} for solar neutrinos and θ_{23} for atmospheric neutrinos, are large. The lack of detected participation of ν_e in oscillations of atmospheric and long-baseline experiments, as well as $\bar{\nu}_e$ in the CHOOZ and Palo Verde reactor experiments, indicates that the third mixing angle (θ_{13}) is small, and that atmospheric and solar neutrino oscillations are nearly decoupled from each other. There are only two independent mass-squared differences, δm^2_{31} for atmospheric neutrinos and δm^2_{21} for solar neutrinos.

We now have a fairly precise knowledge of the solar neutrino oscillation parameters δm^2_{21} (its sign is known from solar matter effects) and θ_{12}, and the atmospheric neutrino oscillation parameters δm^2_{31} (its sign is not known) and θ_{23}. The major challenge before us now is the measurement of θ_{13}, which has been established as nonzero in reactor experiments, and the CP phase, which is completely unknown.

Ongoing reactor experiments (Double Chooz [80] in France, Daya Bay [81] in China, RENO [82] in Korea) will precisely measure the value of $\sin^2 2\theta_{13}$, independently of the CP phase, down to the 1% level or better by measuring $\bar{\nu}_e$ survival. Accelerator based experiments, such as T2K [83] in Japan and NOνA [84] in the USA, may also be able to measure θ_{13} through ν_e appearance in a ν_μ beam.

The accelerator experiments can also test for CP nonconservation associated with the complex phase in the 3×3 neutrino mixing matrix. In order that δ be measurable, both δm^2 scales must contribute to the oscillation [77]. Therefore the size of CP violation in long-baseline experiments also depends on the value of δm^2_{21} in addition to δm^2_{31}. Also, the CP-violating phase enters oscillations via a factor $\sin\theta_{13}e^{-i\delta}$. Nonzero θ_{13} allows us to pursue the measurement of δ and admits interesting matter effects in long-baseline neutrino oscillations. A further complication exists due to an eight-fold oscillation parameter ambiguity [85] that must be resolved by the experiments to obtain a unique solution. If neutrinos are Majorana [86], two further CP-violating phases (ϕ_2, ϕ_3) enter in the calculation of neutrinoless double-beta decay [87] but not oscillations [88].

The anticipated steps in the long-baseline program are off-axis beams [83,89,90], superbeams [91,92], wide-band beams [93–95], and detectors with larger fiducial volumes and sophistication [83, 96, 97]. Beta beams, which utilize $\bar{\nu}_e$ from beta decay, are also under consideration [98]. The ultimate sensitivities can be derived from neutrino factories [99, 100], where the neutrino beams are obtained from the decays of muons that are stored in a ring with straight sections.

Although neutrino oscillations have established that neutrinos have mass, oscillations do not probe the absolute neutrino mass scale. In particle and nuclear physics, the only avenues for this are tritium beta decay and neutrinoless double-beta decay, and the latter works only if neutrinos are Majorana particles. These experiments currently probe the interesting eV scale of neutrino mass. Another route to the absolute mass is the power spectrum of galaxies, which gets modified on small length scales when the sum of neutrino masses is nonzero [101]. Several cosmological analyses of the Cosmic Microwave Background (CMB) and large-scale structure data have already given an upper limit on $\sum m_\nu$ below 1 eV [102]. Big Bang Nucleosynthesis (BBN), at the time scale of a few minutes in the early universe, determines the number of relativistic neutrino degrees of freedom, with results consistent with either $N_\nu = 3$ or 4. Neutrinoless double-beta decay experiments are the only known means of determining the Majorana nature of light neutrinos. An ambitious experimental program is underway to probe below the present upper limits of order 1 eV on the diagonal mass matrix element associated with ν_e using this process.

The theory of neutrino masses and mixings is a wide open area of investigation. If neutrinos are massless, there are 19 free parameters in the Standard Model (SM) Lagrangian: three gauge couplings, six quark masses, three quark mixing angles, and a CP-violating phase in the quark mixing matrix, the strong CP phase, three charged-lepton masses, and the Higgs boson self-coupling and vacuum expectation value. If neutrinos have mass, there are at least seven more: three neutrino masses, and the three mixing angles, and one leptonic CP-violating phase in the leptonic mixing matrix. If neutrinos are Majorana particles, there are also two other CP-violating phases. Therefore understanding neutrino masses is an essential part in the development of any theory of elementary particles.

The starting point in the construction of models is to account for the tribimaximal mixing pattern [103] ($\theta_{23} = 45°$, $\theta_{12} = 35°$, $\theta_{13} = 0°$) that is favored by neutrino data. There are diverse models by which this pattern can be realized, such as having a flavor symmetry, of which the A_4 group is a popular example [104]. With perfect tribimaximal mixing, the angle θ_{13} is zero and all the interesting physical phenomena that would be associated with a nonzero θ_{13} go away. Thus, perturbations from exact tribimaximal mixing are the essence of model constructions and their tests.

The incorporation of neutrinos in the framework of a Grand Unified Theory (GUT) is an attractive possibility [105]. The existence of a right-handed neutrino at the GUT mass scale can provide an explanation of the light neutrino mass scale through the seesaw mechanism [106]. In the simplest form of the seesaw the light neutrinos are predicted to be Majorana particles, hence the importance of the neutrinoless double beta decay experimental program.

There is the potential for fundamental neutrino physics beyond what is now apparent. A dramatic example is a possible environmental dependence of the neutrino masses on the density of the medium in which they propagate. In a model in which neutrinos have a new interaction with a very light scalar field, neutrinos may be connected with the dark energy in the universe [107].

Unexplained deviations from standard three-neutrino expectations that have been reported by several experiments could, if confirmed, be of new physics origin, such as possible CPT violation [108, 109] and the existence of sterile neutrinos that do not couple to SM fields [110]. Sterile neutrinos have also been invoked in explanations of astrophysical phenomena, such as neutron star "kicks" [111]. Although considerable efforts have been devoted to theoretical studies of sterile neutrinos, evidence for their existence is inconclusive.

Neutrino astrophysics is the newest frontier of the field. The advent of the large neutrino telescopes IceCube [112] and ANTARES [113] makes possible the search for neutrinos from astrophysical sources [114], which produce distinctive flavor mixes at the sources that can be inferred from the ratios observed at the earth where their oscillations have averaged [115]. Dark matter capture by the sun followed by annihilations in the solar core could yield neutrinos that may be observed in the DeepCore detector of IceCube and provide an important diagnostic about the nature of dark matter.

Neutrinos were pivotal in testing the Standard Model of particle physics. The SM is a fully renormalizable theory where the charged-current interaction occurs due to the exchange of a W-boson, replacing the effective field theory approach of Fermi's four-fermion description of beta decay. Another new ingredient of the SM was the neutral current mediated by the Z-boson. Accelerator measurements of neutrino scattering cross sections provided many of the tests to show the validity of the SM. Neutrinos are also of unique importance at colliders, since they give rise to missing energy. The discovery of the predicted *W*-boson of the SM [116] utilized a transverse mass variable [117, 118] constructed from the transverse energy of the charged lepton and the transverse missing energy of neutrinos in leptonic *W*-decays, $W \to \ell\nu$. The transverse mass has an upper endpoint of M_W, smeared by the *W*-width, by which the *W*-mass and *W*-width have been measured. In high energy collisions, the decays to neutrinos of produced *Z*-bosons is a source of missing transverse energy that is a background to new physics models that give missing energy through the emission of a stable particle. Many decay processes include transitions to final states with neutrinos, including the decays of the muon, tau lepton, charged pions, charged kaons, etc. Thus, neutrinos are ubiquitous in their presence in high energy physics.

The field of neutrino physics has progressed dramatically over the last decade and a half, and this fruitful era of neutrino exploration continues at a high rate. Our goal is to summarize the present status of the field and to discuss ways that progress will be made in answering the outstanding questions. Anticipating the future is a fragile enterprise and neutrino physics has a long history of unexpected surprises. We can look forward with anticipation to the surprises.

2

Neutrino Basics

2.1 Dirac and Majorana Neutrinos

Massive neutrino fields are constructed from Weyl fields (also called chiral fields) that are obtained from the chirality projections of the Dirac field. The Weyl equation governing the motion of Weyl fermions tells us that Weyl fields are eigenstates of helicity and are therefore massless. Massive neutrinos may be of the Dirac or Majorana type.

We denote the 4-component Dirac spinor by

$$\psi = \begin{pmatrix} \xi \\ \omega \end{pmatrix}, \tag{2.1}$$

where ξ and ω are 2-component Weyl spinors; the left-handed and right-handed projections of the Dirac spinor are $\psi_L = (0\ \omega)^T$ and $\psi_R = (\xi\ 0)^T$, and their charge conjugate counterparts are ψ_L^c and ψ_R^c. By definition, a Majorana field χ must be invariant under charge conjugation (up to a phase), i.e.,

$$\chi^c = e^{-i\eta}\chi. \tag{2.2}$$

It can be checked that $\chi = e^{i\alpha}\psi_L + e^{i(\eta-\alpha)}\psi_L^c$ is a Majorana field. The free Lagrangian for a massive Majorana neutrino is

$$\mathcal{L} = \frac{i}{2}\bar{\chi}\partial\!\!\!/\chi - \frac{1}{2}M\bar{\chi}\chi, \tag{2.3}$$

which can also be written as

$$\mathcal{L} = i\bar{\psi}_L\partial\!\!\!/\psi_L - \frac{m}{2}\bar{\psi}_L^c\psi_L + \text{h.c.}, \tag{2.4}$$

where $m = Me^{-i(\eta-2\alpha)}$. The factors of 1/2 in equation 2.3 ensure canonical normalization of the field.

With two Weyl fields ψ_1 and ψ_2 (taken to be left-handed) it is possible to write a mass term

$$\mathcal{L} = -\frac{m_{ij}}{2}\bar{\psi}_i^c \psi_j + \text{h.c.}, \tag{2.5}$$

where $i = 1, 2$ and $j = 1, 2$. Suppose that the mass matrix m_{ij} is purely off-diagonal so that a conserved charge is admitted. Then, defining $\psi_L = \psi_i$ and $\psi_R = \psi_j^c$ results in the mass term for a Dirac neutrino,

$$\mathcal{L} = -m(\bar{\psi}_L \psi_R + \bar{\psi}_R \psi_L) = -m\bar{\psi}\psi. \tag{2.6}$$

The kinetic piece of the Lagrangian $i(\bar{\psi}_L \not{\partial} \psi_L + \bar{\psi}_R \not{\partial} \psi_R)$ is more compactly written as $i\bar{\psi}\not{\partial}\psi$.

The most general neutrino mass matrix in the (ψ_L, ψ_R^c) basis of left-handed fields is

$$M = \begin{pmatrix} m_L & m_D \\ m_D & m_R \end{pmatrix}, \tag{2.7}$$

where m_L and m_R are the Majorana mass terms for ψ_L and ψ_R^c, respectively, and m_D is the Dirac mass term.

2.2 Neutrino Counting

Studies of e^+e^- annihilation at the Z-resonance pole at the Large Electron Positron (LEP) collider have determined the invisible width of the Z boson. The experimental value $N_\nu = 2.984 \pm 0.008$ is close to the number expected from 3 active light neutrinos, though the value is 2σ low [119].

The standard Big Bang Model with an early inflationary epoch provides a highly successful description of cosmological observations including the primordial abundances of the light elements, D and ^{4}He, the temperature asymmetries of the Cosmic Microwave Background (CMB), the energy budget of the universe and the large scale structure of galaxies. The Cosmic Neutrino Background (CNB) is a firm prediction of early universe cosmology. Neutrinos decoupled from matter a couple of seconds after the Big Bang when the temperature of the universe was approximately 1 MeV. Their degrees of freedom contributed to the energy density, along with photons, in the expansion of the universe in the radiation era. Big Bang Nucleosynthesis (BBN) took place between 3 and 20 minutes after the Big Bang.

CMB anisotropies and BBN probe the effective number of neutrinos N_ν that were present in the early universe. The extra relativistic energy density due to sterile neutrinos, or other possible light particles, is normalized to that of an equivalent neutrino flavor as [120]

$$(N_\nu - 3)\rho_\nu = \frac{7}{8}(N_\nu - 3)\rho_\gamma\,, \tag{2.8}$$

where ρ_γ is the energy density in photons. Sterile neutrinos would contribute to $N_\nu - 3$, but so could other new physics sources. (Since sterile neutrinos can contribute along with the three active neutrinos, there is no conflict with the LEP result.)

The precise WMAP measurements of the CMB have been analyzed in various combinations with other datasets to constrain N_ν [102]. Under the assumption of a ΛCDM universe, measurements of baryon acoustic oscillations from the Sloan Digital Sky Survey (SDSS), and WMAP and Hubble Space Telescope data indicate $N_\nu = 4.34^{+0.86}_{-0.88}$ [102].

BBN is a much better probe of N_ν than the CMB. The prediction of the primordial abundance of ^{4}He depends sensitively on the early expansion rate, while the prediction of the D abundance is most sensitive to the baryon density [121]. A recent determination of the primordial ^{4}He mass fraction, $Y \equiv \frac{4n_{He}/n_H}{1+4n_{He}/n_H} = 0.2565 \pm 0.0060$ [122], suggests that $N_\nu = 4$ is allowed at the 1σ C.L. [123]. Another study that paid special attention to systematic uncertainties in the extraction of the helium abundance found the value $Y = 0.2561 \pm 0.0108$ [124], with a much larger uncertainty. Even more recently, a new method to evaluate the systematic uncertainties was proposed in [125]. Note that a previous measurement of the helium abundance $Y = 0.240 \pm 0.006$ [126] gave a best-fit value $N_\nu = 2.4$. Clearly, the value of Y, and hence N_ν, remains under debate.

Cosmological Relic Neutrinos

BBN probes the cosmological relic neutrinos at early times. Detection of relic neutrinos at the present time is a long range goal of great interest but this is extremely challenging because of the very small neutrino scattering cross sections at low energies [127].

The present temperature of the CMB is $T_\gamma = 2.73\,\text{K}$. The corresponding prediction for the temperature of the CNB today is $T_\nu = 1.95\,\text{K}$. The number density of neutrinos and antineutrinos is

$$n_\nu = \frac{3}{11} n_\gamma = 112\,\text{cm}^{-3}, \tag{2.9}$$

which is comparable to the CMB number density when summed over the 3 neutrino types. The averaged neutrino 3-momentum is

$$p_\nu = 3\left(\frac{4}{11}\right)^{1/3} T_\gamma = 5 \times 10^{-4}\ \text{eV}. \tag{2.10}$$

Thus, at least some of the relic neutrinos are nonrelativistic now. Due to their clustering in the gravitational potential wells of dark matter and baryons, there may be an overdensity in our region of the Milky Way.

Several methods of detection of the cosmological relic neutrinos have been proposed, but the necessary sensitivities have not been reached. One way is the detection of an annual modulation signal from the force due to coherent scattering of the relic neutrinos on a Cavendish-type torsion balance [128]. A second way of detection is through relic neutrino capture on tritium, $\nu_e + {}^3\text{H} \to {}^3\text{He} + e^-$, which would give a spike in the tritium beta decay spectrum beyond the kinematic endpoint [129]. A third way is the detection via an absorption feature from $\nu\bar{\nu} \to Z$ in the energy spectrum of ultra high energy cosmic rays [130, 131] (the Z-burst mechanism discussed in section 11.5).

2.3 Neutrinos from Weak Decays

Since neutrinos are colorless and have no electric charge, they do not participate in the strong or electromagnetic interactions. Therefore the weak interactions are the basis for both neutrino production and detection.

Pion Decays

The charged π mesons decay via the leptonic modes

$$\pi^- \to \ell^- \bar{\nu}_\ell\,, \qquad \pi^+ \to \ell^+ \nu_\ell\,, \qquad \ell = e, \mu\,. \tag{2.11}$$

Because of the V–A coupling of the virtual W-boson to the final state leptons, the decay width is proportional to m_ℓ^2 and thus the electron channel is suppressed relative to the muon channel by a factor of 1.3×10^{-4}. The ν_μ has a line spectrum with energy

$$E(\nu_\mu) = (m_\pi^2 - m_\mu^2)/(2m_\pi) = 29.8 \text{ MeV}\,. \tag{2.12}$$

For the decays of a π-meson in flight, with boost factor $\gamma = E_\pi/m_\pi$, the neutrino energy spectrum is flat with endpoints of the distribution determined by the theta function

$$\theta(1 - m_\mu^2/m_\pi^2 - E_\nu/E_\pi)\,. \tag{2.13}$$

Muon Decays

In the muon rest frame, the distribution of muon antineutrinos (neutrinos) from the decay

$$\mu^\pm \to e^\pm + \nu_e(\bar{\nu}_e) + \bar{\nu}_\mu(\nu_\mu) \tag{2.14}$$

of polarized muons is given by the expression

$$\frac{d^2 N_{\nu_\mu}}{dx d\Omega} = \frac{2x^2}{4\pi}\left[(3 - 2x) \mp (1 - 2x)\cos\theta\right]\,, \tag{2.15}$$

where $x \equiv 2E_\nu/m_\mu$, θ is the angle between the neutrino momentum vector and the muon spin direction, and m_μ is the muon rest mass. The corresponding expression describing the distribution of electron neutrinos (antineutrinos) is

$$\frac{d^2 N_{\nu_e}}{dx d\Omega} = \frac{12x^2}{4\pi}\left[(1 - x) \mp (1 - x)\cos\theta\right]. \tag{2.16}$$

Thus, the neutrino and antineutrino energy- and angular-distributions depend upon the parent muon energy, the decay angle, and the direction of the muon spin vector. The neutrino spectra for an unpolarized muon at rest are shown in figure 2.1.

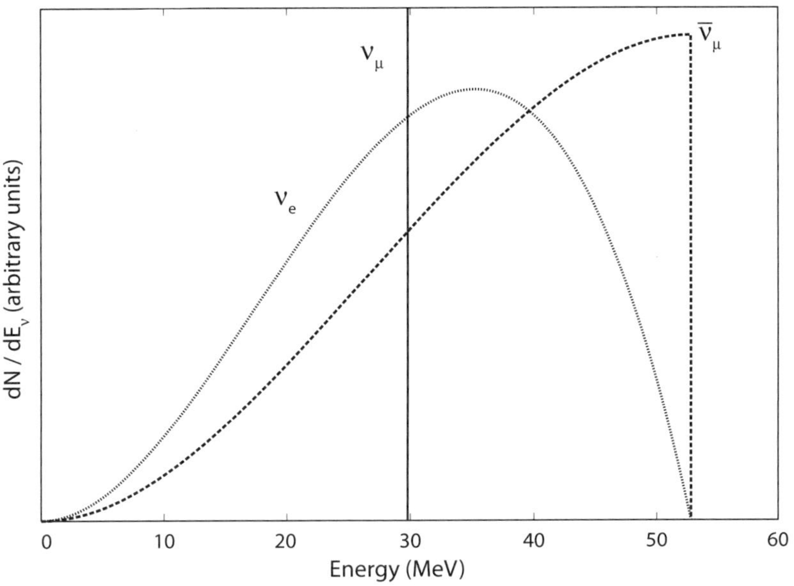

Figure 2.1. Muon neutrino and electron antineutrino spectra from unpolarized muon decay, in the rest frame of the decaying muon. Also shown is the monoenergetic muon neutrino from the decay of a charged pion at rest. Adapted from [132].

Tau Decays

The branching fractions and fragmentation functions of τ decays in flight are given in table 2.1 along with the functional forms of the neutrino decay distributions [133]

$$\frac{dn}{dz} = g_0(z) + P g_1(z). \tag{2.17}$$

Here $z = E_\nu/E_\tau$, P is the polarization of the decaying τ^-, which is $P = -1$ for neutrino $V - A$ production of the τ. Formulas for $g_0(z)$ and $g_1(z)$ are given in tables 2.1 and 2.2, where the smeared distributions f^* are given by a Breit-Wigner approximation,

$$f^* = C \frac{1}{(1 - x_\nu - r_{mes})^2 + r_{mes}^2 \Gamma_{mes}^2 m_{mes}^{-2}}, \tag{2.18}$$

where $x_\nu = E_\nu/m_\tau$, $r_{mes} = m_{\rho,a_1}^2/m_\tau^2$, Γ_{mes} is the decay width of the meson and C is a normalization factor. For the inclusive $\tau \to \nu_\tau X$ mode, $f_X(x)$ is approximated by a generic phase-space decay into four pions and ν_τ. Note that the leptonic decay distributions here also apply to energetic muon decays, with appropriate changes of flavor subscripts.

Neutron Beta Decay

The classic radioactive decay process is that of neutron decay

$$n \to p + e^- + \bar{\nu}_e\,. \tag{2.19}$$

TABLE 2.1
Branching fractions and ν_τ fragmentation functions.

τ^- decay mode*	B_τ	$g_0(z)$	$g_1(z)$
$\nu_\tau \ell \bar{\nu}_\ell$	0.18	$\frac{5}{3} - 3z^2 + \frac{4}{3}z^3$	$\frac{1}{3} + \frac{8}{3}z^3 - 3z^2$
$\nu_\tau \pi$	0.12	$\frac{1}{1-r_\pi}\theta(1 - r_\pi - z)$	$-\frac{2z-1+r_\pi}{(1-r_\pi)^2}\theta(1 - r_\pi - z)$
$\nu_\tau a_1$	0.13	$\int_z^1 f^*(x)x^{-1}dx$	$\int_z^1 (z - 2x) f^*(x)x^{-2}dx$
$\nu_\tau \rho$	0.26	$\int_z^1 f^*(x)x^{-1}dx$	$\int_z^1 (z - 2x) f^*(x)x^{-2}dx$
$\nu_\tau X$	0.13	$\int_z^1 f_X(x)x^{-1}dx$	0

*For various decay modes of the τ^- lepton, with $z = E_{\nu_\tau}/E_\tau$ and $r_j = m_j^2/m_\tau^2$. The distributions f^* and f_X are discussed in the text. From [133–135].

TABLE 2.2
Fragmentation functions for $\bar{\nu}_\ell$.

τ^- decay mode*	B_τ	$g_0(y)$	$g_1(y)$
$\nu_\tau \ell \bar{\nu}_\ell$	0.18	$2 - 6y^2 + 4y^3$	$-2 + 12y - 18y^2 + 8y^3$

In the decay $\tau^- \to \nu_\tau \ell^- \bar{\nu}_\ell$ [134], with $y = (E_{\bar{\nu}_\ell})/E_\tau$.

The neutrino flux from the decay of a nucleus is given by [136]

$$\Phi_{\rm cm}(E_\nu) = bE_\nu^2 E_e \sqrt{E_e^2 - m_e^2}\; F(\pm Z, E_e)\,\Theta(E_e - m_e)\,, \tag{2.20}$$

where the constant $b = \ln(2)/m_e^5 {\rm ft}_{1/2}$, m_e is the electron mass, ${\rm ft}_{1/2}$ is the comparative half-life[1] and Θ is the Heaviside step function. The quantities appearing in equation 2.20 are the energy $E_e = Q - E_\nu$ of the emitted lepton (electron or positron), where Q is the Q-value of the reaction, and the Fermi function $F(\pm Z, E_e)$, which accounts for the Coulomb modification of the spectrum.

2.4 Neutrino Cross Sections

Neutrino-electron Elastic Scattering

The scattering of neutrinos on electrons is free of the complications of the strong interactions and can be used to determine the weak angle θ_w in the Standard Model Lagrangian or test for the existence of new physics. The experimentally accessible reactions are

$$\begin{aligned} \nu_\ell + e^- &\to \nu_\ell + e^-\,, \\ \bar{\nu}_\ell + e^- &\to \bar{\nu}_\ell + e^-\,, \end{aligned} \tag{2.21}$$

[1] The comparative half-life is the product of the half-life $t_{1/2}$ and a factor f that accounts for the effect of the Coulomb field on the emission of the electrons and the energy release; it is significantly bigger for forbidden transitions than for allowed ones [137].

which proceed through Z-boson exchange for $\ell = \mu$ or τ and through Z- and W-boson exchanges for $\ell = e$. For $Q^2 \ll m_W^2$ propagator effects can be neglected and the effective four-fermion interaction used. The cross sections are governed by the Fermi constant, G_F, determined by the muon lifetime to be

$$G_F = \frac{g^2}{\left(4\sqrt{2}m_W^2\right)} = 1.16637(1) \times 10^{-5}\ \text{GeV}^{-2}\,. \tag{2.22}$$

The very small sizes of these cross sections are set by the scale

$$\kappa = \frac{G_F^2 m_e}{(2\pi)} = 4.3 \times 10^{-42}\ \text{cm}^2/\text{GeV}\,, \tag{2.23}$$

and the relative sizes are governed by $x_W = \sin^2\theta_w = 0.23$, where θ_w is the weak mixing angle. At tree level the cross sections in units of κE_ν (GeV) are [138]

$$\sigma(\nu_\mu e) = 1 - 4x_W + \frac{16x_W^2}{3} = 0.362 \tag{2.24}$$

$$\sigma(\bar{\nu}_\mu e) = \frac{1}{3} - \frac{4x_W}{3} + \frac{16}{3}x_W^2 = 0.309 \tag{2.25}$$

$$\sigma(\nu_e e) = 1 + 4x_W + \frac{16x_W^2}{3} = 2.2 \tag{2.26}$$

$$\sigma(\bar{\nu}_e e) = \frac{1}{3} + \frac{4x_W}{3} + \frac{16x_W^2}{3} = 0.922. \tag{2.27}$$

The current precision of x_W from measurements of these processes is at the $\pm$ 3.5% level.

Many extensions of the Standard Model have extra neutral weak bosons (usually called Z') that couple to neutrinos and charged leptons. In a process that is measured at low energies, the Z' exchange can be approximated as an effective four-fermion interaction. In neutrino electron scattering, a Z' exchange will add extra contributions to the SM cross section that are proportional to $(M_Z/M_{Z'})^2$. An $\mathcal{O}(0.1\%)$ measurement of $\sigma(\nu_\mu e)$ at low energies could probe Z' masses up to 2–3 TeV [138].

Elastic and Quasielastic Scattering of Neutrinos on Nucleons

In the limit that the charged-lepton mass can be ignored, the differential cross section for the processes

$$\nu, \bar{\nu}(k_1) + N(p_1) \rightarrow \ell(k_2) + N'(p_2) \tag{2.28}$$

can be written in terms of the neutrino energy E and the momentum transfer $Q^2 = -(k_2 - k_1)^2$ as [139]

$$\frac{d\sigma^{\nu,\bar{\nu}}}{dQ^2} = \frac{G_F^2}{2\pi} \{G_E^2 \left[(1 - \tau M_N/E)^2(1+\tau)^{-1} - \tau M_N^2/E^2\right]$$
$$+G_M^2\tau \left[(1 - \tau M_N/E)^2(1+\tau)^{-1} + \tau M_N^2/E^2\right] + G_A^2 \left[(1 - \tau M_N/E)^2\right.$$
$$\left. + \tau(1+\tau)M_N^2/E^2\right] \pm 4G_A G_M \tau \left[(1 - \tau M_N/E)M_N/E\right]\} \ , \qquad (2.29)$$

where $\tau \equiv Q^2/(4M_N^2)$, M_N is the nucleon mass and the electric and magnetic form factors are effectively given by the dipole parameterizations

$$G_E(Q^2)/G_E(0) = G_M(Q^2)/G_M(0) = (1 + Q^2/M_V^2)^{-2}, \qquad (2.30)$$

$$G_A(Q^2)/G_A(0) = (1 + Q^2/M_A^2)^{-2}, \qquad (2.31)$$

with $M_V^2 \simeq 0.71\,\text{GeV}^2$ and $M_A^2 \simeq 1.0\,\text{GeV}^2$ [140].[2] For the quasielastic reactions $\nu_\mu n \to \mu^- p$ and $\bar{\nu}_\mu p \to \mu^+ n$, the $Q^2 = 0$ values are

$$\begin{aligned} G_E(0) &= \cos\theta_C \ , \\ G_M(0) &= \cos\theta_C(1 + \kappa_p - \kappa_n), \\ G_A(0) &= g_A \cos\theta_C, \end{aligned} \qquad (2.32)$$

where θ_c is the Cabibbo angle ($\cos\theta_c = 0.974$), $\kappa_p = 1.79$ and $\kappa_n = -1.91$ are anomalous magnetic moments of the nucleons, and $g_A = 1.23$ is the axial coupling constant. There are nuclear corrections to quasielastic scattering, the most important arising from the Pauli exclusion principle [140]. The integrated CC cross sections are compared to data in figure 2.2.

For the neutral current reactions the $Q^2 = 0$ values of the proton (neutron) electric and magnetic NC form factors are

$$G_A^{p,n}(0) = \pm\frac{g_A}{2} \ . \qquad (2.33)$$

Single Pion (Δ Resonance) Production

For neutrino energies of order 1 GeV, the neutrino cross section is dominated by the production of the $\Delta(J = 3/2, I = 3/2)$ resonance [144]

$$\nu_\mu + p \to \mu^- + \Delta^{++}(1232\ \text{MeV}) \to \mu^- + p + \pi^+ \ . \qquad (2.34)$$

[2] A recent MiniBooNE measurement [141] suggests a larger value for M_A, but a detailed analysis [142] including multinucleon effects indicates that the MiniBooNE data is compatible with the world average from other experiments.

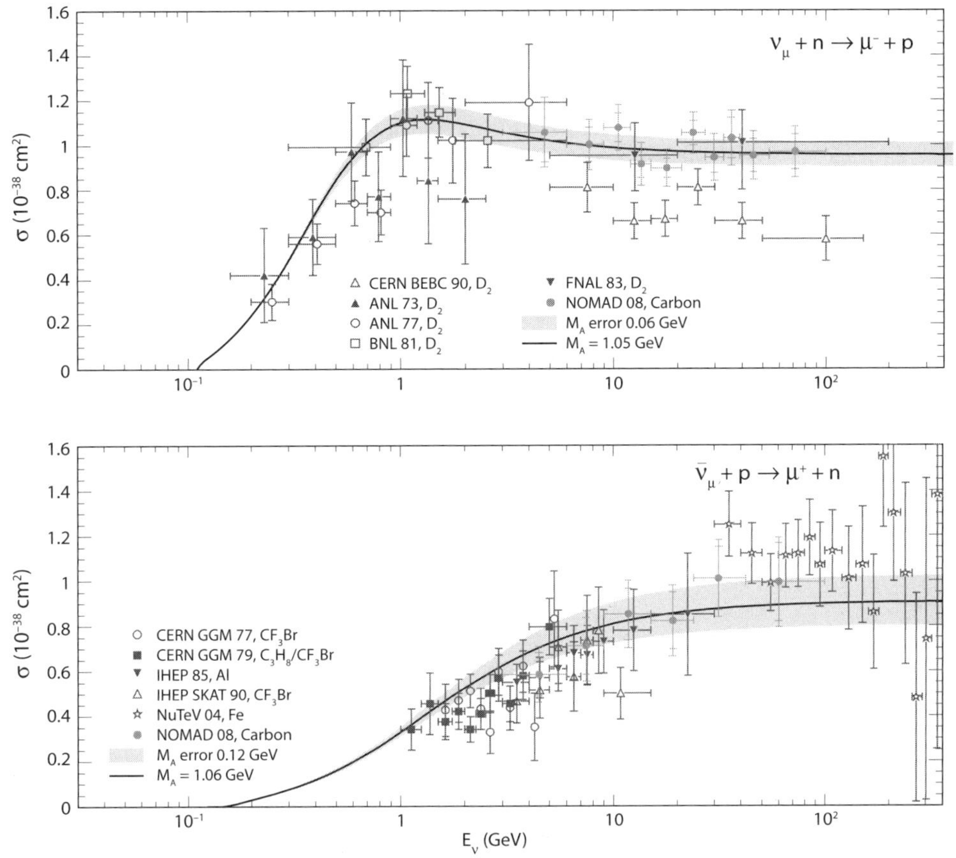

Figure 2.2. Quasielastic CC cross section for ν_μ and $\bar{\nu}_\mu$ scattering. The solid lines and bands show the calculated values and error bars, including nuclear effects, for the best-fit value of $M_A = 1.05 \pm 0.06$ GeV. From [143].

The contribution to this reaction from the conserved vector current (CVC) is dominated by the magnetic dipole form factor and can be inferred from electroproduction data, $e + p \to e + p + \pi^0 (e + n + \pi^+)$. The contribution from the partially conserved axial vector current (PCAC) is calculated using an empirically modified dipole Q^2 dependence. This simple phenomenological formalism reproduces the measured differential cross sections for charged single pion production by neutrinos.

Deep Inelastic Scattering (DIS)

At high Q^2, the neutrino scattering cross sections are well described by the quark parton model when the quantum chromodynamic (QCD) corrections to the lowest-order calculations are included. The leading order cross sections are expressed in terms of the scaling variables

$$x = Q^2/(2M_N\nu)\,, \qquad y = \frac{E_\nu - E_\ell}{E_\nu} = \frac{E_{\text{hadron}}}{E_\nu}\,, \tag{2.35}$$

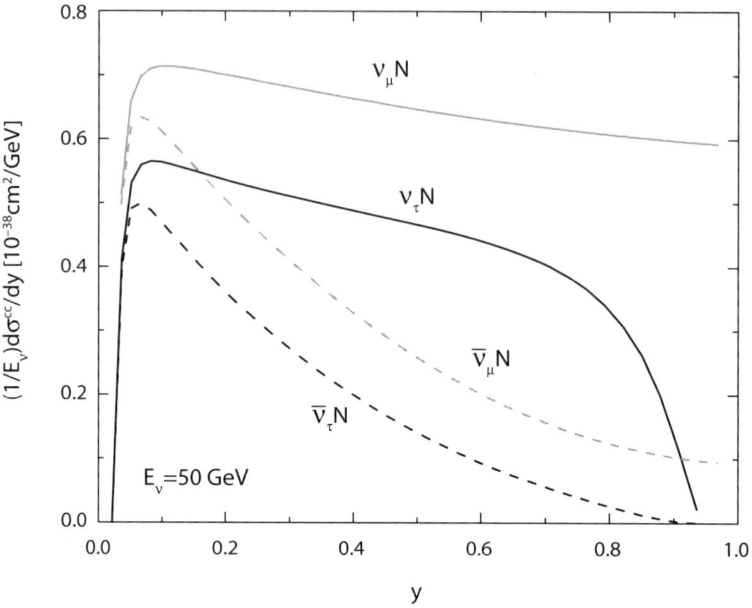

Figure 2.3. The differential cross section $(1/E_\nu)d\sigma/dy$ for DIS CC scattering of a 50 GeV neutrino on an isoscalar nucleon target. Adapted from [147].

where $\nu = E_\nu - E_\ell$ with E_ν the incident neutrino energy and E_ℓ the energy of the charged lepton produced in the charged-current (CC) interactions. If the charged-lepton mass is ignored, the y-dependence of differential neutrino and antineutrino cross sections on an isoscalar nuclear target (equal number of protons and neutrons), integrated over x, is approximately parameterized by [118, 145]

$$\frac{d\sigma}{dy}(\nu N) = \sigma_0 \left[1 + r(1-y)^2\right] I_1, \tag{2.36}$$

$$\frac{d\sigma}{dy}(\bar{\nu} N) = \sigma_0 \left[(1-y)^2 + r\right] I_1, \tag{2.37}$$

where $I_1 \equiv \int_0^1 \frac{1}{2}[u(x) + d(x)]\, dx \approx 0.212$, $r \approx 1/5$ is the relative size of the sea-quark contribution compared to the valence quarks, and the scale of the DIS cross sections is

$$\sigma_0 = \frac{2G_F^2 M_N E_\nu}{\pi} = \left(3.16 \times 10^{-38}\ \text{cm}^2/\text{GeV}\right) E_\nu. \tag{2.38}$$

The $(1-y)^2$ dependence is a direct consequence of the V–A interaction. These parameterizations neglect the Q^2 dependences of the structure functions. Exact expressions using the proton structure functions can be found, e.g., in [146]. The y-dependences are illustrated in figure 2.3 for $E_\nu = 50$ GeV.

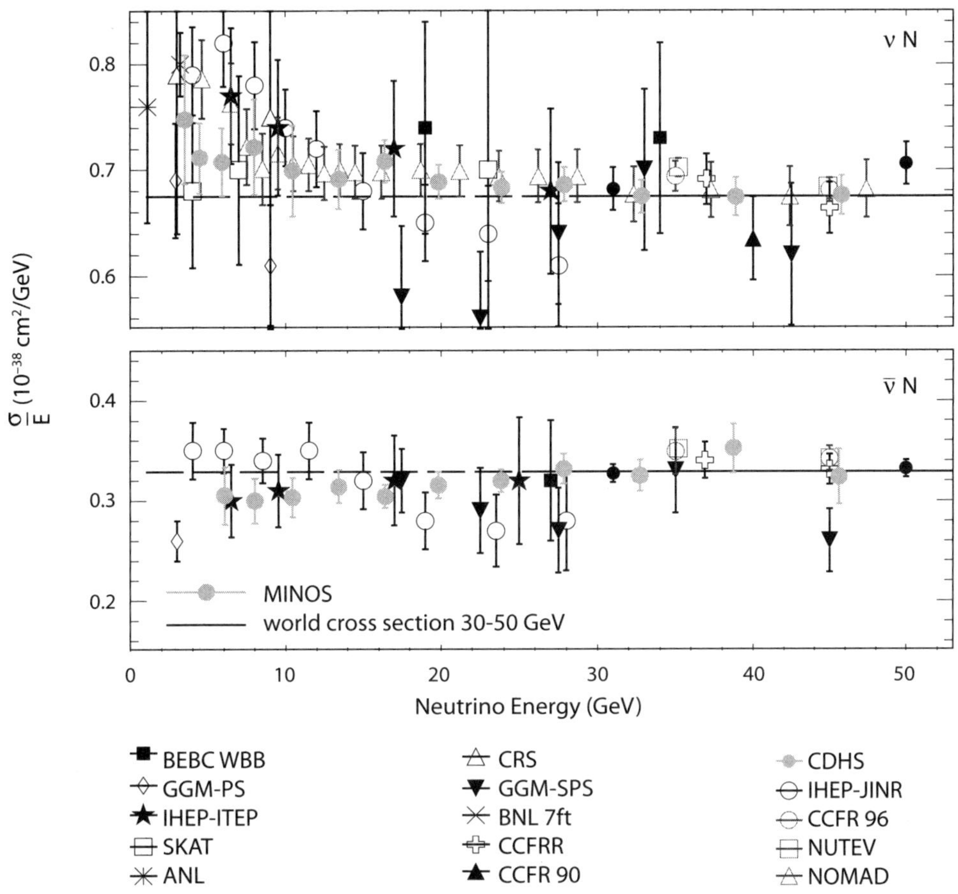

Figure 2.4. Total CC cross section for for ν_μ and $\bar{\nu}_\mu$ scattering on an isoscalar target. The dashed lines are the average values from equation 2.39. From [149].

After integration over y the total DIS CC cross sections are [148]

$$\begin{aligned}
\sigma(\nu n) &= 0.881 \times 10^{-38}\text{cm}^2 \left(\frac{E_\nu}{\text{GeV}}\right), \\
\sigma(\nu p) &= 0.451 \times 10^{-38}\text{cm}^2 \left(\frac{E_\nu}{\text{GeV}}\right), \\
\sigma(\bar{\nu} n) &= 0.250 \times 10^{-38}\text{cm}^2 \left(\frac{E_\nu}{\text{GeV}}\right), \\
\sigma(\bar{\nu} p) &= 0.399 \times 10^{-38}\text{cm}^2 \left(\frac{E_\nu}{\text{GeV}}\right).
\end{aligned} \tag{2.39}$$

For an isoscalar target, one should take the average of the neutron and proton cross sections. The scaling of the CC cross sections with energy are shown in figure 2.4. A measurement of the tau neutrino cross section [150] is consistent with equation 2.39.

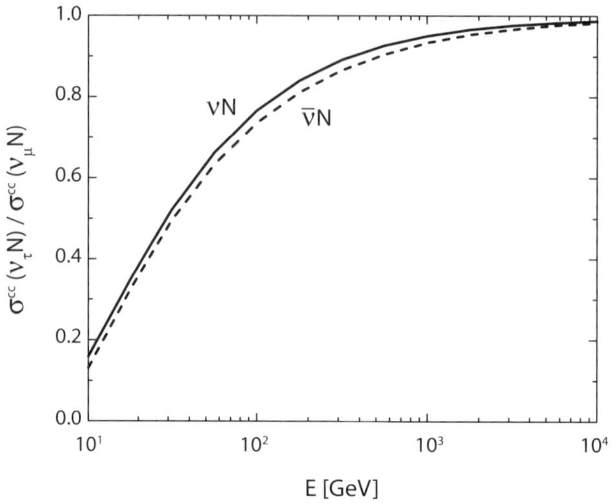

Figure 2.5. The ratio $\sigma(\nu_\tau N)/\sigma(\nu_\mu N)$ for DIS CC scattering of neutrinos (solid curve) and antineutrinos (dashed) on an isoscalar target. From [147].

At higher energies, at which there are higher Q^2 values, the W-propagator modification $1/(Q^2 + m_W^2)^2$ of the differential cross sections becomes important and the growth of the total cross sections with energy slows. Also, the kinematic suppression due to the tau mass, evident in figure 2.3, must be taken into account for tau production at lower energies [147]. The combined effects of the tau mass and W propagator on the $\nu_\tau N$ cross section are shown in figure 2.5.

The corresponding results for the neutral current (NC) differential cross sections are [118]

$$\frac{d\sigma}{dy}(\nu N) = \sigma_0 \left\{ \left[\frac{1}{2} - x_W + \frac{5}{9}x_W^2\right] \left[1 + r(1-y)^2\right] + \frac{5}{9}x_W^2 \left[(1-y)^2 + r\right] \right\} I_1, \quad (2.40)$$

$$\frac{d\sigma}{dy}(\bar{\nu} N) = \sigma_0 \left\{ \left[\frac{1}{2} - x_W + \frac{5}{9}x_W^2\right] \left[(1-y)^2 + r\right] + \frac{5}{9}x_W^2 \left[1 + r(1-y)^2\right] \right\} I_1, \quad (2.41)$$

where N is an isoscalar nucleon target. After integration over y the total DIS NC cross sections for an isoscalar target are [148]

$$\sigma(\nu N) = 0.209 \times 10^{-38}\mathrm{cm}^2 \left(\frac{E_\nu}{\mathrm{GeV}}\right),$$
$$\sigma(\bar{\nu} N) = 0.115 \times 10^{-38}\mathrm{cm}^2 \left(\frac{E_\nu}{\mathrm{GeV}}\right). \quad (2.42)$$

The ratios $R = \sigma^{\mathrm{NC}}/\sigma^{\mathrm{CC}}$ of neutral current to charged-current cross sections in neutrino DIS are approximately

$$R(\nu N) = \frac{1}{2} - x_W + \frac{20}{9}\left(\frac{1+r}{3+r}\right) x_W^2, \quad (2.43)$$

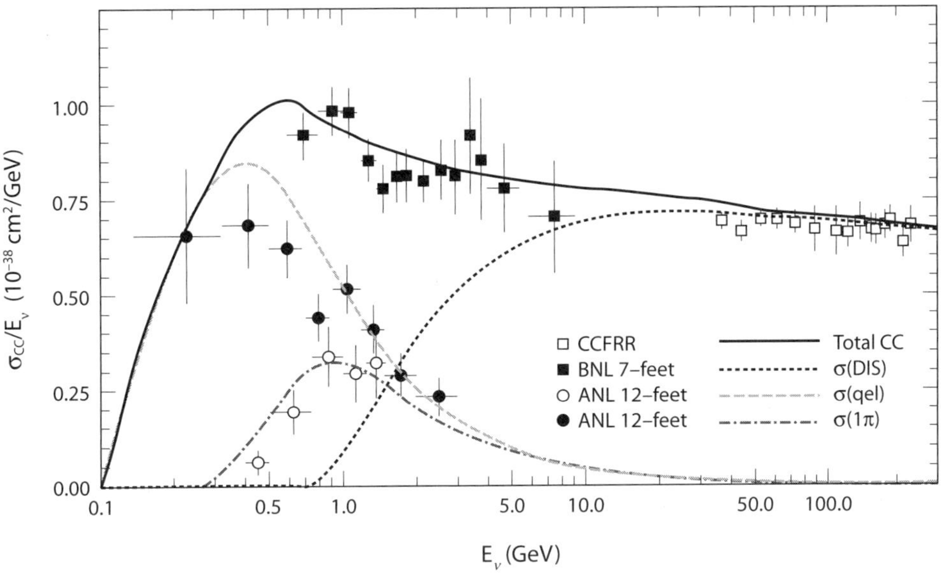

Figure 2.6. Summary of contributions to the neutrino-nucleon charged-current cross sections in the few GeV region, including deep inelastic scattering (dotted curve), quasielastic scattering (dashed), and single pion production from the Δ resonance (dash-dotted). From [159].

$$R(\bar{\nu}N) = \frac{1}{2} - x_W + \frac{20}{9}\left(\frac{1+r}{1+3r}\right)x_W^2. \tag{2.44}$$

In the valence quark approximation $r = 0$. Measurements of the two R values determine x_W.

GeV Neutrino Cross Sections

The existing neutrino oscillation experiments require knowledge of the cross sections in the few GeV region where there are non-negligible contributions from quasielastic, resonance, and deep-inelastic processes, but there are still sizeable uncertainties on the measured values. A summary of the situation is given in figure 2.6 versus energy. The SciBooNE, [151] MINERνA [152], and ArgoNeuT [153] experiments in progress are designed to improve the precision of the neutrino cross sections in this energy region; proposed experiments include FINeSSE [154] and MicroBooNE [155]. Near detectors at short- and long-baseline neutrino experiments such as MiniBooNE [156, 157], K2K [158], MINOS [149], T2K [83], and NOνA [84] can also provide neutrino cross section measurements.

Neutrino Capture

Another class of neutrino reactions is neutrino capture, (A, Z) (ν_e, e^-) $(A, Z+1)$. A precise calculation of these cross sections requires knowledge of the appropriate nuclear matrix elements, which in most cases is not available. Two reactions for which the matrix elements are sufficiently known are ^{37}Cl (ν_e, e^-) ^{37}Ar, with threshold 0.814 MeV, and ^{71}Ga (ν_e, e^-) ^{71}Ge, with threshold 0.233 MeV. Just above

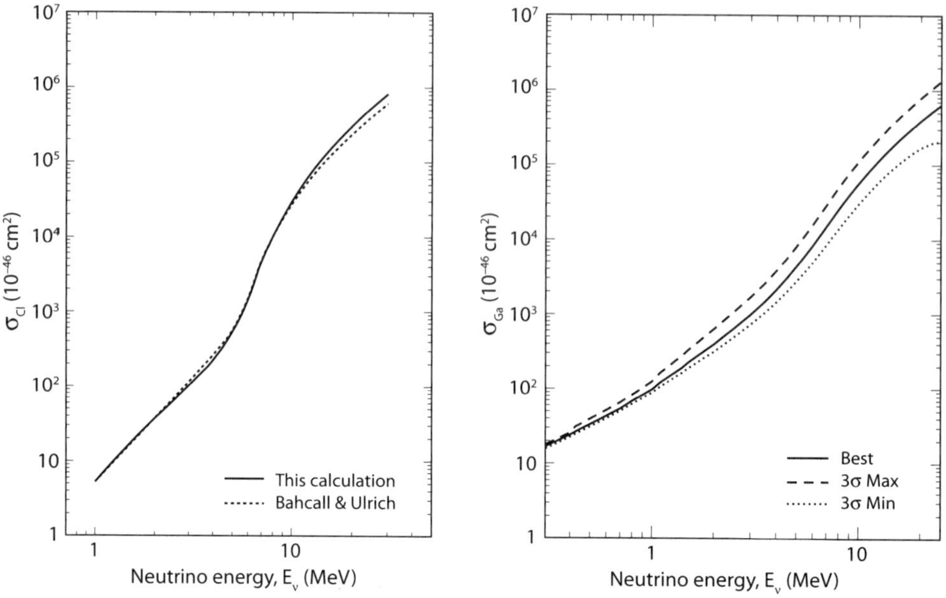

Figure 2.7. Left panel: electron neutrino capture cross section on ^{37}Cl, from [160] (solid line) and [161] (dashed). Right panel: electron neutrino capture cross section on ^{71}Ga from [162].

threshold only the ground-state to ground-state transition is possible and the cross sections are well-known. At higher energies many excited states also contribute; the cross section rises dramatically and is less well-known. Cross sections for neutrino capture on ^{37}Cl and ^{71}Ga are shown in figure 2.7.

2.5 Neutrino Detectors

The design of a neutrino detector involves many factors: (i) the interaction cross section of the target material at the relevant neutrino energies; (ii) the trade-off between the granularity of the detector, i.e., the ability to make precise position measurements, and the size, to which the event rate is proportional; (iii) the sensitivity to a particular final state; and (iv) the size of the background relative to the signal. In this section we outline the principal classes of neutrino detectors.

Radiochemical

Radiochemical neutrino detectors are sensitive only to electron neutrinos via a neutrino capture reaction on the target nucleus. The first solar neutrino experiments were radiochemical; the Homestake mine experiment using the neutrino capture reaction ^{37}Cl (ν_e, e^-) ^{37}Ar, and the SAGE, GALLEX, and GNO experiments, which used the reaction ^{71}Ga (ν_e, e^-) ^{71}Ge. The resulting radioactive final-state atoms are then chemically flushed from the system and their decays observed later.

Radiochemical detectors have a relatively low threshold (of order 1 MeV) and the neutrino capture cross sections rise very rapidly with neutrino energy, so they are especially well-suited for measuring the solar neutrino flux, which has copious

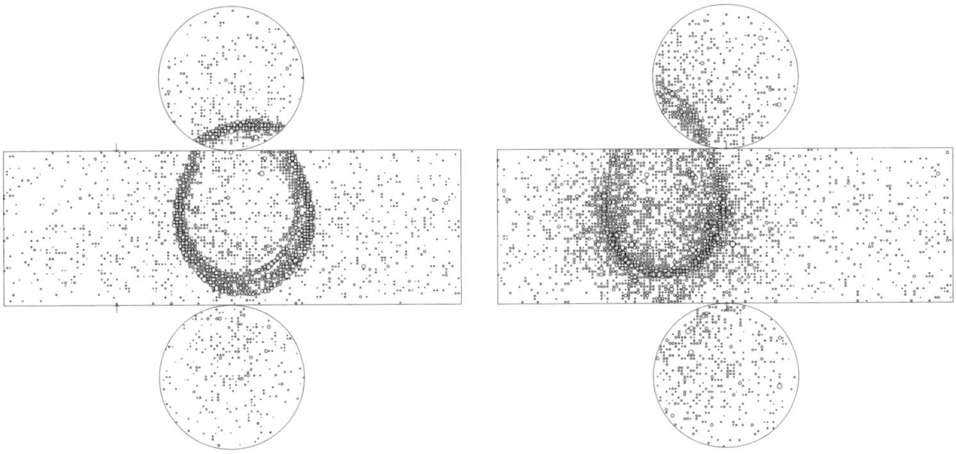

Figure 2.8. Ring patterns for ν_μ and ν_e CC interactions in a Cherenkov detector. The muon from the ν_μ (left) creates a sharper ring than the electron from the ν_e (right). From [163].

numbers of lower-energy neutrinos and fewer (by orders of magnitude) higher-energy neutrinos. However, there is no measurement of the incident neutrino energy, so only the total rate is determined. Radiochemical detectors are generally placed far underground to avoid background from cosmic rays.

Water Cherenkov

In water Cherenkov detectors, charged particles created by neutrino scattering emit Cherenkov radiation if they move faster than the speed of light in the medium ($\approx c/1.33$ in water), e.g., electron energy above 0.77 MeV and muon energy above 160 MeV. This radiation, which has a characteristic conical shape, forms a ring pattern on the outer surface of the detector, where it is detected by photomultiplier tubes (PMTs) surrounding the detector volume. The opening angle of the cone (measured from the electron or muon direction) is $\theta = \cos^{-1}(c/nv)$, where n is the index of refraction, so in principal both the direction and energy of the charged particle may be determined.

At solar neutrino energies the relevant reaction is $\nu e \to \nu e$. Because much of the Cherenkov radiation is scattered or absorbed before it is detected, the minimum threshold for detection is 5 MeV or above, depending on the efficiency and coverage of the PMTs.

Above the muon threshold, the dominant reaction is CC neutrino scattering off nuclei. Electron rings are much more diffuse than muon rings due to multiple Coulomb scattering and electromagnetic showers, which allows incident ν_e to be distinguished from ν_μ with a high degree of accuracy (see figure 2.8). Below 500 MeV, quasielastic reactions are dominant. For 500 MeV $\lesssim E_\nu \lesssim$ 2 GeV, Δ resonance production becomes important, while at higher energies DIS provides the largest contribution to the neutrino cross section, as discussed in section 2.4.

Neutral current events such as $\nu N \to \nu N' \pi^0$ also occur. The π^0 will decay to two photons, each of which will create a diffuse ring pattern in the PMTs due to electromagnetic showering. At pion energies below about 1 GeV the photons are sufficiently separated that these NC events may be distinguished from CC ν_e events. At higher energies, however, the two photons are more collinear, and the NC events may be misidentified as CC ν_e events. NC background rejection in the few GeV region is therefore very important [164].

Early water Cherenkov detectors include the Irvine-Michigan-Brookhaven (IMB) detector [61], which measured atmospheric neutrinos, and the original Kamioka Nucleon Decay Experiment (Kamiokande) detector, which measured both solar and atmospheric neutrinos [60]. Super-Kamiokande [30, 63] is a large (50 kiloton (kt)) water Cherenkov detector that has been used to measure both higher-energy solar neutrinos and atmospheric and long-baseline neutrinos. Much larger Cherenkov detectors–300 kt and up–are being considered for the future, such as UNO [96], Hyper-Kamiokande [165], and MEMPHYS [166].

Liquid Scintillator

When charged particles pass through matter, their electric field excites atoms along their path, which leads to ionization energy loss. In organic scintillating materials some of this energy is emitted as photons, which may then be detected by PMTs. Compared to Cherenkov radiation, scintillation has a relatively high yield (emitted radiation compared to energy lost by the particle) and is especially effective for detecting neutrinos with $E_\nu < 5\,\text{MeV}$. Because of this, experiments attempting to detect reactor and low-energy solar neutrinos use scintillating materials.

At very low energies, $E_\nu \lesssim 1\,\text{MeV}$, scintillators have a large background from cosmogenic radioactivity (due to radioactive nuclei produced by cosmic ray muons); placing the detector underground can help eliminate such events. These radioactive nuclei have a short lifetime, so this background may be further reduced by rejecting events that are coincident with a muon. Natural radioactivity due to trace amounts of unstable nuclei is also a problem at low neutrino energies; in most cases this may be reduced by purification of the scintillating material. However, some ^{14}C is present naturally and cannot be removed, making the background below about 200 keV insurmountable.

The primary reaction for detecting neutrinos is νe elastic scattering (at low energies) and quasielastic scattering off nuclei, while antineutrinos undergo inverse beta decay, $\bar{\nu}_e p \to n e^+$. In each case the reaction can be identified by detection of the final-state electron. For inverse beta decay, the outgoing neutron can also be detected (see the discussion of reactor neutrino experiments in (section 8.2).

KamLAND [54] and LVD [167] are currently the largest liquid scintillator detectors with a mass of 1 kt; Hanohano [168] and NOνA [84] are planned to be 10 kt and 15 kt, respectively, while the proposed LENA [169] detector will be of order 50 kt.

Liquid Argon TPC

In a liquid argon time projection chamber (TPC), charged particles ionize the material, creating electron-ion pairs. If an electric field is applied across the material, the ionization electrons will drift to the edges of the detector, where they are collected and measured. From the position and timing of the ionization signals, a full 3D picture of the track is obtained. Charged particle interactions with liquid argon give scintillation radiation and, if the charged particle is sufficiently relativistic, will

TABLE 2.3
Comparison of physics capabilities of three detector types being considered for a future long-baseline experiment.

Source	Mode	*50* kt Scint	*100* kt Water	*20* kt Argon
Long baseline	$\nu_\mu \rightarrow \nu_e$	Yes	Yes	Yes
	e^+/e^- discrimination	No(?)*	No	Yes
Reactor	$\bar{\nu}_e$	Yes	with Gd	No
Solar	ν_e from ^{8}B	Yes	Yes(?)*	Yes
	ν_e from *hep*	Yes	Yes	Yes
	$\bar{\nu}_e$	Yes	No	No
Other	Geoneutrinos	Yes	No	No
	Atmospheric ν	Yes	Yes	Yes
	Supernova ν	Yes	Yes	Yes
	Relic supernova ν	Yes	Yes	No
	DM annihilation ν	Yes	Yes	Yes

*Question marks indicate uncertainty. Adapted from [173].

lead to Cherenkov radiation. The dominant neutrino reaction is neutrino-nucleon scattering. Liquid argon (LAr) detectors have low backgrounds and are an excellent choice for long-baseline experiments (see the discussion of long-baseline experiments in section 8.1); the main issue is whether a viable detector can be made large enough to provide a significant number of events. The ICARUS [170] detector is currently the largest LAr detector (800 tons); much larger LAr detectors, from 20 kt [171] to as large as 100 kt [172], are being planned.

Table 2.3 compares the physics capabilities of water Cherenkov, liquid scintillator, and liquid argon detectors for measuring a variety of physics processes.

Iron Calorimeter

An iron calorimeter has interleaved layers of iron and active detector material. Muons from CC ν_μ interactions may be discriminated from CC electrons and NC hadrons since muons travel much further through the iron; the different shapes of hadronic and electromagnetic showers distinguish CC electrons from NC interactions. Magnetizing the iron allows the determination of the sign of the particle charge. The 5.4 kt MINOS [71] uses solid scintillator for the active detection technology, while the proposed 50 kt INO detector [174] has resistive plate chambers.

Emulsion Cloud Chamber

An emulsion cloud chamber consists of alternating thin (of order 50 μm) emulsion layers and thicker (of order 1 mm) metal plates. The emulsions provide precision tracking that can resolve the displaced vertices of tau decays, which make this

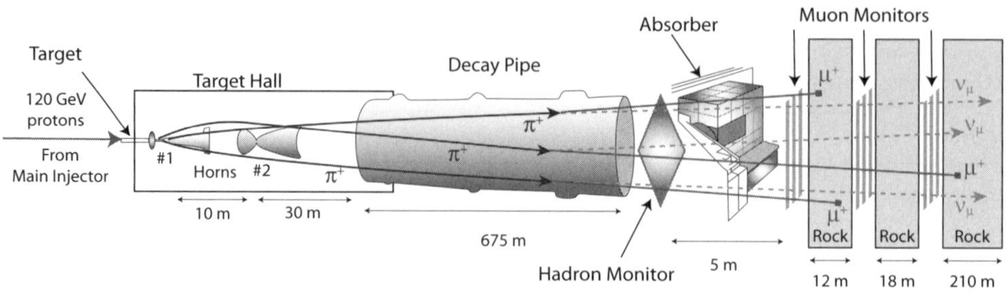

Figure 2.9. The basic elements employed in a conventional neutrino beam. From [175].

detector ideal for detecting ν_τ CC interactions. The metal layers provide a large target mass. A separate electronic tracker is used as a trigger for the neutrino interaction and to locate the region in the emulsion where the event occurs; the emulsion in the region of the interaction is removed and scanned to reconstruct the event.

An emulsion cloud chamber was used in the DONUT experiment that first detected tau neutrinos [21], and is part of the OPERA detector at CNGS (CERN to Gran Sasso); OPERA has a target mass of 1.8 kt.

2.6 Neutrino Beams

Conventional Neutrino Beam

The traditional way to create a neutrino beam is by first smashing protons into a fixed target (i.e., into particles at rest in the lab frame). This produces a large number of high-energy pions and kaons, which are then focused magnetically and allowed to decay in a long decay tunnel; the decay products of these particles also have high energy due to the Lorentz boost. At the end of the decay tunnel neutrinos are then selected by shielding that filters out strongly- and electromagnetically-interacting particles. Detectors that measure the secondary hadron and muon flux may also be employed (see figure 2.9). Because the parent pion production mechanism is complicated, a neutrino detector beyond the shielding but still near the source is often used to monitor the normalization of the neutrino flux.

If the decaying particle is a π^+, the primary decay is $\pi^+ \to \mu^+\nu_\mu$; for antineutrinos, the decay $\pi^- \to \mu^-\bar{\nu}_\mu$ is used. In the rest frame of the pion the decay is isotropic and the neutrino has energy $E_\nu^* = (m_\pi^2 - m_\mu^2)/(2m_\pi) \approx 30\,\text{MeV}$. If α is the angle of the neutrino from the beam axis (pion direction) and $\gamma = E_\pi/m_\pi$ is the Lorentz boost factor of the pion, then the neutrino energy in the lab frame is

$$E_\nu = \frac{E_\nu^*}{\gamma}\frac{1}{1-\beta\cos\alpha} \approx \frac{2\gamma E_\nu^*}{1+\gamma^2\alpha^2}, \tag{2.45}$$

where the approximation is for small α, i.e., close to the beam axis, and for large γ.

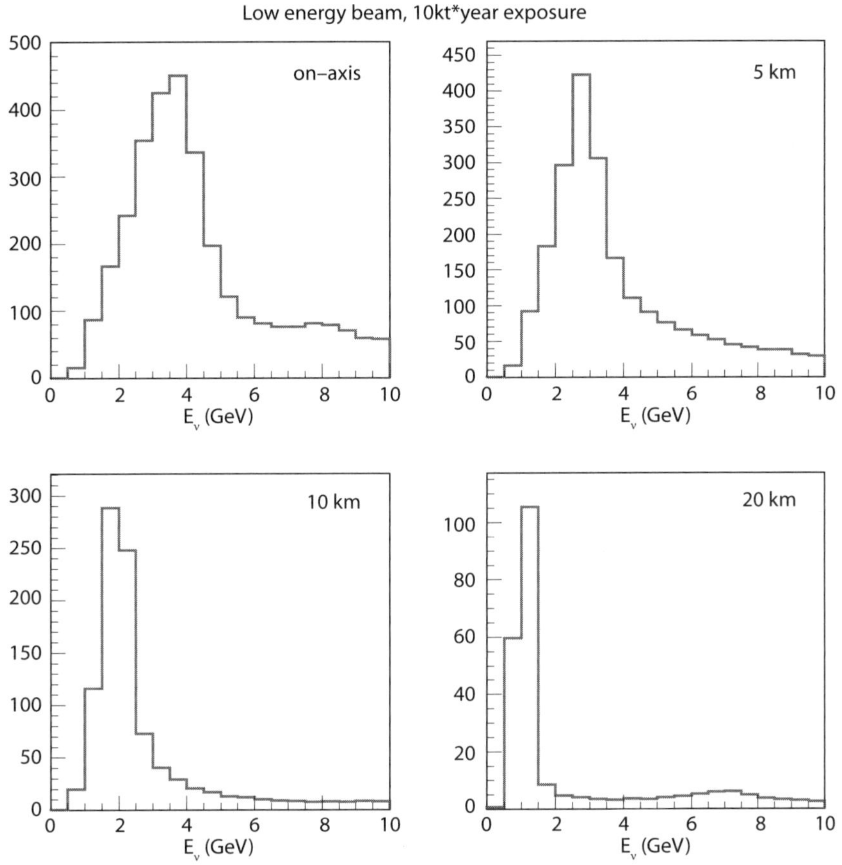

Figure 2.10. Neutrino spectrum on and off axis for the low-energy NuMI (Neutrino Main Injector at Fermilab) neutrino beam. The four spectra are for detectors 730 km from the beam source and on axis, 5 km (0.4°) off axis, 10 km (0.8°) off axis, and 20 km (1.6°) off axis. From [90].

The neutrino flux in the lab frame is [90]

$$\Phi \equiv \frac{dN_\nu}{dAdt} = \frac{n_0}{4\pi L^2 \gamma^2 (1 - \beta \cos\alpha)^2} \approx \frac{n_0}{4\pi L^2} \left(\frac{2\gamma}{1 + \gamma^2 \alpha^2} \right)^2, \tag{2.46}$$

where n_0 is the number of neutrinos from π decays per unit time, dA is the differential area at the detector and L is the distance from the pion decay to the detector. On the beam axis the neutrino energy is boosted by a factor 2γ and the flux by a factor $4\gamma^2$ in the lab frame.

For a monoenergetic meson source, the neutrino energy is also monoenergetic at a given angle from the beam axis. In practice the pions have some energy spread, which results in an energy spread for the neutrino beam (although the pions can be energy-selected to some extent to narrow their range of energies). From equation 2.45 one can see that for an off-axis beam, i.e., $\alpha \neq 0$, higher- (lower-) energy pions have a larger (smaller) energy-suppression factor for a given off-axis

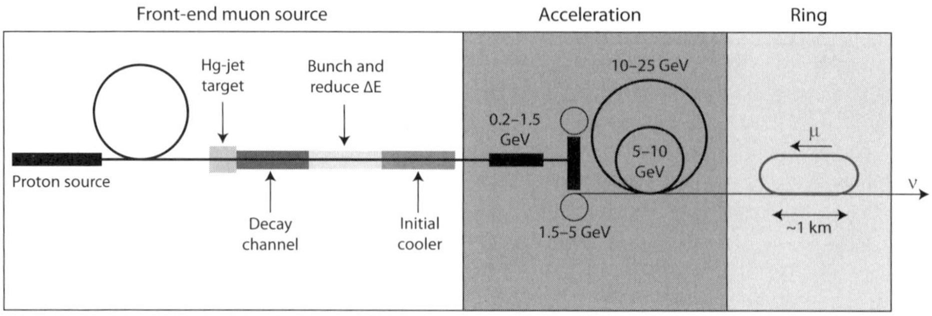

Figure 2.11. Sample schematic of a neutrino factory. First target material is placed in a high-intensity proton beam. Proton-nucleus interactions then lead to the creation of pions, which subsequently decay to muons. Next the muons are phase-rotated to a smaller energy spread, and cooled to a smaller transverse momentum spread. Then they are accelerated by a combination of linear accelerators (linacs), recirculating linear accelerators (RLAs), and fixed-field alternating gradient (FFAG) accelerators. Finally the muons are injected into a muon storage ring with a race-track design, and the muons decaying in the long straight sections provide a highly-collimated neutrino beam. From [176].

angle, so that the energy distribution of the neutrinos is narrower than it is on the beam axis. This effect is more pronounced as the off-axis angle increases (see figure 2.10). Furthermore, the flux of lower-energy neutrinos does not fall off as fast with angle as for higher-energy neutrinos, so that off axis there is a contribution to lower-energy neutrinos both from higher- and lower-energy pions. Therefore placing a detector off the beam axis can be an effective method to produce a narrower neutrino energy spread for lower-energy neutrinos without a large loss in intensity. This technique is used in the T2K [83] and NOνA [84] experiments.

Conventional neutrino beams have an intrinsic contamination of electron neutrinos due to K and μ decays, typically at a little below the 1% level. In principle, the decay $\pi^+ \to e^+\nu_e$ also contributes to this contamination, but the branching ratio for this decay is only of order 10^{-4}.

Neutrino Factory

In a neutrino factory (NuFact) [99], high-energy muons are accumulated in a storage ring, where they eventually decay. If the storage ring has one or more long straight sections, the muon decays to neutrinos in those sections will give a collimated neutrino beam (see figure 2.11). A decaying μ^+ (μ^-) in the ring yields both $\bar{\nu}_\mu$ and ν_e (ν_μ and $\bar{\nu}_e$). Because muon decay is three-body, the neutrino does not have a unique energy; in the muon rest frame the distribution of $\bar{\nu}_\mu$ (ν_μ) from the decay of unpolarized μ^+ (μ^-) is

$$\frac{d^2 N_{\nu_\mu}}{dx\, dt} = 2x^2(3 - 2x)n_0\,, \tag{2.47}$$

where $x \equiv 2E_\nu^*/m_\mu$, E_ν^* is the neutrino energy in the muon rest frame, and n_0 is the number of decaying muons per unit time. In the lab frame on the beam axis, $E_\nu = 2\gamma E_\nu^* = 2E_\mu E_\nu^*/m_\mu$, so $x = E_\nu/E_\mu$. The corresponding expression describing

TABLE 2.4
Signals for oscillation channels assuming a decaying μ^- in a neutrino factory.

Channel	*Detect*	*Nomenclature*
$\nu_\mu \to \nu_\mu$	μ^-	right–sign μ survival
$\nu_\mu \to \nu_e$	e^-	right–sign e appearance
$\nu_\mu \to \nu_\tau$	τ^-	right–sign τ appearance
$\bar{\nu}_e \to \bar{\nu}_e$	e^+	wrong–sign e survival
$\bar{\nu}_e \to \bar{\nu}_\mu$	μ^+	wrong–sign μ appearance
$\bar{\nu}_e \to \bar{\nu}_\tau$	τ^+	wrong–sign τ appearance

the distribution of ν_e ($\bar{\nu}_e$) is

$$\frac{d^2 N_{\nu_e}}{dx\, dt} = 12x^2(1-x)n_0 \,. \tag{2.48}$$

For highly relativistic muons, the differential flux along the beam direction is

$$\frac{d\Phi_\nu}{dE_\nu} = \frac{\gamma^2}{\pi L^2} \frac{d^2 N_\nu}{dx\, dt}\,, \tag{2.49}$$

where $\gamma = E_\mu / m_\mu$.

As with conventional neutrino beams, the flux in a neutrino factory grows quadratically with the decaying particle energy. Unlike neutrino beams from π and K decays, neutrinos from a neutrino factory have a broad spread of energies for monoenergetic muons, from 0 to E_μ, due to the three-body decay. For ν_μ (ν_e) the peak in the spectrum occurs at E_μ ($2E_\mu/3$) and the average neutrino energy is 0.7 E_μ ($0.6 E_\mu$).

In a sense, the beam in a neutrino factory is maximally contaminated; a decaying μ^- in the ring yields both ν_μ and $\bar{\nu}_e$. Detection of the charge of the final state lepton in a charged-current event allows one to determine the initial neutrino flavor. Table 2.4 shows the six oscillation channels possible with stored μ^-; the six charge-conjugate channels can be tested using μ^+ decays.

Beta Beam

A beta beam [98] is a beam of pure electron neutrinos (or antineutrinos) that are produced in the decay of radioactive ions circulating in a storage ring. Since beta decay is typically three-body, there is a broad spectrum like a muon-based neutrino factory. A nuclear decay has a different energy spectrum in the parent particle rest frame, given by

$$\frac{dN_\nu}{dE} \propto (Q - E_\nu)\sqrt{(Q - E_\nu)^2 - m_e^2}\, E_\nu^2\, F(\pm Z, Q - E_\nu), \tag{2.50}$$

where Q is the available energy in the decay and F is the Fermi function that accounts for the modification of the spectrum due to the Coulomb interaction between the nucleus and the electron. The typical average energy for the neutrino is about $0.55Q$. Boosting to the lab frame occurs as in a conventional neutrino beam or a neutrino factory. Because beta beams are formed from nuclear beta decay, which involves only electrons, there is no intrinsic flavor contamination.

A novel twist on the beta beam concept is possible for some high-Z nuclei that have fast electron capture decays, such as $e^- + {}^{150}\mathrm{Dy} \rightarrow {}^{150}\mathrm{Tb} + \nu_e$. Since the decay is two-body, the neutrino is monoenergetic in the decaying nucleus rest frame. If the nuclei are accelerated to high energy and accumulated in a storage ring, a high-energy monoenergetic ν_e beam will result [177]. The energy and angular distributions in the lab frame are given by equations 2.45 and 2.46.

✵ 3 ✵

Neutrino Mixing and Oscillations

3.1 Vacuum Oscillations

The dramatic increase in our knowledge of neutrino properties has come from observational evidence of neutrino oscillations. These neutrino flavor changes require that the neutrino flavor states, ν_α, are not the same as the neutrino mass eigenstates, ν_i. The eigenstates are related by a unitary matrix V [18],

$$\nu_\alpha = \sum V_{\alpha i}^* \nu_i. \tag{3.1}$$

V is often denoted as V_{MNS}, where MNS represents the authors of [18].[1] For 3 neutrinos, the mixing matrix V is specified by three rotation angles θ_{23}, θ_{13}, θ_{12} $(0 \le \theta_i \le \pi/2)$ and three CP-violating phases δ, ϕ_2 and ϕ_3 $(0 \le \delta, \phi_i \le 2\pi)$. V can be conveniently written as the matrix product

$$V = \begin{bmatrix} 1 & 0 & 0 \\ 0 & c_{23} & s_{23} \\ 0 & -s_{23} & c_{23} \end{bmatrix} \begin{bmatrix} c_{13} & 0 & s_{13}e^{-i\delta} \\ 0 & 1 & 0 \\ -s_{13}e^{i\delta} & 0 & c_{13} \end{bmatrix} \begin{bmatrix} c_{12} & s_{12} & 0 \\ -s_{12} & c_{12} & 0 \\ 0 & 0 & 1 \end{bmatrix} \begin{bmatrix} 1 & 0 & 0 \\ 0 & e^{i\phi_2/2} & 0 \\ 0 & 0 & e^{i\phi_3/2} \end{bmatrix} \tag{3.2}$$

where c_{jk} denotes $\cos\theta_{jk}$ and s_{jk} denotes $\sin\theta_{jk}$. The angle θ_{23} governs the oscillations of atmospheric neutrinos, the angle θ_{12} describes solar neutrino oscillations, and the angle θ_{13} has been measured by reactor neutrino experiments at short distances ($L \simeq 1$ km). The oscillation probabilities are independent of the Majorana phases ϕ_2 and ϕ_3, and depend only on the Dirac phase δ. Setting $\phi_2 = \phi_3 = 0$ we

[1] Sometimes it is denoted as V_{PMNS} or V_{MNSP} to acknowledge the contributions of Pontecorvo.

obtain the standard form of the mixing matrix for oscillations

$$V = \begin{bmatrix} c_{13}c_{12} & c_{13}s_{12} & s_{13}e^{-i\delta} \\ -s_{12}c_{23} - c_{12}s_{23}s_{13}e^{i\delta} & c_{12}c_{23} - s_{12}s_{23}s_{13}e^{i\delta} & s_{23}c_{13} \\ s_{12}s_{23} - c_{12}c_{23}s_{13}e^{i\delta} & -c_{12}s_{23} - s_{12}c_{23}s_{13}e^{i\delta} & c_{23}c_{13} \end{bmatrix}. \tag{3.3}$$

The propagation of the neutrino wavefunction is determined by the Hamiltonian via $i\, d\Psi/dt = H\Psi$. In vacuum for highly relativistic neutrinos $H \approx p + m^2/2E$; after removing the common phase e^{-ipt} (which has no effect on the physical observables), $i\, d\psi_j/dt = (m_j^2/2E)\psi_j$ in the mass eigenstate basis.[2] For a neutrino initially created in the charged-current eigenstate ν_α, the probability for oscillating into ν_β is

$$P(\nu_\alpha \to \nu_\beta) = \left| \sum_j V_{\beta j} e^{-\frac{i m_j^2 L}{2E_\nu}} V_{\alpha j}^* \right|^2 = \sum_{j,k} V_{\alpha j}^* V_{\beta j} V_{\alpha k} V_{\beta k}^* e^{-\frac{i \delta m_{jk}^2 L}{2E_\nu}}, \tag{3.4}$$

where the m_j are the neutrino eigenmasses and $\delta m_{jk}^2 = m_j^2 - m_k^2$. The oscillation probabilities depend only on differences of squared neutrino masses. Defining the oscillation arguments for the atmospheric and solar phenomena as

$$\Delta_{31} \equiv \frac{\delta m_{31}^2 L}{4E_\nu}, \qquad \Delta_{21} \equiv \frac{\delta m_{21}^2 L}{4E_\nu}, \tag{3.5}$$

respectively, where

$$\delta m_{31}^2 = m_3^2 - m_1^2, \qquad \delta m_{21}^2 = m_2^2 - m_1^2, \tag{3.6}$$

the vacuum oscillation probabilities are then [182]

$$\begin{aligned} P(\nu_e \to \nu_e) = 1 &- \sin^2 2\theta_{13} \sin^2 \Delta_{31} - (c_{13}^4 \sin^2 2\theta_{12} + s_{12}^2 \sin^2 2\theta_{13}) \sin^2 \Delta_{21} \\ &+ s_{12}^2 \sin^2 2\theta_{13} \left(\frac{1}{2} \sin 2\Delta_{21} \sin 2\Delta_{31} + 2 \sin^2 \Delta_{31} \sin^2 \Delta_{21} \right), \end{aligned} \tag{3.7}$$

$$\begin{aligned} P(\nu_e \to \nu_\mu) = {} & s_{23}^2 \sin^2 2\theta_{13} \sin^2 \Delta_{31} + 4J (\sin 2\Delta_{21} \sin^2 \Delta_{31} - \sin 2\Delta_{31} \sin^2 \Delta_{21}) \\ & -(s_{23}^2 s_{12}^2 \sin^2 2\theta_{13} - 4K) \left[\frac{1}{2} \sin 2\Delta_{21} \sin 2\Delta_{31} + 2 \sin^2 \Delta_{21} \sin^2 \Delta_{31} \right] \\ & + [c_{13}^2 (c_{23}^2 - s_{13}^2 s_{23}^2) \sin^2 2\theta_{12} + s_{23}^2 s_{12}^2 \sin^2 2\theta_{13} - 8K s_{12}^2] \sin^2 \Delta_{21}, \end{aligned} \tag{3.8}$$

[2] This treatment assumes that the different mass eigenstates have equal momentum, which is technically incorrect. A more careful analysis using neutrino wave packets accounts for differing neutrino momenta; see [178] for a recent discussion. Alternative derivations of the oscillation formulas can be found with equal momentum [179] and equal energy [180] assumptions, or describing the neutrino as a state entangled with its production partner [181]. Under the ordinary conditions found in all oscillation experiments conducted to date, all of these treatments give the same oscillation formulas; see section 3.6 for the conditions under which the oscillation formulas deviate from the standard results shown here.

$$\begin{aligned}P(\nu_\mu \to \nu_\mu) = 1 &- (c_{13}^4 \sin^2 2\theta_{23} + s_{23}^2 \sin^2 2\theta_{13}) \sin^2 \Delta_{31} \\ &+[c_{13}^2(c_{12}^2 - s_{13}^2 s_{12}^2) \sin^2 2\theta_{23} + s_{12}^2 s_{23}^2 \sin^2 2\theta_{13} - 8K s_{23}^2] \\ &\times \left[\frac{1}{2} \sin 2\Delta_{21} \sin 2\Delta_{31} + 2 \sin^2 \Delta_{21} \sin^2 \Delta_{31}\right] \\ &-[\sin^2 2\theta_{12}(c_{23}^2 - s_{13}^2 s_{23}^2)^2 + s_{13}^2 \sin^2 2\theta_{23}(1 - c_\delta^2 \sin^2 2\theta_{12}) \\ &+2s_{13} \sin 2\theta_{12} \cos 2\theta_{12} \sin 2\theta_{23} \cos 2\theta_{23} c_\delta - 16K s_{23}^2 s_{12}^2 \\ &+ \sin^2 2\theta_{23} c_{13}^2 (c_{12}^2 - s_{13}^2 s_{12}^2) + s_{12}^2 s_{23}^2 \sin^2 2\theta_{13}] \sin^2 \Delta_{21}, \end{aligned} \tag{3.9}$$

$$\begin{aligned}P(\nu_\mu \to \nu_\tau) = c_{13}^4 &\sin^2 2\theta_{23} \sin^2 \Delta_{31} + 4J (\sin 2\Delta_{21} \sin^2 \Delta_{31} - \sin 2\Delta_{31} \sin^2 \Delta_{21}) \\ &-[c_{13}^2 \sin^2 2\theta_{23}(c_{12}^2 - s_{13}^2 s_{12}^2) + 4K \cos 2\theta_{23}] \\ &\times \left[\frac{1}{2} \sin 2\Delta_{21} \sin 2\Delta_{31} + 2 \sin^2 \Delta_{21} \sin^2 \Delta_{31}\right] \\ &+[\sin^2 2\theta_{23}(c_{12}^2 - s_{13}^2 s_{12}^2)^2 + s_{13}^2 \sin^2 2\theta_{12}(1 - \sin^2 2\theta_{23} c_\delta^2) + 4K \cos 2\theta_{23} \\ &+s_{13} \sin 2\theta_{12} \cos 2\theta_{12} \sin 2\theta_{23} \cos 2\theta_{23}(1 + s_{13}^2) c_\delta] \sin^2 \Delta_{21}, \end{aligned} \tag{3.10}$$

where the CP-violating Jarlskog invariant J [183] is

$$J = \frac{1}{8} c_{13} \sin 2\theta_{13} \sin 2\theta_{23} \sin 2\theta_{12} s_\delta\,, \tag{3.11}$$

and for convenience we have defined

$$K = \frac{1}{8} c_{13} \sin 2\theta_{13} \sin 2\theta_{23} \sin 2\theta_{12} c_\delta\,. \tag{3.12}$$

Oscillation probabilities for other neutrino channels may be obtained by probability conservation, i.e., $\sum_\alpha P(\nu_\alpha \to \nu_\beta) = \sum_\beta P(\nu_\alpha \to \nu_\beta) = 1$. Probabilities for antineutrino channels are obtained by CP conjugation, under which δ is replaced by $-\delta$ in the corresponding neutrino formulas. If CP is conserved, then $P(\bar\nu_\alpha \to \bar\nu_\beta) = P(\nu_\alpha \to \nu_\beta)$. Also, since CPT is conserved for ordinary neutrino oscillations, the antineutrino probabilities are given by $P(\bar\nu_\alpha \to \bar\nu_\beta) = P(\nu_\beta \to \nu_\alpha)$, and $P(\nu_\alpha \to \nu_\beta)$ is obtained from $P(\nu_\beta \to \nu_\alpha)$ by replacing δ by $-\delta$.

For oscillations of atmospheric and long-baseline neutrinos, the oscillation argument Δ_{31} is dominant, and the vacuum oscillation probabilities in the leading oscillation approximation (where only the δm_{31}^2 oscillations are appreciable) are

$$P(\nu_e \to \nu_e) \simeq 1 - \sin^2 2\theta_{13} \sin^2 \Delta_{31} \tag{3.13}$$

$$P(\nu_e \to \nu_\mu) \simeq s_{23}^2 \sin^2 2\theta_{13} \sin^2 \Delta_{31} \tag{3.14}$$

$$P(\nu_\mu \to \nu_\mu) \simeq 1 - (c_{13}^4 \sin^2 2\theta_{23} + s_{23}^2 \sin^2 2\theta_{13}) \sin^2 \Delta_{31} \tag{3.15}$$

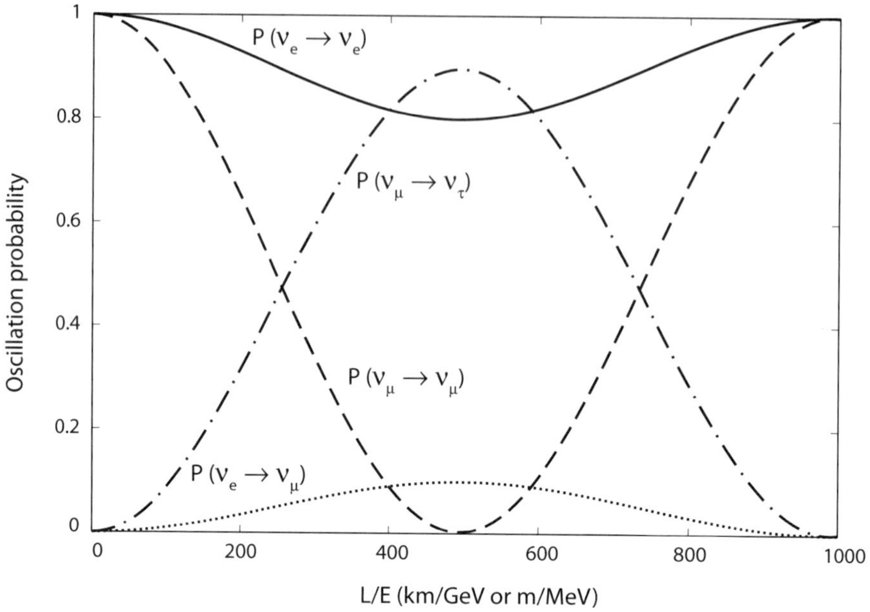

Figure 3.1. Probabilities for the leading oscillation of $\nu_e \to \nu_e$ (solid curve), $\nu_\mu \to \nu_\mu$ (dashed), $\nu_e \to \nu_\mu$ (dotted), and $\nu_\mu \to \nu_\tau$ (dash-dotted) for the oscillation parameters $\sin^2 2\theta_{23} = 1$, $\delta m^2_{31} = 2.5 \times 10^{-3}$ eV^2, and $\sin^2 2\theta_{13} = 0.2$.

$$P(\nu_\mu \to \nu_\tau) \simeq c_{13}^4 \sin^2 2\theta_{23} \sin^2 \Delta_{31}. \tag{3.16}$$

Figure 3.1 shows representative oscillation probabilities for the oscillation channels in equations 3.13–3.16.

For vacuum oscillations of solar or reactor neutrinos, where $|\Delta_{31}| \gg 1$, the terms involving Δ_{31} approach their average values over a complete cycle and

$$P(\nu_e \to \nu_e) \simeq 1 - \frac{1}{2} \sin^2 2\theta_{13} - c_{13}^4 \sin^2 2\theta_{12} \sin^2 \Delta_{21}\,. \tag{3.17}$$

Note that in the limit $\theta_{13} \to 0$, $\nu_\mu \to \nu_\tau$ oscillations of atmospheric neutrinos and $\nu_e \to \nu_e$ oscillations of solar neutrinos completely decouple, i.e., they are determined by independent parameters, and each has the form of two-neutrino oscillations. Also, the ν_μ survival probability is insensitive to the quadrant of θ_{23} and the sign of δm^2_{31} in the leading oscillation in the $\theta_{13} \to 0$ limit.

3.2 Matter Effects on Oscillations

The scattering of ν_e on electrons in matter can modify the vacuum oscillation probabilities [35,36,184]. For the two-neutrino case, with mixing angle θ and mass-squared difference $m_2^2 - m_1^2 \equiv \delta m^2$, the equation for two-neutrino propagation in matter is

$$i\frac{d}{dt}\begin{pmatrix}\nu_e\\ \nu_\mu\end{pmatrix} = \frac{1}{4E}\begin{pmatrix}4\sqrt{2}G_F E N_e & \delta m^2 \sin 2\theta\\ \delta m^2 \sin 2\theta & 2\delta m^2 \cos 2\theta\end{pmatrix}\begin{pmatrix}\nu_e\\ \nu_\mu\end{pmatrix}, \tag{3.18}$$

where N_e is the electron number density (the product of the electron fraction Y_e and matter density ρ) and the common phase $m_1^2 \cos^2\theta + m_2^2 \sin^2\theta$ has been removed. For antineutrinos the matter term (proportional to N_e) changes sign.

Let the angle θ_m represent the mixing between the flavor eigenstates ν_α and the instantaneous eigenstates in matter ν_{im}:

$$\begin{pmatrix}\nu_{1m}\\ \nu_{2m}\end{pmatrix} = \begin{pmatrix}\cos\theta_m & -\sin\theta_m\\ \sin\theta_m & \cos\theta_m\end{pmatrix}\begin{pmatrix}\nu_e\\ \nu_\mu\end{pmatrix}. \tag{3.19}$$

Then the neutrino propagation in the instantaneous eigenstate basis is given by

$$i\frac{d}{dt}\begin{pmatrix}\nu_{1m}\\ \nu_{2m}\end{pmatrix} = \begin{pmatrix}0 & -id\theta_m/dt\\ id\theta_m/dt & \delta m_m^2/2E\end{pmatrix}\begin{pmatrix}\nu_{1m}\\ \nu_{2m}\end{pmatrix}, \tag{3.20}$$

where a common phase has again been removed and [36]

$$\delta m_m^2 = \delta m^2\sqrt{\left(\frac{A}{\delta m^2} - \cos 2\theta\right)^2 + \sin^2 2\theta}. \tag{3.21}$$

For matter of constant electron number density, $d\theta_m/dt = 0$ and the oscillation probability amplitude $\sin^2 2\theta_m$ in matter is [36]

$$\sin^2 2\theta_m = \frac{\sin^2 2\theta}{\left(\frac{A}{\delta m^2} - \cos 2\theta\right)^2 + \sin^2 2\theta}, \tag{3.22}$$

where

$$A = 2\sqrt{2}\, G_F\, N_e\, E_\nu = 1.54 \times 10^{-4}\ \mathrm{eV}^2\ Y_e\, \rho(\mathrm{g/cm}^3)\, E_\nu(\mathrm{GeV}). \tag{3.23}$$

The oscillation amplitude for neutrinos in matter is enhanced if $\delta m^2 \cos 2\theta > 0$, and a resonance occurs (i.e., the amplitude reaches its maximal value of unity) at the critical density $N_e^c = \delta m^2 \cos 2\theta/(2\sqrt{2}\, G_F\, E_\nu)$ (see figure 3.2). For antineutrinos, $A \to -A$ in equation 3.22, and the oscillation amplitude in matter is enhanced if $\delta m^2 \cos 2\theta < 0$. For N_e much larger than the critical density ($A \gg \delta m^2$), the oscillation amplitude in matter is strongly suppressed for both neutrinos and antineutrinos. The oscillation wavelength, which is inversely proportional to δm_m^2, is also changed in matter.

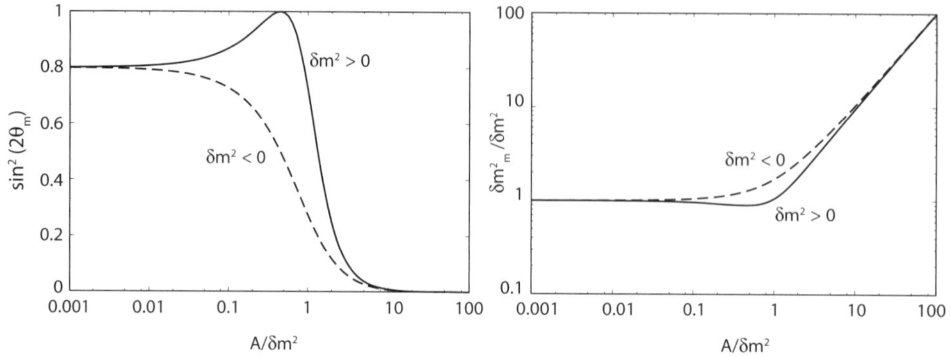

Figure 3.2. Effective oscillation amplitude, $\sin^2 2\theta_m$, and $\delta m_m^2/\delta m^2$ in matter versus $A/\delta m^2$ for two-neutrino oscillations involving ν_e with vacuum mixing $\sin^2 2\theta = 0.8$. The maximum amplitude ($\theta_m = 45^\circ$) occurs for the critical density, where $A/\delta m^2 = \cos 2\theta$. For antineutrinos, the $\delta m^2 > 0$ and $\delta m^2 < 0$ curves are exchanged.

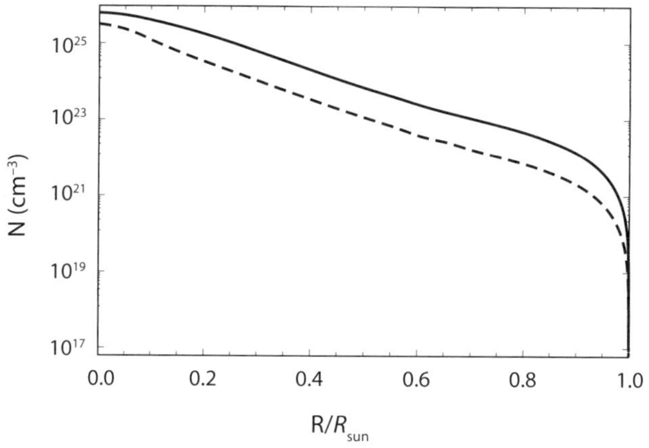

Figure 3.3. Electron density (solid line) and neutron density (dashed line) in the sun. From [185].

3.3 Solar Neutrino Oscillations

For matter of varying density, such as for neutrinos propagating through the sun, the instantaneous eigenstates change as the neutrinos propagate. If the electron density is N_e^0 when the neutrino is created, the initial value of θ_m is given by

$$\cos 2\theta_m^0 = -\frac{\frac{A^0}{\delta m^2} - \cos 2\theta}{\sqrt{(\frac{A^0}{\delta m^2} - \cos 2\theta)^2 + \sin^2 2\theta}}, \tag{3.24}$$

where $A^0 = 2\sqrt{2}\, G_F\, N_e^0\, E_\nu$. The electron density profile in the sun is shown in figure 3.3. A neutrino originally created as a ν_e can be expressed in terms of the lower and upper eigenstates as $\nu_e = \cos\theta_m^0\, \nu_{1m} + \sin\theta_m^0\, \nu_{2m}$. Once a solar neutrino reaches the vacuum, the lower eigenstate is $\nu_1 = \cos\theta\, \nu_e - \sin\theta\, \nu_\mu$ and the upper eigenstate is $\nu_2 = \sin\theta\, \nu_e + \cos\theta\, \nu_\mu$.

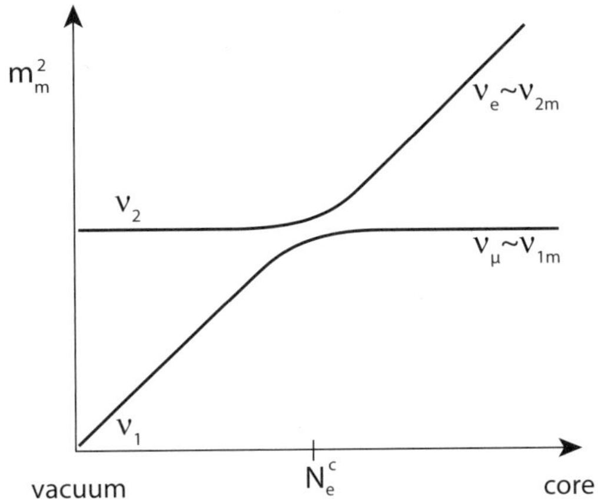

Figure 3.4. The level-crossing diagram for solar neutrinos showing the effective m^2 in matter versus electron number density.

A ν_e created far above the resonance density is predominantly in the upper eigenstate. If N_e changes slowly enough (adiabatic propagation), then $d\theta_m/dt \approx 0$ (i.e., small compared to the difference of the diagonal elements of the Hamiltonian) and the neutrino will remain in the upper eigenstate (see equation 3.20). Then if θ is small, it will be predominantly ν_μ once it reaches the vacuum [38,42] (see figure 3.4). For nonadiabatic propagation, if the probability of jumping from one eigenstate to another is P_x, then averaging over the oscillations gives [40]

$$\langle P(\nu_e \to \nu_e)\rangle = \frac{1}{2}\left[1 + (1 - 2P_x)\cos 2\theta \cos 2\theta_m^0\right], \tag{3.25}$$

where [186]

$$P_x = \frac{\exp\left(-\frac{\pi}{2}\gamma\,\mathrm{F}\right) - \exp\left(-\frac{\pi}{2}\gamma\,\frac{\mathrm{F}}{\sin^2\theta}\right)}{1 - \exp\left(-\frac{\pi}{2}\gamma\,\frac{\mathrm{F}}{\sin^2\theta}\right)}, \tag{3.26}$$

is the transition probability with $F = 1 - \tan^2\theta$ for the exponentially varying matter density in the sun, and [40,42]

$$\gamma = \frac{(\delta m^2)^2 \sin^2 2\theta}{4\sqrt{2} G_F E_\nu^2\, |dN_e/dL|_c}, \tag{3.27}$$

measures the degree of adiabaticity of the transition. In equation 3.27, $|dN_e/dL|_c$ is the density gradient at the critical density. This process is analogous to level crossings in atoms [187].

For an electron neutrino that undergoes a perfectly adiabatic transition ($\gamma \to \infty$, $P_x \to 0$), the oscillation probability is

$$\langle P(\nu_e \to \nu_e)\rangle = \frac{1}{2}\left[1 + \cos 2\theta \cos 2\theta_m^0\right] = \cos^2\theta \cos^2\theta_m^0 + \sin^2\theta \sin^2\theta_m^0. \tag{3.28}$$

If the the neutrino is created well above the resonance density ($\theta_m^0 \simeq \pi/2$) and undergoes an adiabatic transition, the oscillation probability is $\langle P(\nu_e \to \nu_e)\rangle = \sin^2\theta$.

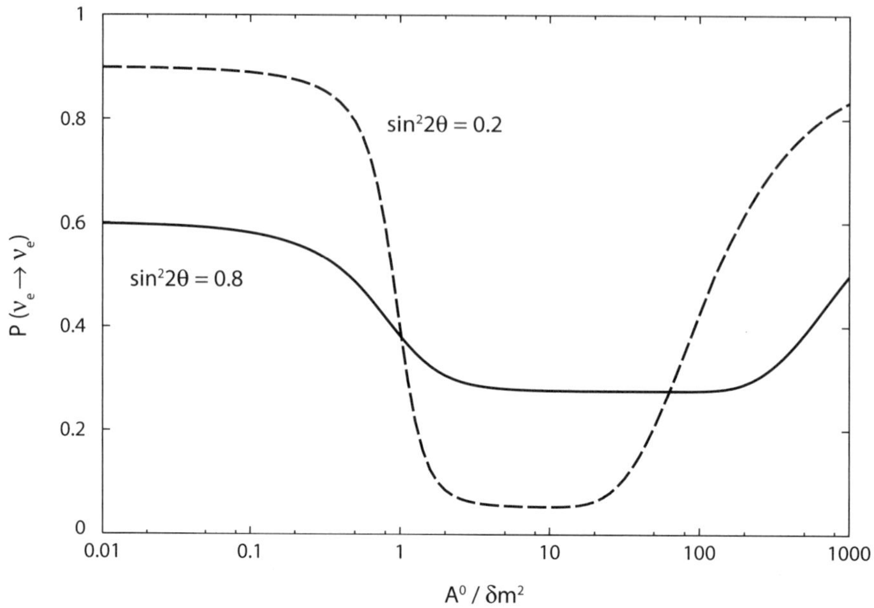

Figure 3.5. Survival probability for solar neutrinos versus $A^0/\delta m^2$ for oscillation amplitude 0.8 (solid curve) and 0.2 (dashed), where $A^0 = 2\sqrt{2}G_F N_e^0 E_\nu$ and $\delta m^2 = 8 \times 10^{-5}$ eV^2.

Thus a very large depletion of solar ν_e's is possible even for small vacuum mixing angles. This is known as the MSW effect, and was first studied numerically in [37]. In the extreme nonadiabatic limit ($\gamma \to 0$) [41], $P_x \to \cos^2\theta$ and the oscillation probability for a neutrino created well above resonance approaches $1 - \frac{1}{2}\sin^2 2\theta$, the expected value for two-neutrino vacuum oscillations averaged over the oscillations.

For neutrino energies below $\delta m^2 \cos 2\theta/(2\sqrt{2}G_F N_e^0)$, the initial density is below the critical density and the neutrino starts with a large fraction in the lower eigenstate. For much larger neutrino energies the transition becomes very nonadiabatic and the neutrino has a high probability of hopping from the upper eigenstate to the lower eigenstate. In either case, a large component of the neutrino ends up in the lower eigenstate, in which case the survival probability is greater than $\frac{1}{2}$ (see figure 3.5). For $N_e^0 = 100\, N_A/\text{cm}^3$ (the approximate number density at the center of the sun), neutrinos with energies $E_\nu \gtrsim 2\,\text{MeV}$ will start above resonance, assuming the best-fit oscillation parameters of the LMA solution ($\delta m^2_{21} = 8 \times 10^{-5}\,\text{eV}^2$ and $\sin^2 2\theta_{12} = 0.85$).

Exact formulas that include cases when the neutrino is created near resonance are presented in [188]. More discussions of exact formulas for the transition probability are given in [189]. A semi-classical treatment for an arbitrary density profile is given in [190]. Formulas for the MSW effect in the three-neutrino case are presented in [191]. For adiabatic propagation among n neutrino species, the oscillation probability for solar neutrinos is

$$\langle P(\nu_e \to \nu_e)\rangle = \sum_{j=1}^{n} |V_{ej}|^2 |(V_m^0)_{ej}|^2, \tag{3.29}$$

where V and V_m^0 are the mixing matrices that relate the flavor eigenstates to the mass eigenstates in vacuum and at the creation point in the sun, respectively, as in equation 3.1.

3.4 Long-baseline Oscillations through the Earth

Oscillations of long-baseline neutrinos are affected by electrons in the earth if the path length is an appreciable fraction of the earth's diameter. The full propagation equations for three neutrinos in matter are

$$i\frac{d\nu_\alpha}{dL} = \frac{1}{2E_\nu}\sum_\beta \left(A\delta_{\alpha e}\delta_{\beta e} + \sum_i V^*_{\beta i}\delta m^2_{i1} V_{\alpha i} \right) \nu_\beta. \tag{3.30}$$

A constant density approximation often provides a good representation of the neutrino propagation over long baselines through the earth. Since $\delta m^2_{21} \ll |\delta m^2_{31}|$ and θ_{13} is small, the probabilities can be expanded to second order in terms of the small parameters θ_{13} and $|\delta m^2_{21}/\delta m^2_{31}|$ [192, 193]. The following useful approximations for $\delta m^2_{31} > 0$ are obtained [85]:

$$P(\nu_\mu \to \nu_e) = x^2 f^2 + 2xyfg(\cos\delta\cos\Delta - \sin\delta\sin\Delta) + y^2 g^2, \tag{3.31}$$

$$P(\bar{\nu}_\mu \to \bar{\nu}_e) = x^2 \bar{f}^2 + 2xy\bar{f}g(\cos\delta\cos\Delta + \sin\delta\sin\Delta) + y^2 g^2, \tag{3.32}$$

where

$$x = \sin\theta_{23}\sin 2\theta_{13}, \tag{3.33}$$

$$y = \alpha\cos\theta_{23}\sin 2\theta_{12}, \tag{3.34}$$

$$f, \bar{f} = \sin\left[(1 \mp \hat{A})\Delta\right] / (1 \mp \hat{A}), \tag{3.35}$$

$$g = \sin(\hat{A}\Delta)/\hat{A}, \tag{3.36}$$

and

$$\Delta = |\Delta_{31}|, \qquad \hat{A} = |A/\delta m^2_{31}|, \qquad \alpha = |\delta m^2_{21}/\delta m^2_{31}|. \tag{3.37}$$

For $\delta m^2_{31} < 0$, the corresponding formulas are

$$P(\nu_\mu \to \nu_e) = x^2 \bar{f}^2 - 2xy\bar{f}g(\cos\delta\cos\Delta + \sin\delta\sin\Delta) + y^2 g^2, \tag{3.38}$$

$$P(\bar{\nu}_\mu \to \bar{\nu}_e) = x^2 f^2 - 2xyfg(\cos\delta\cos\Delta - \sin\delta\sin\Delta) + y^2 g^2. \tag{3.39}$$

Figure 3.6 shows the matter effect on $\nu_\mu \to \nu_e$ oscillations for a representative choice of parameters.

Oscillation probabilities for an initial ν_e and final ν_μ can be found by changing the sign of the $\sin\delta$ term in equations 3.31, 3.32, 3.38, and 3.39. These expansions

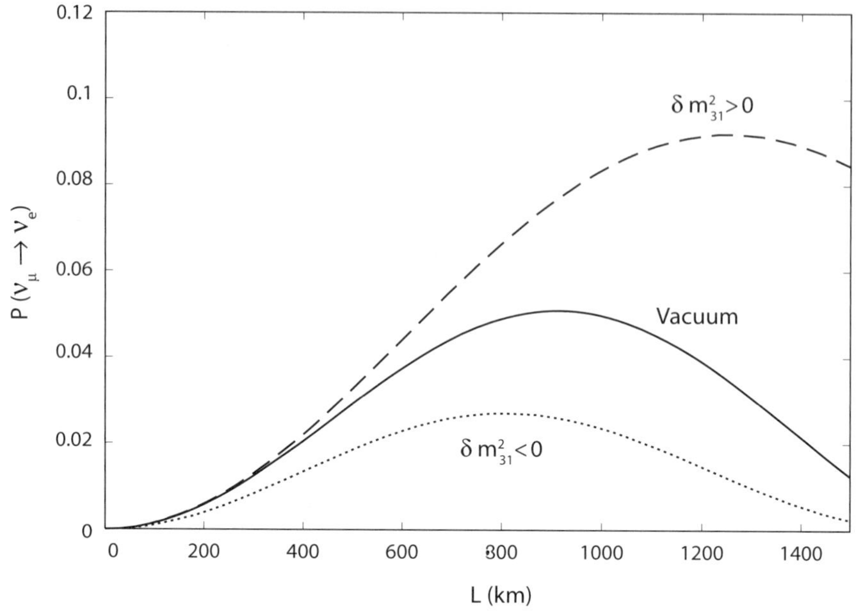

Figure 3.6. Oscillation probability for $\nu_\mu \rightarrow \nu_e$ in vacuum (solid curve), for $\delta m_{31}^2 > 0$ (dashed), and $\delta m_{31}^2 < 0$ (dotted), with $N_e^0 = 2.5\ N_A/\text{cm}^3$, $|\delta m_{31}^2| = 2.5 \times 10^{-3}\ \text{eV}^2$, $|\delta m_{21}^2| = 8 \times 10^{-5}\ \text{eV}^2$, $\sin^2 2\theta_{23} = 1$, $\sin^2 2\theta_{12} = 0.87$, $\sin^2 2\theta_{13} = 0.1$, and $\delta = 0$. For antineutrinos, oscillations in matter are suppressed (enhanced) relative to the vacuum for $\delta m_{31}^2 < 0$ ($\delta m_{31}^2 > 0$).

are nearly exact for distances less than 4000 km when $E_\nu \gtrsim 0.5$ GeV [85]. For more accurate results at longer distances, equation 3.30 may be integrated numerically over the density profile of the neutrino path.

Other approximate solutions can also be useful in certain situations. At relatively short distances where the matter effect is not as large, an expansion can be made in α and $A/\delta m_{31}^2$ for the constant density solution [194]. Relationships between the vacuum and matter oscillation parameters for three-neutrino oscillations are given in [195]. Exact results for the three-neutrino case with constant density are given in [36, 196]. Several properties of the general three-neutrino solution with a nonconstant density profile are discussed in [197]. Consequences of random density fluctuations are discussed in [198]; they are not expected to play an important role in most situations.

3.5 Matter Effects for Sterile Neutrinos

The evolution in equation 3.30 is modified if a sterile neutrino ν_s, is involved [199]:

$$A\delta_{\alpha e}\delta_{\beta e} \rightarrow 2\sqrt{2}G_F E_\nu \left[N_e\delta_{\alpha e}\delta_{\beta e} - \frac{1}{2}N_n(\delta_{\alpha\beta} - \delta_{\alpha s}\delta_{\beta s})\right], \qquad (3.40)$$

where N_n is the neutron number density. In two-neutrino oscillations between ν_e and a sterile neutrino, the electron number density N_e is changed to an effective

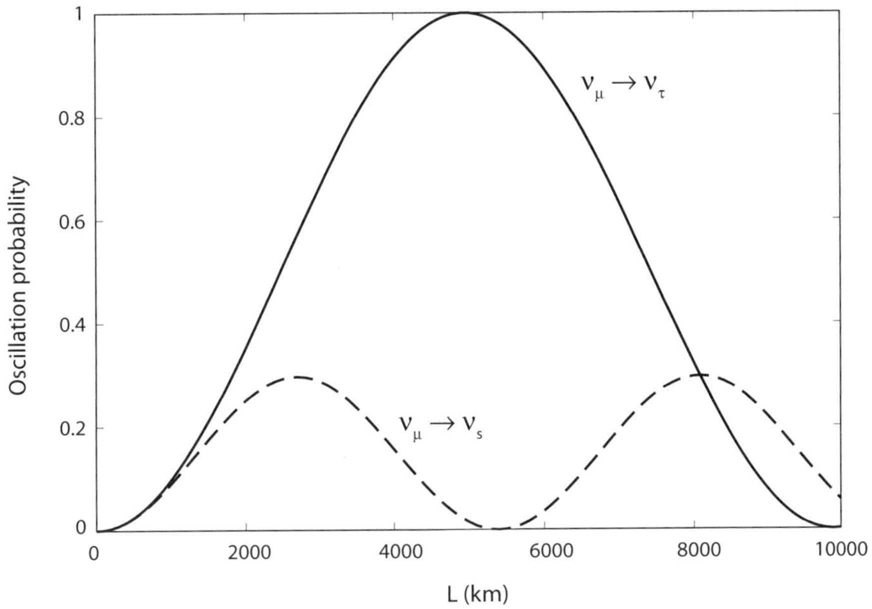

Figure 3.7. Two-neutrino oscillation probabilities versus distance for $\nu_\mu \to \nu_\tau$ (solid curve) and $\nu_\mu \to \nu_s$ (dashed) with $E_\nu = 10$ GeV, $N_n = 5\ N_A/\mathrm{cm}^3$, $\sin^2 2\theta_{23} = 1$ and $\delta m^2_{31} = 2.5 \times 10^{-3}$ eV^2. In the two-neutrino limit, $\nu_\mu \to \nu_\tau$ oscillations in matter are the same as in vacuum, $P(\bar{\nu}_\mu \to \bar{\nu}_\tau) = P(\nu_\mu \to \nu_\tau)$, and the $\nu_\mu \to \nu_s$ probabilities for $\delta m^2_{31} < 0$ and/or for antineutrinos are identical to these for maximal vacuum mixing.

number density $N_{eff} = N_e - \frac{1}{2}N_n$, which changes the critical density for a resonance in oscillations of solar neutrinos. Also, in two-neutrino oscillations between a ν_μ or ν_τ and a sterile neutrino, $N_{eff} = -\frac{1}{2}N_n$; consequently there can be substantial matter effects in $\nu_\mu \to \nu_s$ oscillations of atmospheric neutrinos (see figure 3.7).

For solar neutrino oscillations to sterile neutrinos, A^0 in equation 3.24 is given by $2\sqrt{2}G_F(N_e^0 - \frac{1}{2}N_n^0)E_\nu$. The neutron density in the sun is shown in figure 3.3.

3.6 Decoherence

The expression for the oscillation probabilities in equation 3.4 is derived assuming plane-wave neutrino states with fixed momentum. However, neutrino states are more accurately described by wave packets with momentum and spatial uncertainty; different mass eigenstates will propagate with different group velocities, and when the wavepackets of different mass eigenstates no longer overlap coherence is lost. The subject of neutrino decoherence has been extensively studied in the literature [200].

The widths of initial-state wave packets are determined by the size of the production region, σ_{xP}. Likewise, the widths of final-state wave-packets are determined by the size of the detection region, σ_{xD}. The degree of coherence depends on the size of both the production and detection processes, and, assuming Gaussian wave packets, the effective wave packet width is given by

$$\sigma_x^2 = \sigma_{xP}^2 + \sigma_{xD}^2. \tag{3.41}$$

In the relativistic limit the transition probability in a wave packet treatment is then [201]

$$P(\nu_\alpha \to \nu_\beta) \simeq \sum_{j,k} V_{\alpha j}^* V_{\beta j} V_{\alpha k} V_{\beta k}^* \exp\left[-2\pi i \frac{L}{L_{jk}^{osc}} - \left(\frac{L}{L_{jk}^{coh}}\right)^2\right], \tag{3.42}$$

where $L_{jk}^{osc} = 4\pi E/\delta m_{jk}^2$ are the usual vacuum oscillation lengths (defined by setting $\Delta_j = \pi$ in equation 3.5) and $L_{jk}^{coh} = 4\sqrt{2}E^2\sigma_x/\delta m_{jk}^2$ are the decoherence lengths.[3]

When $L \gg L_{jk}^{coh}$, such as for neutrinos from astrophysical sources, then only terms with $j = k$ in equation 3.42 are non-negligible and the neutrino state becomes an incoherent mixture of mass eigenstates, with

$$P(\nu_\alpha \to \nu_\beta) = \sum_j |V_{\alpha j}|^2 |V_{\beta j}|^2. \tag{3.43}$$

For example, for a 1 MeV neutrino with $\sigma_x = 1$ Å, decoherence effects will be observed for $\delta m^2 \gtrsim 4 \times 10^{-9}\,\text{eV}^2$. Even if the baseline L is much longer than the coherence length due to the production process, which occurs if $L \gg 4\sqrt{2}E^2\sigma_{xP}/\delta m^2$, coherence can be recovered if the detection region is not sufficiently localized, i.e., $L \ll 4\sqrt{2}E^2\sigma_{xD}/\delta m^2$ [202].

So far there is no evidence for decoherence effects in experiments with a positive oscillation signal: for terrestrial experiments the baseline is generally too short. It has been suggested that decoherence might be observed in a reactor neutrino experiment with a liquid scintillator detector a long distance from the source [203]. As pointed out in [203], the observation of oscillations in the KamLAND experiment puts a lower bound on σ_x for reactor neutrinos of about 10^{-3} Å.

Another aspect of neutrino oscillations not accounted for in the standard treatment is the quantum entanglement of the neutrino with other decay products in the neutrino production process [204]. Energy-momentum conservation implies that the momentum of a neutrino is correlated with the momentum of other daughter particles and in principle an Einstein-Podolski-Rosen (EPR) paradox [205] may result. However, it has been shown [206] that if the neutrino is created in a charged-current process and the coherence condition is satisfied at production, correlation effects are negligible. On the other hand, if the neutrino is created in a neutral-current process such as $Z \to \nu\bar{\nu}$, then the neutrino and antineutrino are entangled and EPR-like effects are possible [206].

[3] There is an additional term not included in equation 3.42 because of the delocalization of the source and/or detector. For example, if the oscillation wavelength is shorter than the size of the production region, only an incoherent mixture of mass eigenstates is observed; this is the case for MSW oscillations of solar neutrinos, where the exact production point for a given neutrino is not known.

4

Solar Neutrinos

4.1 Origin of Solar Neutrinos

Solar neutrinos are created by chains of fusion reactions in the sun. In the so-called *pp* chain, the primary reaction (which takes place with frequency 86%) is

$$p + p \to (^2\mathrm{H})^+ + e^+ + \nu_e\,, \tag{4.1}$$

where the neutrino can have kinetic energy in the range 0 to 0.42 MeV. A secondary branch includes the reaction (frequency 14%)

$$(^7\mathrm{Be})^+ + e^- \to\ ^7\mathrm{Li} + \nu_e\,, \tag{4.2}$$

where the neutrino has energy 0.86 MeV or 0.38 MeV, depending on whether the ^{7}Li is in the ground state (90% of the time) or an excited state (10%). A tertiary branch (frequency 0.11%) includes the reaction

$$(^8\mathrm{B})^+ \to\ ^8\mathrm{Be} + e^+ + \nu_e + \gamma\,, \tag{4.3}$$

where the neutrino has energy 0–14.06 MeV. Much rarer than the *pp* reaction (by a factor of about 400) is the *pep* reaction

$$p + e^- + p \to (^2\mathrm{H})^+ + \nu_e\,, \tag{4.4}$$

which is simply a crossed version of the *pp* reaction and has a monoenergetic neutrino with energy 1.44 MeV. The extremely rare *hep* reaction

$$^3\mathrm{He} + p \to\ ^4\mathrm{He} + e^+ + \nu_e\,, \tag{4.5}$$

emits a neutrino with energy in the range 0–18.8 MeV.

In the CNO cycle, neutrinos are created in the following reactions

$$(^{13}\mathrm{N})^+ \to\ ^{13}\mathrm{C} + e^+ + \nu_e\,, \tag{4.6}$$

TABLE 4.1
Calculations of solar neutrino fluxes.

Process	High metallicity	Low metallicity
pp	$(5.97 \pm 0.04) \times 10^{10}$	$(6.04 \pm 0.03) \times 10^{10}$
^{7}Be	$(5.07 \pm 0.30) \times 10^{9}$	$(4.55 \pm 0.27) \times 10^{9}$
pep	$(1.41 \pm 0.16) \times 10^{8}$	$(1.45 \pm 0.15) \times 10^{8}$
^{8}B	$(5.94 \pm 0.65) \times 10^{6}$	$(4.72 \pm 0.52) \times 10^{6}$
hep	$(7.90 \pm 1.19) \times 10^{3}$	$(8.22 \pm 1.23) \times 10^{3}$
^{13}N	$(2.88 \pm 0.43) \times 10^{8}$	$(1.89 \pm 0.26) \times 10^{8}$
^{15}O	$(2.15 \pm 0.35) \times 10^{8}$	$(1.34 \pm 0.21) \times 10^{8}$
^{17}F	$(5.82 \pm 1.05) \times 10^{6}$	$(3.25 \pm 0.50) \times 10^{6}$

Note: At the earth (in $cm^{-2}\ s^{-1}$) in two versions of the Standard Solar Model [208].

$$(^{15}\mathrm{O})^+ \rightarrow {}^{15}\mathrm{N} + e^+ + \nu_e\,, \tag{4.7}$$

$$(^{17}\mathrm{F})^+ \rightarrow {}^{17}\mathrm{O} + e^+ + \nu_e\,, \tag{4.8}$$

where the neutrinos have energies 0–2.22 MeV, 0–2.75 MeV, and 0–2.76 MeV, respectively.

The net effect of each of these chain reactions is

$$4p \rightarrow (^4\mathrm{He})^{++} + 2\nu_e + n\gamma\,, \tag{4.9}$$

where n depends on the particular reaction. The CNO cycle uses carbon, nitrogen, and oxygen as catalysts (none of these elements are created or destroyed in the process) and produces about 1.7% of the ^{4}He created in these fusion processes.

There have been many attempts to calculate solar neutrino fluxes based on these reactions [207]. Solar models are updated frequently as cross section values and other inputs to the model are improved; models developed by Bahcall and collaborators [208] are generally considered the most comprehensive and are called the Standard Solar Model (SSM). Presently there is some difficulty in reconciling helioseismological measurements with the observed heavy-element abundances (metallicity) in solar models. There appears to be a slight preference for models with high metallicity [209]. Recent best calculations of neutrino fluxes for both high and low metallicity are shown in table 4.1. Although model differences for the pp neutrino flux are small, they are much larger for ^{8}B and CNO neutrinos.

4.2 Solar Neutrino Experiments

Decades of study of neutrinos from the sun have convincingly established that neutrino oscillations are the cause of the deficits of 1/3 to 1/2 in the measured ν_e flux relative to the SSM. The water Cherenkov experiments of Super-K [31,32,210]

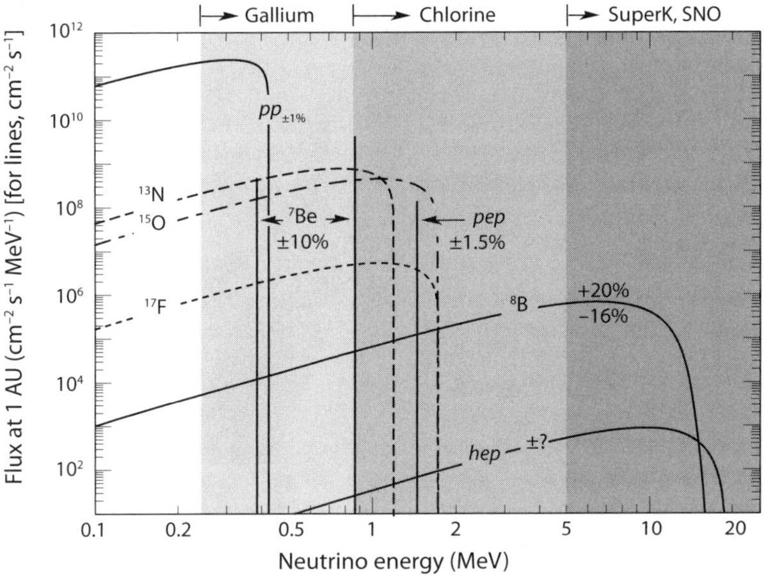

Figure 4.1. The neutrino flux predictions of the Standard Solar Model [208]. The energy thresholds for several solar neutrino experiments are also shown. From [79].

and SNO [33, 34, 48] measure the high-energy neutrinos ($E \gtrsim 5$ MeV) from the ^{8}B chain, the ^{37}Cl experiment [211] also detects the intermediate-energy neutrinos from ^{7}Be and *pep* line sources and from the CNO cycle, and the ^{71}Ga experiments, SAGE [27], GALLEX [28], and GNO [29], have dominant contributions from the *pp* neutrinos; see figure 4.1.

The early inferences that there was a deficiency of solar neutrinos depended on comparisons with the SSM predictions of the flux. With the SNO experiment, which directly measured the total active neutrino flux via neutral currents, the evidence for flavor conversion became robust. The SNO experiment utilized a heavy water target and measured the following processes:

$$\text{Charged Current (CC):} \quad \nu_e + d \to e^- + p + p \tag{4.10}$$

$$\text{Neutral Current (NC):} \quad \nu_x + d \to \nu_x + n + p \tag{4.11}$$

$$\text{Elastic Scattering (ES):} \quad \nu_x + e^- \to \nu_x + e^-. \tag{4.12}$$

The CC/NC ratio,

$$\frac{\text{CC}}{\text{NC}} = \frac{\text{flux}(\nu_e)}{\text{flux}(\nu_e + \nu_\mu + \nu_\tau)}, \tag{4.13}$$

established the oscillations of ν_e to ν_μ and ν_τ flavors. Only ν_e are produced in the sun; the ν_μ and ν_τ fluxes are a consequence of oscillations. Assuming an undistorted energy spectrum, the inferred fluxes for the three reactions were measured in SNO

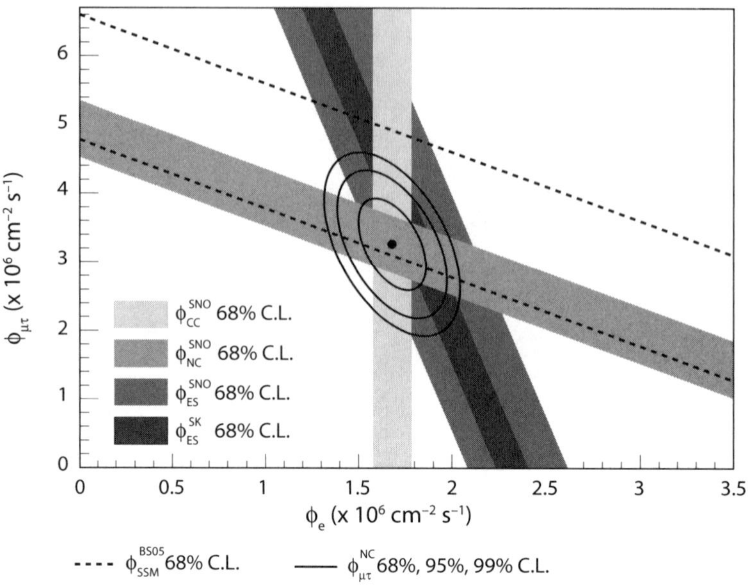

Figure 4.2. Measurement of solar neutrino fluxes $\phi_e = \phi(\nu_e)$ and $\phi_{\mu\tau} = \phi(\nu_\mu)+\phi(\nu_\tau)$ in SNO. The gray bands show the 1σ allowed region from the CC, ES, and NC measurements. Also shown are the best-fit point, the 1σ, 2σ, and 3σ combined fit to the fluxes (ellipses), the Super-K ES measurement [31] (black band), and the SSM prediction [208] for the total flux (dashed lines). Adapted from [48].

to be [34]

$$\phi_{CC} = 1.67^{+0.05}_{-0.04}(\text{stat.})^{+0.07}_{-0.08}(\text{syst.}) \times 10^6 \text{ cm}^{-2} \text{ s}^{-1} \tag{4.14}$$

$$\phi_{NC} = 5.54^{+0.33}_{-0.31}(\text{stat.})^{+0.36}_{-0.34}(\text{syst.}) \times 10^6 \text{ cm}^{-2} \text{ s}^{-1} \tag{4.15}$$

$$\phi_{ES} = 1.77^{+0.24}_{-0.21}(\text{stat.})^{+0.09}_{-0.10}(\text{syst.}) \times 10^6 \text{ cm}^{-2} \text{ s}^{-1}, \tag{4.16}$$

where the SSM flux is $(5.69 \pm 0.93) \times 10^6 \text{ cm}^{-2} \text{ s}^{-1}$ [208]. The measured CC flux, which consists of only ν_e, is about 5σ below the NC flux. Thus, the oscillations of ν_e to ν_μ and/or ν_τ is definitively established. Combining the results of the ES process, in which ν_μ and ν_τ cross sections are about 17% of the ν_e cross sections for ^{8}B neutrinos, and the NC process, in which all active neutrinos have the same cross section, the combined $\nu_\mu + \nu_\tau$ flux may be determined; see figure 4.2. The result,

$$\phi(\nu_\mu) + \phi(\nu_\tau) = 3.41^{+0.45}_{-0.45}(\text{stat.})^{+0.48}_{-0.45}(\text{syst.}) \times 10^6 \text{ cm}^{-2} \text{ s}^{-1}, \tag{4.17}$$

shows that $\phi(\nu_\mu) + \phi(\nu_\tau)$ is 5.3σ away from zero, a clear indication of neutrino oscillations. The constraints of the SNO CC are also shown in figure 4.2. Super-K made a more precise measurement of the elastic scattering flux, $\phi_{ES} = (2.36 \pm 0.07) \times 10^6 \text{ cm}^{-2} \text{ s}^{-1}$ [32, 210]. SNO subsequently extended their measurement to lower energies, with an electron kinetic energy threshold of 3.5 MeV [212]. This improved their statistics by 30% in the CC and ES channels, and by 70% in the NC channel. Including the lower-energy electrons, the NC flux is determined to be

$$\phi_{NC} = 5.140^{+0.160}_{-0.158}(\text{stat.})^{+0.132}_{-0.117}(\text{syst.}) \times 10^6 \text{ cm}^{-2} \text{ s}^{-1} . \tag{4.18}$$

The day and night energy spectra of charged-current events are potentially sensitive to matter effects on oscillations that occur when the neutrinos travel through the earth [213], though a statistically significant day-night asymmetry is yet to be established. The most precise measurement of the day minus night ν_e rate relative to the average rate was made by the first phase of Super-K, with value $(-2.1 \pm 2.0^{+1.3}_{-1.2})\%$ [47], compared to the MSW oscillation prediction of about 2%. Subsequent measurements include the second and third phases of Super-K, with values $(-6.3 \pm 4.2 \pm 3.7)\%$ [210] and $(-5.7 \pm 3.1 \pm 1.3)\%$ [214], respectively, and SNO, which reported $(-3.2 \pm 4.0)\%$ [212].

4.3 KamLAND

The KamLAND experiment [54] measures the electron antineutrino flux at the Kamiokande site from surrounding reactors. The dominant reactor is at $L = 160$ km and the average distance from the sources is $L_0 \sim 180$ km. The measured reaction is $\bar{\nu}_e + p \to e^+ + n$. If CPT invariance holds, which is expected in a local quantum field theory, then $P(\bar{\nu}_e \to \bar{\nu}_e) = P(\nu_e \to \nu_e)$. Therefore, for the LMA solar solution, reactor antineutrinos should also disappear due to oscillations. For any other solar oscillation solution, no $\bar{\nu}_e$ disappearance would be observed at KamLAND.

The KamLAND results show significant depletion [55] and spectral distortion [56, 57] in the final-state positron energy spectrum due to oscillations. A suppressed positron spectrum with no distortion from oscillations is excluded at more than 5σ [57]. When compared to the expected rates, the inferred neutrino survival probability shows spectacular evidence of oscillatory behavior (see figure 4.3). The best-fit oscillation parameters to the KamLAND data, assuming two-neutrino oscillations, are [57]

$$\delta m^2_{21} = 7.58^{+0.14}_{-0.13}\ (\text{stat.})^{+0.15}_{-0.15}\ (\text{syst.}) \times 10^{-5}\ \text{eV}^2 \tag{4.19}$$

$$\tan^2\theta_{12} = 0.56^{+0.10}_{-0.07}\ (\text{stat.})^{+0.10}_{-0.06}\ (\text{syst.}). \tag{4.20}$$

4.4 Solar/Reactor Neutrino Parameters

SNO data alone yields three distinct regions in a two-neutrino parameter space: the large-mixing angle (LMA), small mixing angle (SMA), and low-δm^2 (LOW) solutions; see figure 4.4. The combination of Chlorine, Gallium, Super-K, and SNO solar neutrino data eliminates all solar solutions but LMA; a combined analysis of solar and KamLAND data select the LMA solution uniquely at the 4–5σ C.L. [57].

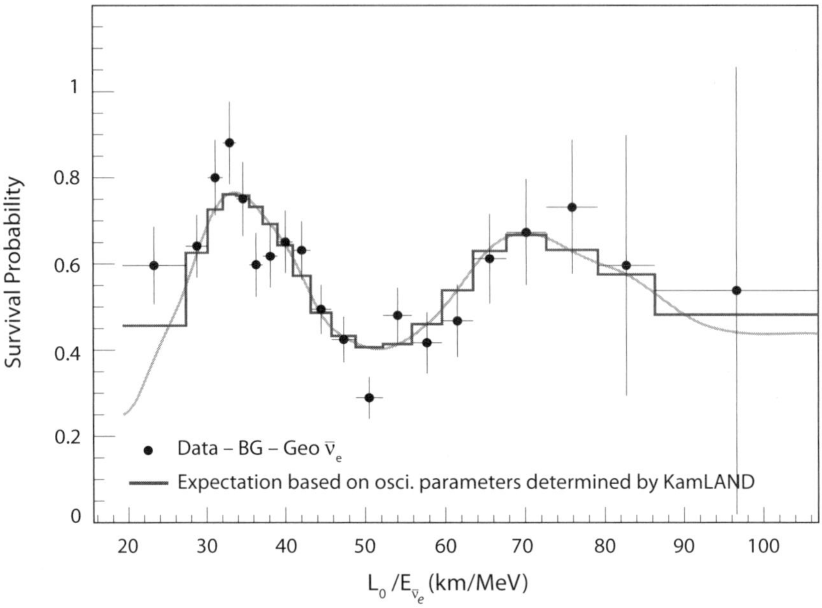

Figure 4.3. Reactor neutrino survival probability versus $L_0/E_{\bar{\nu}_e}$, where L_0 is the weighted average distance from the reactors and $E_{\bar{\nu}_e}$ is the $\bar{\nu}_e$ energy. The data error bars are for statistical uncertainties only; the histogram and curve indicate the best fit with oscillations. From [57].

Maximal mixing is excluded at greater than the 5σ level. KamLAND provides a very good measurement of δm^2_{21}, while the combined solar neutrino experiments better determine θ_{12}. The complementary allowed regions of the solar and KamLAND data are shown separately in figure 4.5a; the overall fit to solar plus KamLAND data is shown in figure 4.5b. The neutrino oscillation fit to the combined KamLAND and solar data, including the measurement of the low-energy electrons in SNO, gives the parameters [212]

$$\delta m^2_{21} = 7.59^{+0.20}_{-0.21} \times 10^{-5}\ \text{eV}^2\,, \tag{4.21}$$

$$\tan^2\theta_{12} = 0.457^{+0.041}_{-0.028}\,. \tag{4.22}$$

KamLAND has not only eliminated all oscillation solutions other than LMA, but has relegated nonoscillation solutions to the solar neutrino problem to be at most subleading effects. Non-standard interactions (NSI) of neutrinos (see section 12.7), which lead to energy-independent conversion probabilities, were consistent [215] with the flat energy spectra seen by Super-K and SNO. With KamLAND data, neutrino NSI are rejected as the primary cause of $\bar{\nu}_e$-disappearance at about the 3σ C.L. [216]. Also, the resonant and nonresonant spin-flavor precession solutions (see section 12.9) are allowed only at the 99.86% and 99.88% C.L., respectively [217].

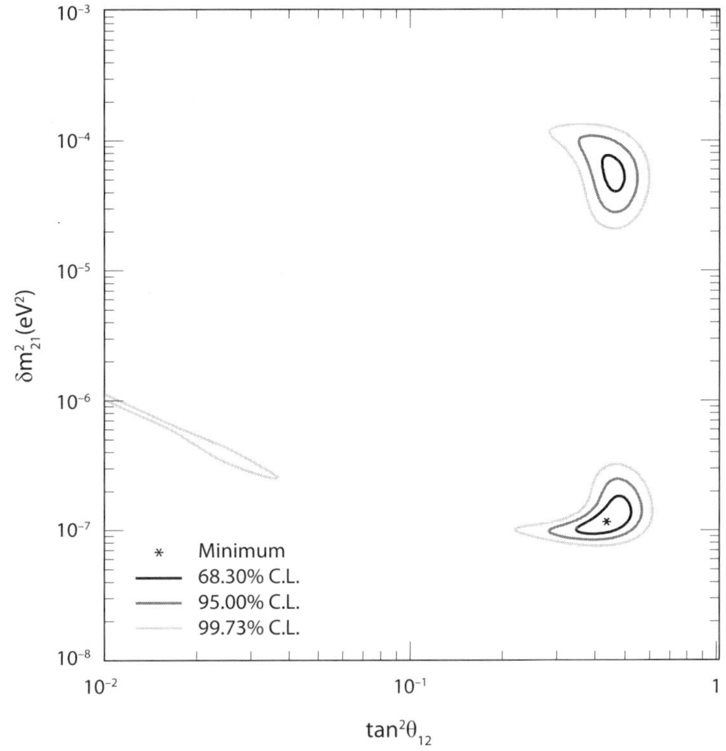

Figure 4.4. Allowed regions for the two-neutrino oscillation parameters from SNO: LMA (upper right), LOW (lower right), and SMA (left). Adapted from [212].

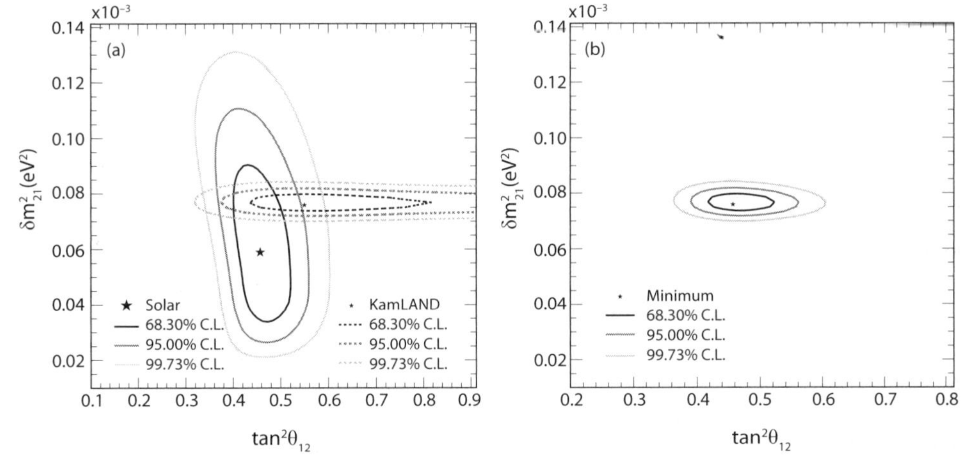

Figure 4.5. Allowed regions for the two-neutrino oscillation parameters from (a) separate fits to KamLAND and the global solar neutrino data, and (b) a combined fit to KamLAND and solar neutrino data. Adapted from [212].

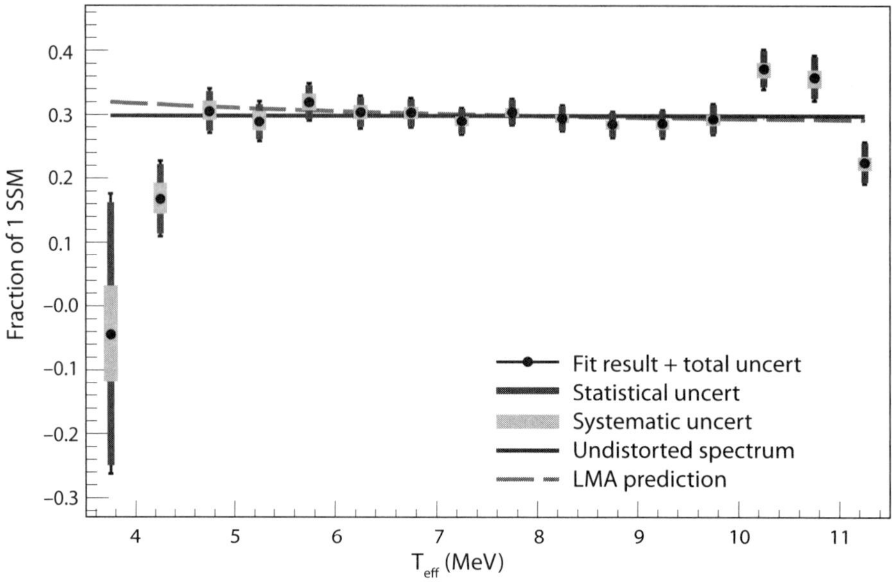

Figure 4.6. CC electron spectrum as a fraction of the SSM prediction in SNO. Also shown are the LMA prediction (dashed curve) and an undistorted spectrum that agrees with the overall rate (dash-dotted). From [212].

If LMA is the dominant neutrino oscillation mechanism, it is unlikely that the effects of spin-flavor precession can be seen in solar neutrinos [218]. Yet another excluded alternative invokes the violation of the equivalence principle to induce oscillations even for massless neutrinos [219]. These three solutions fail because the KamLAND baseline is too short for any significant disappearance to occur.

The LMA solution predicts an almost uniform suppression of ^{8}B neutrinos in the Super-K and SNO experiments. The measured survival probability versus final-state electron energy in the SNO CC process is shown in figure 4.6. The survival probability is very flat above 5 MeV, consistent with the LMA prediction. Below 5 MeV there is a sharp drop, contrary to LMA, although the uncertainties are much larger there.

The Chlorine, Super-K, Gallium, and SNO solar neutrino experiments did not provide a good measurement of the intermediate-energy neutrinos (^{7}Be, *pep*, CNO). The liquid scintillator detector Borexino [49] has a low detection threshold of 200 keV, and therefore can detect intermediate-energy neutrinos as well as some *pp* neutrinos via elastic scattering off electrons (see figure 4.7). The Borexino measurement [51,52] of 0.862 MeV ^{7}Be neutrinos indicates an oscillation probability of $P(\nu_e \to \nu_e) = 0.52^{+0.07}_{-0.06}$, consistent with the prediction $P(\nu_e \to \nu_e) = 0.541 \pm 0.017$ from the standard solar model and the best fit oscillation parameters to other solar data and KamLAND. Borexino also measured the day-night asymmetry of ^{7}Be neutrinos to be $A_{dn} = 0.001 \pm 0.014$, which is consistent only with the LMA solution at 90% C.L. [220]. Borexino also has measured ^{8}B neutrinos with a threshold of 3 MeV and found $P(\nu_e \to \nu_e) = 0.29 \pm 0.10$ [53], consistent with the Super-K and SNO results.

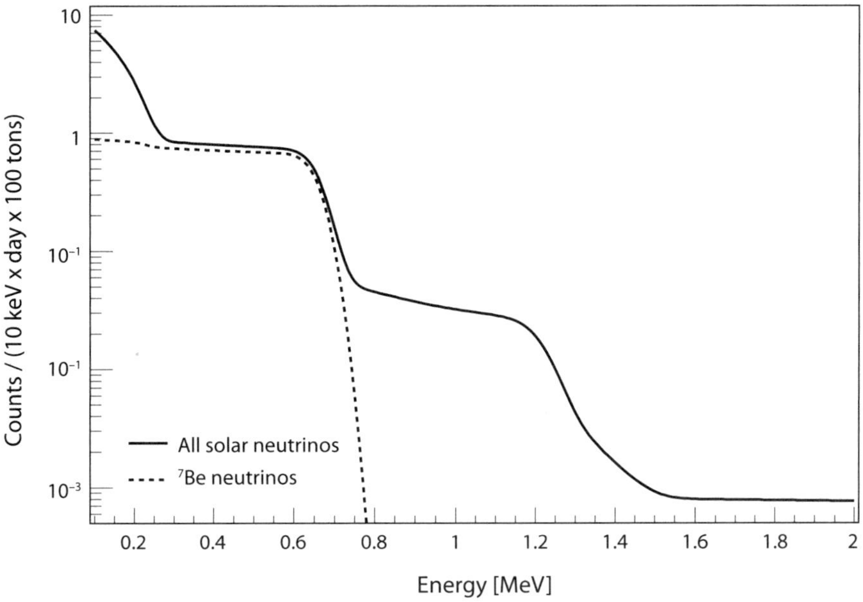

Figure 4.7. Recoil electron spectrum for neutrino elastic scattering off of electrons in Borexino. The ^{7}Be contribution, shown separately by the lower curve, dominates in the region just below 665 keV. The bump near 1.2 MeV is due to *pep* neutrinos ($E_\nu = 1.44\,\text{MeV}$). From [50].

4.5 Flux-independent Tests

Before Borexino, there was a tension in the LMA fits to the different solar neutrino data sets. Since the intermediate energy solar neutrino flux was not well-measured by either the Chlorine or Gallium experiments, it had to be inferred from a combination of all the data. The oscillation probabilities in different neutrino energy bands can be determined following the analysis of [221]. If R is defined as the ratio of the measured rate to the SSM prediction for a given experiment, β is a flux normalization relative to the SSM, and P_L, P_I, and P_H are average survival probabilities of low-energy (*pp*), intermediate-energy (^{7}Be, *pep*, ^{15}O, ^{13}N), and high-energy (^{8}B, *hep*) neutrinos, respectively, then for three active neutrinos the relative rates are given by

$$R_{Ga} = 0.109\beta_H P_H + 0.335\beta_I P_I + 0.556\beta_L P_L, \tag{4.23}$$

$$R_{Cl} = 0.803\beta_H P_H + 0.197\beta_I P_I, \tag{4.24}$$

$$R_{SNO}^{CC} = \beta_H P_H, \tag{4.25}$$

$$R_{SNO}^{NC} = \beta_H. \tag{4.26}$$

TABLE 4.2
Current solar neutrino measurements.

Measurement	Value	Source
R_{Ga}	0.54 ± 0.03	SAGE [27], GALLEX [28], GNO [29]
R_{Cl}	0.31 ± 0.03	Homestake [211]
R_{SNO}^{CC}	0.293 ± 0.016	SNO [34]
R_{SNO}^{NC}	$0.97^{+0.09}_{-0.08}$	SNO [34]
$R_{Borexino}^{ES}$	0.63 ± 0.03	Borexino [52]

Note: where R is the ratio of the measured rate to the SSM expectation. Only the experimental uncertainties are included.

TABLE 4.3
Inferred solar neutrino oscillation probabilities.

Measurement	Before Borexino	After Borexino
P_L	0.68 ± 0.12	0.60 ± 0.06
P_I	0.38 ± 0.17	0.52 ± 0.06
P_H	0.30 ± 0.03	0.30 ± 0.03

Here, CC and NC refer to charged-current and neutral-current measurements, respectively. From equations 4.23–4.26 it is evident that no experiment measures the intermediate-energy neutrinos well. The measured R values are summarized in table 4.2.

The SNO NC data provides a direct measurement of β_H, but there is no equivalent measurement of β_L and β_I. However, we can use the SSM as an additional input, the so-called flux constraint, which predicts

$$\beta_L^{SSM} = 1.00 \pm 0.01, \tag{4.27}$$

$$\beta_I^{SSM} = 1.00 \pm 0.06. \tag{4.28}$$

Then the flux normalizations and average oscillation probabilities may be determined from equations 4.23–4.28; the results are given in table 4.3, without and with the Borexino data.

Since the Borexino result measures electron scattering, which has a neutral-current component, we have

$$R_{Borexino}^{ES} = \beta_I P_I + \beta_I (1 - P_I) r_I, \tag{4.29}$$

where $r_I = 0.23$ is the ratio of the NC cross section to the charged-current cross section at the ^{7}Be neutrino energy.

There are two ways to make use of the Borexino data in the probability analysis. Since the ES process measured in the Borexino experiment has an NC

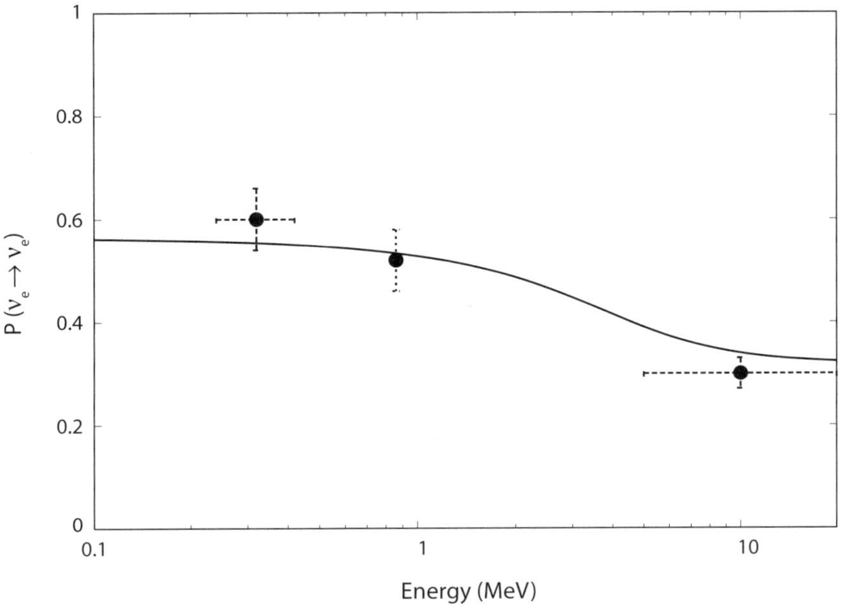

Figure 4.8. Solar neutrino survival probablities versus energy. The rightmost data point indicates the oscillation probability of ^8B neutrinos measured by SNO [34], the central data point is for ^7Be neutrinos measured by Borexino [51], and the leftmost data point is inferred from the ^{71}Ga data combined with that of SNO and Borexino. The horizontal error bars indicate the range of neutrino energies covered by each experiment. The analysis is an updated version of the one in [221]. The curve is the best-fit LMA prediction for $\delta m^2_{21} = 7.6 \times 10^{-5}\ \mathrm{eV}^2$ and $\sin^2 2\theta_{12} = 0.87$.

component, in principle it can be used to help determine the intermediate-energy flux normalization β_I without using the SSM flux constraint, replacing equation 4.28 with equation 4.29. This gives an oscillation probability that is too low for the LMA solution ($P_I = 0.26 \pm 0.22$) and a flux normalization that is very high ($\beta_I = 1.47 \pm 0.58$), although the uncertainties are so large as to render the determinations largely useless.

Since Borexino can separate out the ^7Be contribution, it does a much better job measuring the intermediate-energy neutrino flux than the ^{37}Cl data. An alternate analysis is to keep the flux constraint (which for the ^7Be neutrinos is $\beta_I = 1.00 \pm 0.06$), and use the Borexino result to replace the Chlorine data, i.e., use equation 4.29 instead of equation 4.24. With this method, the inferred oscillation probabilities before and after Borexino are shown in table 4.3. Clearly, before Borexino P_I was initially close to P_H, albeit with a large uncertainty, but with Borexino data included P_I is close to the value of P_L, in good agreement with the LMA prediction (see figure 4.8). Since R_{Cl} is no longer an input, one obtains a prediction for it that is slightly higher than the experimental value.

When combined with other solar neutrino data, and assuming the LMA oscillation probabilities, Borexino can also provide constraints on the *pp* flux normalization,

$$\beta_L = 1.04^{+0.13}_{-0.19}. \tag{4.30}$$

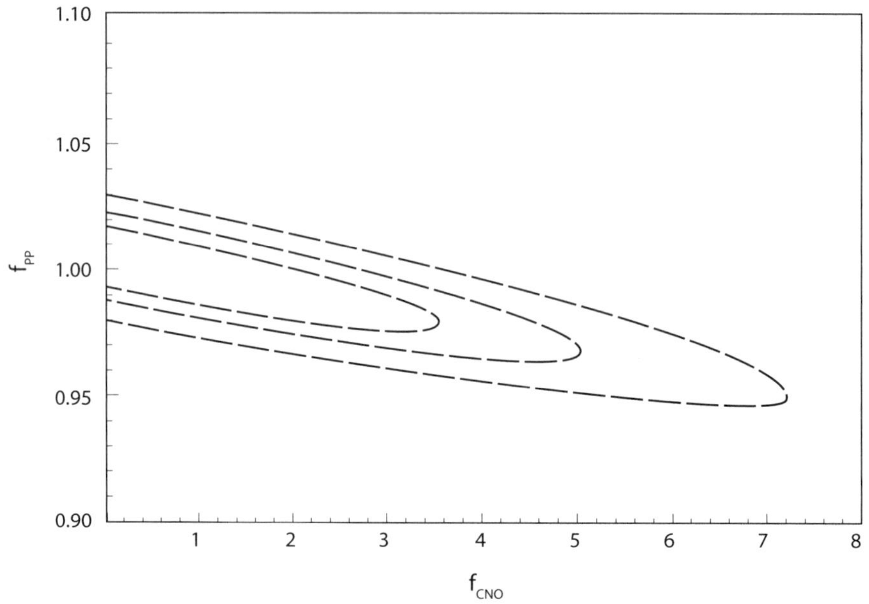

Figure 4.9. Borexino constraints (at 68%, 90%, and 99% C.L., respectively) on the *pp* and CNO solar neutrino flux normalizations, $f_{pp} = \beta_L$ and $f_{CNO} = \beta_I$, respectively, relative to the SSM predictions. Other solar neutrino data, the LMA oscillation probabilities, and the SSM flux constraints are used as inputs. From [51].

With the same inputs Borexino can also place a limit on the flux normalization of the CNO neutrinos; figure 4.9 shows the current bounds on the *pp* and CNO flux normalizations when the SSM flux constraints are imposed. A Borexino measurement of ^{8}B solar neutrinos [222] is in very good agreement with the LMA prediction, albeit with much larger uncertainties than SNO.

One can also make a simultaneous fit to the flux normalizations and oscillation parameters; the result [223] shows that current solar neutrino data provide a better determnation of the ^{8}B neutrino flux than the uncertainties of the SSM, but that new, high-precision measurements of the neutrino energy spectrum below 1.5 GeV are needed to make flux determinations that are independent of the SSM. These improved measurements might also help differentiate between different solar models and provide a better measurement of neutrinos from the CNO and *hep* neutrinos (see figure 4.1).

4.6 Future Experiments

In order to make a model-independent determination of the oscillation probabilities, the flux normalizations and oscillation probabilities must be determined separately in each energy regime. For the high-energy solar neutrinos this has been done by the SNO CC and NC measurements, from which both β_H and P_H may be determined. Current data do not allow a good determination of β_I or β_L. High-precision measurements of both the CC and electron scattering cross sections for low- and intermediate-energy neutrinos are required to determine the oscillation

TABLE 4.4
Solar neutrino experiments for measuring intermediate- and low-energy solar neutrinos.

Experiment	Process	Fluxes measured
Borexino [49]	ES	^{7}Be, *pep*, CNO
KamLAND [225]	ES	^{7}Be, *pep*, CNO
SNO+ [226]	ES	*pep*, CNO
LENS [227]	CC	*pp*, ^{7}Be, *pep*
MOON [228]	CC	*pp*, ^{7}Be, *pep*
SIREN [229]	CC	*pp*, ^{7}Be, *pep*
XMASS [230]	ES	*pp*
CLEAN [231]	ES	*pp*
XAX [232]	ES	*pp*

probabilities independent of the SSM flux constraints. More precise measurements of solar neutrinos can also lead to a better determination of θ_{12}. However, Borexino or any other ^{7}Be solar neutrino experiment will not improve on the accuracy of KamLAND's determination of θ_{12} [224]. A future *pp* solar neutrino experiment with better than 3% precision can lead to significant improvement. A summary of planned solar neutrino experiments that can improve the determinations of the low- and intermediate-energy fluxes is given in table 4.4. Better measurements of ^{7}Be, *pep*, and CNO neutrinos can also help differentiate between solar models.

KamLAND may be able to reduce backgrounds sufficiently that it will be sensitive to ^{7}Be, CNO, and *pep* solar neutrinos [225]. Borexino, in addition to their ^{7}Be measurements, also plans to measure *pep* and CNO neutrinos, as does a liquid scintillator detector at the SNO site (SNO+) [226]. Proposals for measuring *pp* neutrinos include LENS [227], MOON [228], and SIREN [229], which measure ν_e via CC reactions with ^{115}In, ^{100}Mo, and ^{160}Gd targets, respectively, and XMASS [230], CLEAN [231], and XAX [232], which measure all active neutrino types via electron scattering. Another avenue for an improved θ_{12} measurement is a lithium-based radiochemical detector that detects electron neutrinos from the CNO cycle [233].

A reactor neutrino experiment with baseline such that the measured survival probability is a minimum can also lead to a more precise measurement of θ_{12}. Since the reactor neutrino spectrum is accurately known and KamLAND will determine δm^2_{21} precisely, such an experiment is conceivable. For $\delta m^2_{21} = 7 \times 10^{-5}\,\mathrm{eV}^2$, the required baseline is about 70 km [234].

4.7 Geoneutrinos

Geoneutrinos are electron antineutrinos from the radioactive decay of ^{238}U, ^{232}Th, and ^{40}K in the earth [235, 236]. They provide a background for reactor neutrino experiments such as KamLAND, and also can provide information about the make-up of the earth.

Although the amount of radioactive material in the earth's upper crust is fairly well known, abundances in the lower crust, mantle and core are not. Estimates from geochemical arguments indicate that the abundance of radioactive material in the mantle is roughly three orders of magnitude smaller than in the crust, although the mantle contains about as much radioactive material as the crust due to its much larger volume [237]. The standard geophysical prediction is that there is no radioactive material in the core, although the possibility of a large amount of ^{40}K content has been suggested [238].

The earth's surface radiates about 40 TW of heat. Radioactive processes contribute a large fraction of this energy; typical estimates suggest that at least 40% of geothermal heat comes from radioactive decays in the earth [237]. Therefore a measurement of geoneutrinos can provide a means of measuring the radiogenic contribution to the earth's heat flow.

Liquid scintillator detectors provide the best means for measuring the low-energy neutrinos from these radioactive decays via the inverse beta decay reaction $\bar{\nu}_e + p \rightarrow n + e^+$ [239, 240]. Detectors at different sites receive different relative geoneutrino fluxes from the crust and mantle [240], which may help to isolate the contributions from each source [237, 242].

KamLAND has already detected geoneutrinos from U and Th decays (antineutrinos from K decay are below the threshold for detection); the measured number of events is 73 ± 27 [243], where the expected number was 69.7 after accounting for oscillation effects. While there is basic agreement between the prediction and the experimental result, KamLAND alone will not be able to discriminate between competing models for the source of geothermal heat [244, 245]. However, if a reasonable estimate of the Th/U ratio in the earth is made, geoneutrinos can provide additional constraints on the solar neutrino oscillation parameters [244].

The Borexino detector can provide an independent measurement of geoneutrinos without a large background of reactor neutrinos. Its geoneutrino signal will be dominated by radioactive sources in the continental crust. They have seen a signal at more than 3σ C.L. with a limited number of events [246]. The combination of KamLAND and Borexino data confirms a geoneutrino signal at the 5σ level [247].

The Hawaii Antineutrino Observatory (Hanohano) is a proposed 10 kt deep ocean liquid scintillator detector [168]. Since the oceanic crust is much thinner than the continental crust, it will receive a much higher percentage of geoneutrinos from the mantle. If situated approxomately 50 km offshore from a large nuclear reactor, Hanohano can also be used to make precision measurements of θ_{12} and θ_{13}, including a determination of the neutrino mass hierarchy without using matter effects, and it can also serve as a supernova neutrino detector.

Other possible sites for a geoneutrino detector include the Sudbury Neutrino Observatory (SNO) and the Baksan neutrino detector facility in the Caucasus mountains in Russia [248]. The proposed 50 kt LENA (Low Energy Neutrino Astronomy) detector [169] can also detect geoneutrinos.

5

Atmospheric Neutrinos

5.1 Atmospheric Neutrino Experiments

The first compelling evidence for neutrino oscillations came from the measurement of atmospheric neutrinos. Interactions of cosmic rays with the atmosphere produce pions and kaons that decay to muon neutrinos, electron neutrinos, and their antineutrinos:

$$\pi^+, K^+ \to \nu_\mu \mu^+ \to \nu_\mu e^+ \nu_e \bar{\nu}_\mu, \tag{5.1}$$

$$\pi^-, K^- \to \bar{\nu}_\mu \mu^- \to \bar{\nu}_\mu e^- \bar{\nu}_e \nu_\mu. \tag{5.2}$$

On average there are twice as many ν_μ as ν_e at energies of about 1 GeV, although the ν_e tend to be at somewhat lower energies since they are produced only in a secondary decay. The atmospheric neutrino flux is well understood: the normalizations are known to 20% or better (10% or better for neutrino energies below 10 GeV) and ratios of fluxes are known to 5% [249]. The flux falls off rapidly with neutrino energy for $E_\nu \gtrsim 1$ GeV. For very large neutrino energies, $E \gtrsim 10^5$ GeV, one must also include the prompt neutrino flux from charm decays [250].

Neutrinos observed at different zenith angles have path lengths that vary from $L \sim 10$–30 km for downward neutrinos to $L \sim 10^4$ km for upward neutrinos, as illustrated in figure 5.1. The ratio of observed to expected neutrino events provides a sensitive measure of neutrino oscillations, especially since different values of neutrino baselines and energies can be studied. Because neutrinos coming directly down from the atmosphere travel a much shorter distance than upward neutrinos that traverse long distances through the earth, there will be a significant up-down asymmetry in the neutrino survival probability with oscillations.

Initial evidence for atmospheric neutrino oscillations was an overall depletion of ν_μ [60, 61] compared to the theoretical expectation. The Super-K experiment [63] has studied CC events in four categories: fully contained ($E_\nu \sim 1$ GeV), partially

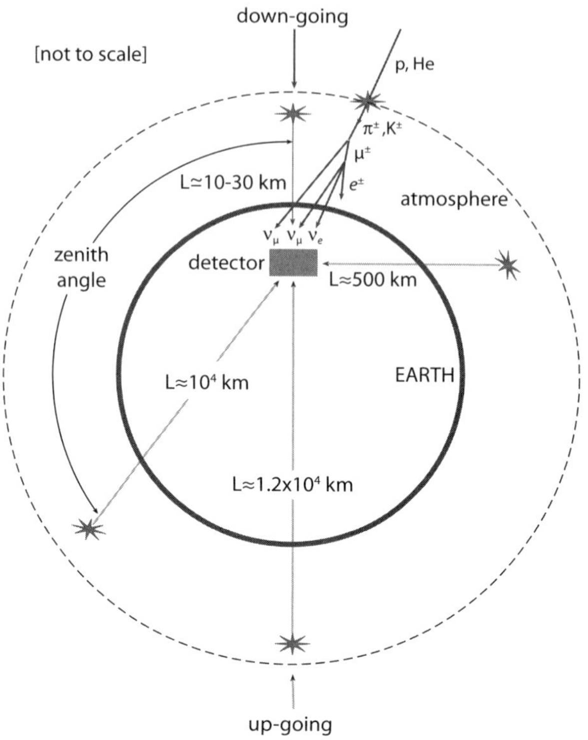

Figure 5.1. A schematic view (not to scale) of the different zenith angles of atmospheric neutrinos and distances they travel before detection. From [182].

contained ($E_\nu \sim 10\,\text{GeV}$), upward-going stopped ($E_\nu \sim 10\,\text{GeV}$), and through-going ($E_\nu \sim 100\,\text{GeV}$). Super-K does not distinguish between atmospheric neutrino and antineutrino events. The contained events have the highest statistics and give the most precise measurement, but all of the data samples are fully consistent with the same oscillation parameters. Zenith angle distributions for the e-like and μ-like contained events are shown in figure 5.2 along with the no-oscillation expectation and the best fit assuming oscillations. The survival probability $P(\nu_\mu \to \nu_\mu)$ versus L/E is shown in figure 5.3.

Atmospheric neutrinos are primarily sensitive to the leading oscillation that involves θ_{23} and θ_{13}. Assuming $\theta_{13} = 0$ in equations (3.9) and (3.10) (i.e., atmospheric ν_μ oscillate exclusively to ν_τ), an L/E analysis of high-resolution events by the Super-K collaboration yields best-fit values $\sin^2 2\theta_{23} = 1.00$ (maximal mixing) and $\delta m^2_{31} = 2.4 \times 10^{-3}\,\text{eV}^2$ [251].[1] The 90% C.L. ranges for these oscillation parameters are $\sin^2 2\theta_{23} \gtrsim 0.90$ and $\delta m^2_{31} \simeq (1.9\text{–}3.0) \times 10^{-3}\,\text{eV}^2$. A more inclusive zenith angle analysis [163] finds best-fit values $\sin^2 2\theta_{23} = 1.00$ and $\delta m^2_{31} = 2.1 \times 10^{-3}\,\text{eV}^2$, with 90% C.L. ranges $\sin^2 2\theta_{23} \gtrsim 0.92$ and $\delta m^2_{31} \simeq (1.5\text{–}3.4) \times 10^{-3}\,\text{eV}^2$. The L/E analysis gives a slightly better determination of δm^2_{31}, while the zenith angle analysis

[1] Although δm^2_{31} values are often quoted without an absolute value sign, the sign of δm^2_{31} has not yet been determined.

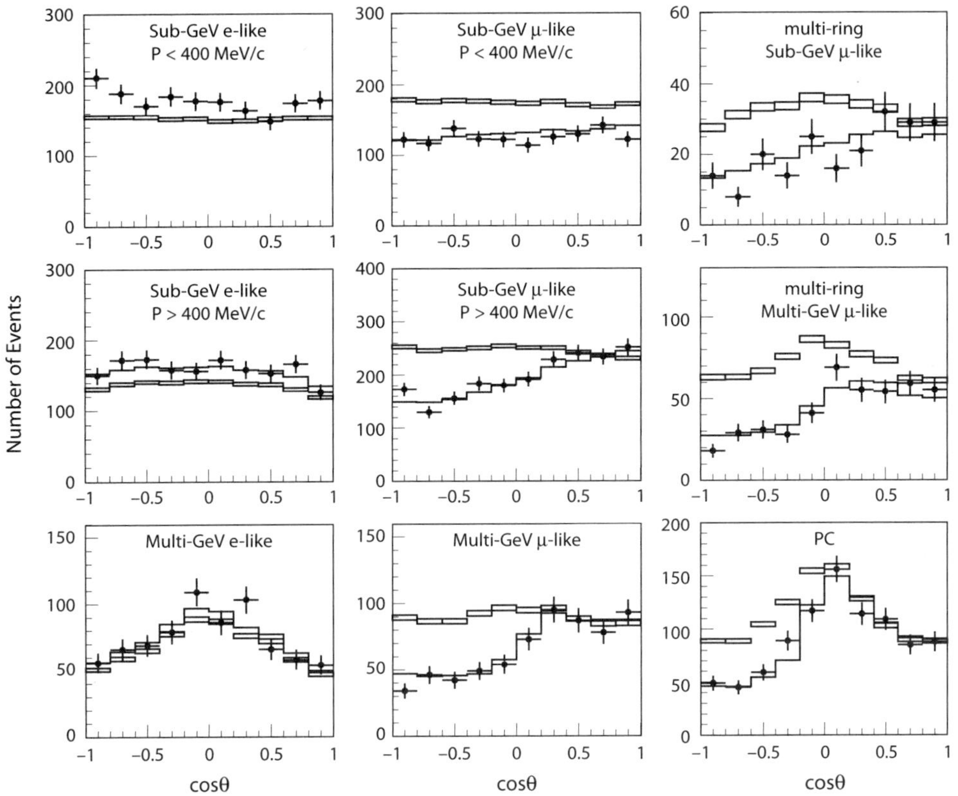

Figure 5.2. Zenith angle distributions for e-like and μ-like atmospheric neutrino events in Super-K, where $\cos\theta = 1$ corresponds to downward events with $L \sim 15$ km and $\cos\theta = -1$ corresponds to upward events with $L \sim 13000$ km. The lines show the best fits ($\delta m^2_{31} = 2.1 \times 10^{-3}\,\text{eV}^2$ and $\sin^2 2\theta_{23} = 1.00$) to the zenith angle distributions with oscillations; the box histograms show the non-oscillated Monte Carlo prediction. The fit was constrained to the region $\sin^2 2\theta \leq 1$. From [163].

does better on $\sin^2 2\theta_{23}$ because of higher statistics in the up-down asymmetry. The allowed regions in $\sin^2 2\theta_{23} - \delta m^2_{31}$ parameter space assuming $\nu_\mu \to \nu_\tau$ oscillations are shown in figure 5.4.

The Soudan-2 [68,252] and MACRO [69,253] experiments have also measured atmospheric neutrinos and find allowed regions consistent with the Super-K result. The MINOS experiment [71] can detect atmospheric neutrinos via the ν_μ and $\bar{\nu}_\mu$ charged-current reactions *and* can determine the signs of the resulting charged leptons with a magnetic field, thereby separately testing oscillations of $\nu_\mu \to \nu_\mu$ and $\bar{\nu}_\mu \to \bar{\nu}_\mu$.

Since $\theta_{13} \neq 0$, reactor $\bar{\nu}_e$ fluxes should exhibit disappearance due to the leading oscillation when $L/E_\nu \gtrsim 40$ m/MeV. Data from the CHOOZ reactor experiment [72] ($L \sim 1000$ m, $E_\nu \sim 3$ MeV) place upper limits on θ_{13} for the values of δm^2_{31} indicated by the atmospheric neutrino data; similar limits have been obtained from the Palo Verde reactor experiment [73]. Another consequence of $\theta_{13} \neq 0$ is that ν_e participates in the oscillations of atmospheric neutrinos with amplitude $\sin^2 2\theta_{13}$. In the Super-K data the number of observed electron neutrinos is consistent with

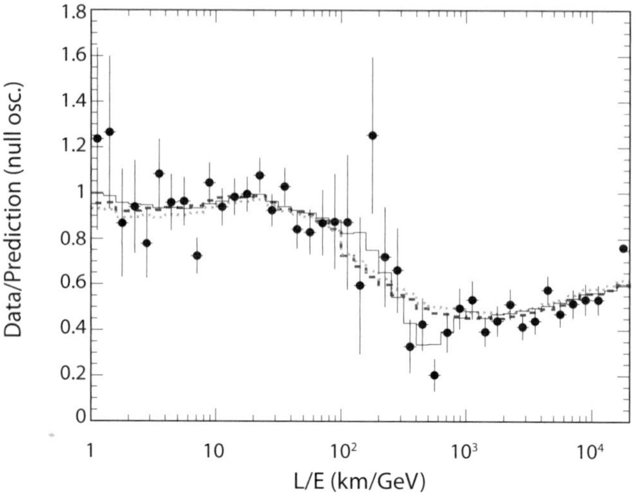

Figure 5.3. Muon neutrino survival probability versus L/E for atmospheric neutrinos in the Super-K experiment. Also shown are the best fits for two-neutrino oscillations (solid curve), neutrino decay (dashed), and neutrino decoherence (dotted). From [251].

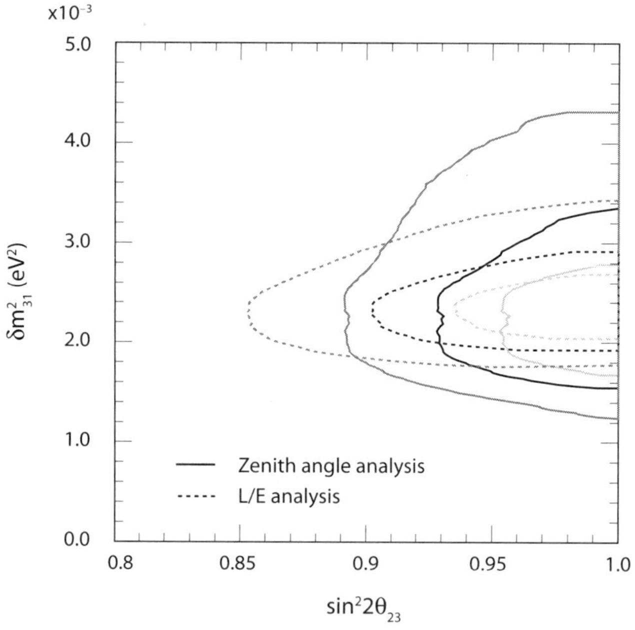

Figure 5.4. The 68%, 90%, and 99% C.L. allowed regions for $\nu_\mu \to \nu_\tau$ oscillations of atmospheric neutrinos from Super-K data from their zenith angle (solid curves) and L/E (dotted) analyses. Adapted from [163].

$\theta_{13} = 0$ (see figure 5.2); a strong enhancement would be expected if there were sizeable $\nu_\mu \to \nu_e$ oscillations (due to the 2 : 1 ratio of ν_μ to ν_e in the flux). Furthermore, the zenith angle distribution of the Super-K muon sample is inconsistent with large oscillations involving ν_e.

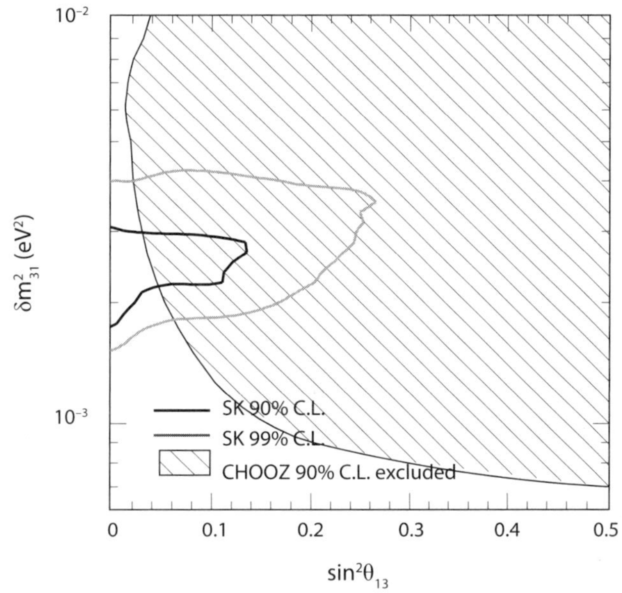

Figure 5.5. Allowed regions in the $\sin^2\theta_{13} - \delta m^2_{31}$ plane for three-neutrino oscillations from Super-K atmospheric data. Also shown is the 90% C.L. exclusion bound from CHOOZ. Adapted from [254].

Figure 5.5 shows the Super-K allowed region in the $\sin^2\theta_{13} - \delta m^2_{31}$ plane, along with the CHOOZ bound. For the δm^2_{31} values obtained from the Super-K collaboration's two-neutrino analysis, bounds on θ_{13} are sensitive to the value of δm^2_{31}. Thus, it is necessary to specify the δm^2_{31} for which a bound on θ_{13} is quoted. For $\delta m^2_{31} = 2.0 \times 10^{-3}$ eV^2 the angle θ_{13} is constrained to be smaller than 13° at 90% C.L. There is also slightly greater than 2σ evidence in the Super-K data of hadronic showers from τ decays [255], consistent with the hypothesis that the primary oscillation of atmospheric neutrinos is $\nu_\mu \to \nu_\tau$. A large Liquid Argon Time Projection Chamber could in principle detect atmospheric ν_τ's [256].

5.2 Matter Effects for Atmospheric Neutrinos

Since upward-going atmospheric neutrinos traverse a large fraction of the earth's diameter, matter effects could be relevant. The dominant oscillation of atmospheric neutrinos appears to be $\nu_\mu \to \nu_\tau$; for two-neutrino $\nu_\mu \to \nu_\tau$ oscillations there would be no matter effects. However, probability conservation in three-neutrino oscillations combined with large matter effects in oscillations involving ν_e lead to small changes the $\nu_\mu \to \nu_\tau$ oscillation probability.

Specific relationships between the changes in the oscillation phase and the matter density can lead to an enhancement of the oscillation probability, analogous to the classical phenomenon of parametric resonance [257]. Also, constructive quantum mechanical interference between the probability amplitudes for different density layers can give total neutrino flavor conversion [258]. Such enhancement phenomena generally require passage of the neutrino through the earth's core.

Conditions for observing matter effects in the earth's mantle and core in atmospheric neutrino experiments are discussed in [259]. A comprehensive discussion of the effects of δm^2_{21} and δ on the earth matter effects is given in [260].

Future atmospheric neutrino oscillation experiments can in principle determine the octant of θ_{23} (i.e., whether θ_{23} is greater than or less than $\pi/4$) through subdominant effects due to δm^2_{21} [261]. Since θ_{13} is large enough ($\sin^2 2\theta_{13} \sim 0.10$), then matter effects can help determine $\mathrm{sgn}(\delta m^2_{31})$ [262]. Atmospheric neutrino detectors that can distinguish neutrino- from antineutrino-induced events (e.g., detectors with a magnetic field) may be able to determine the sign of δm^2_{31} due to the different matter effects on neutrinos and antineutrinos [263–265].

The combination of long-baseline and atmospheric experiments can resolve parameter degeneracies (see chapter 8) in the determination of the oscillation parameters [266]. The oscillations of atmospheric neutrinos can be used to place strong constraints on the flavor content of the matter-neutrino interaction [267], although not on its strength [268]. Also, the absorption of very high-energy atmospheric neutrinos (of order 10 TeV) in kilometer-sized neutrino detectors may be used to probe the earth's density distribution [269]. Finally, the ATLAS detector at the Large Hadron Collider (LHC) at CERN could be used to detect atmospheric neutrinos when the colliding beams are not turned on [270].

5.3 Long-baseline Neutrino Experiments

Long-baseline neutrino experiments can provide an independent measurement of neutrino oscillations seen in atmospheric neutrino experiments. The K2K experiment [158], in which ν_μ with energies of approximately ~ 1 GeV are directed from KEK to Super-K ($L = 250$ km), has measured a ν_μ survival probability consistent with the atmospheric neutrino results, with best-fit values [70] of $\delta m^2_{31} = 2.8 \times 10^{-3}\,\mathrm{eV}^2$ and $\sin^2 2\theta_{23} = 1.00$. The K2K allowed region, from the number of events and the spectrum shape combined, is consistent with the allowed region from the atmospheric neutrino data (see figure 5.6). K2K also searched for $\nu_\mu \to \nu_e$ oscillations, and placed a 90% C.L. upper bound [271] of $\sin^2 2\theta_{13} \sin^2\theta_{23} < 0.13$ for $\delta m^2_{31} = 2.8 \times 10^{-3}\,\mathrm{eV}^2$, somewhat weaker than the CHOOZ exclusion bound.

The MINOS experiment [71], using a ν_μ beam from NuMI at Fermilab to the Soudan mine in Minnesota ($L = 735$ km), has measured the survival probability for ν_μ energies mostly in the range 1–5 GeV, although some data was taken at higher energies (5–10 GeV). The MINOS far/near flux ratio clearly shows oscillation characteristics, including a sharp dip in the probability at a neutrino energy around 1 GeV, as would be expected from the oscillation parameters favored by the Super-K atmospheric neutrino data. Alternative explanations of the MINOS data by neutrino decay (see section 12.4) or quantum decoherence (see section 12.5) are excluded at 7σ and 9σ, respectively [272]. The MINOS allowed regions for the combined neutrino and antineutrino data are shown in figure 5.7 along with those from Super-K. The best-fit two-neutrino oscillation parameters to the MINOS data are [272]

$$\delta m^2_{31} = (2.32^{+0.12}_{-0.08}) \times 10^{-3}\,\mathrm{eV}^2 \tag{5.3}$$

$$\sin^2 2\theta_{23} = 1.00\,, \tag{5.4}$$

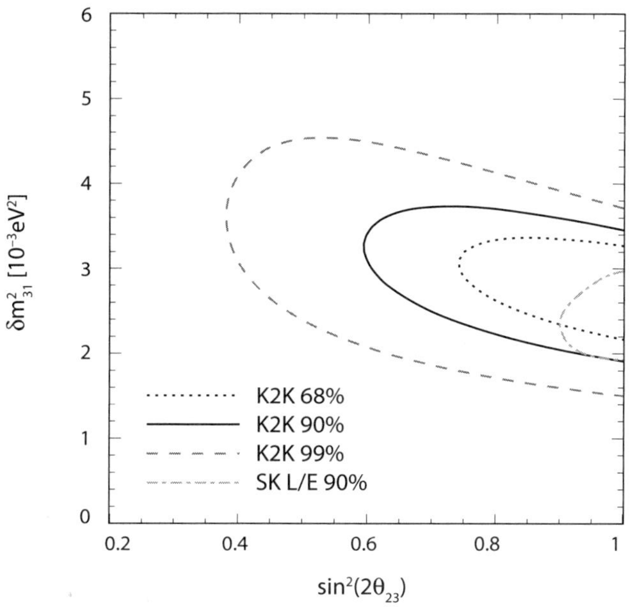

Figure 5.6. Allowed regions in the $\sin^2 2\theta_{23} - \delta m^2_{31}$ plane from K2K, compared with the allowed region from the Super-K atmospheric L/E analysis. Adapted from [70].

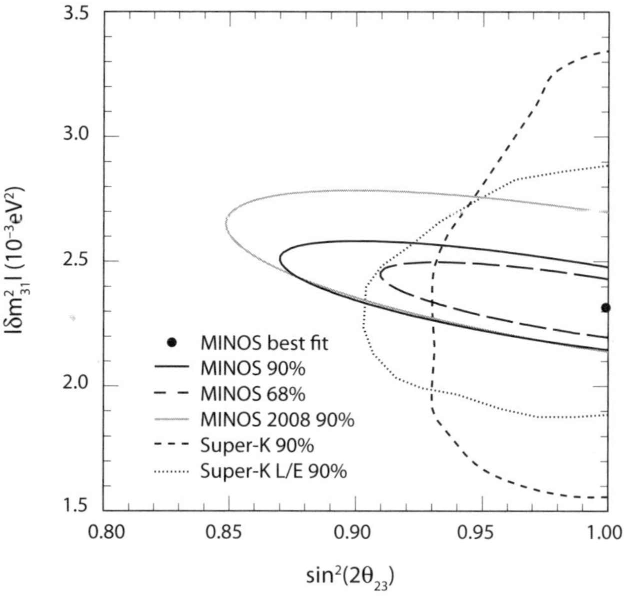

Figure 5.7. Allowed regions in the $\sin^2 2\theta_{23} - \delta m^2_{31}$ plane from ν_μ survival in MINOS, compared with the allowed regions from the Super-K experiment. Adapted from [272].

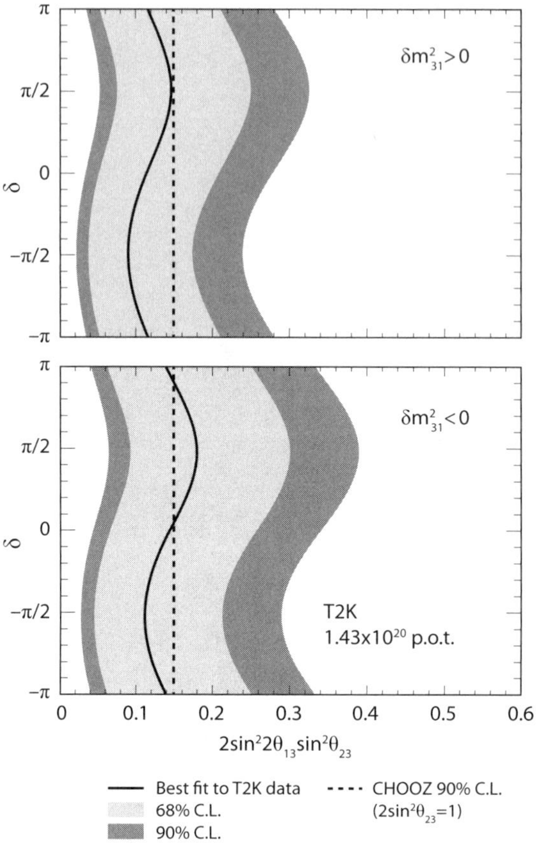

Figure 5.8. Allowed regions from the T2K $\nu_\mu \rightarrow \nu_e$ search for the CP phase δ versus $2\sin^2 2\theta_{13}\sin^2\theta_{23}$ for the normal ($\delta m^2_{31} > 0$) and inverted ($\delta m^2_{31} < 0$) neutrino mass hierarchies. The CHOOZ bound is shown for $\delta m^2_{31} = 2.4 \times 10^{-3}\,\text{eV}^2$ and $\sin^2 2\theta_{23} = 1$. Adapted from [275].

with a 90% C.L. lower bound on $\sin^2 2\theta_{23}$ of 0.90. The MINOS allowed region is consistent with that of Super-K and complementary to it; MINOS has a more precise measurement of δm^2_{31}, while Super-K has a smaller uncertainty for $\sin^2 2\theta_{23}$. The fits of MINOS and Super-K both prefer maximal mixing.

MINOS has made measurements of $\bar{\nu}_\mu$ survival for an antineutrino beam [273], although they are less precise than the neutrino measurements due to less beam time and smaller detection cross section. MINOS finds strong evidence for oscillations in measurements of the $\bar{\nu}_\mu$ flux, with no oscillations excluded at more than the 6σ level. The best-fit point is $\delta m^2_{31} \approx 3.36 \times 10^{-3}\,\text{eV}^2$ and $\sin^2 2\theta_{23} \approx 0.86$, and the 90% C.L. allowed region for antineutrinos does not include the best-fit point for neutrinos, although the 90% C.L. regions overlap. This discrepancy is not statistically compelling, but is nonetheless intriguing since a difference in the neutrino and antineutrino oscillation parameters (after accounting for matter effects) would require CPT violation (see section 12.2).

MINOS has also searched for $\nu_\mu \rightarrow \nu_e$ oscillations and found 54 events [274], consistent with the expected background of $49.1 \pm 7.0(\text{stat.}) \pm 2.7(\text{syst.})$. From this measurement they quote an upper limit of $2\sin^2 2\theta_{13}\sin^2\theta_{23} < 0.12$ (0.20) at 90% C.L. for the normal (inverted) mass hierarchy when the CP violating phase is $\delta = 0$.

The T2K experiment has seen a signal for $\nu_\mu \rightarrow \nu_e$ appearance at the 2.5σ level [275], with best fit value $\sin^2 2\theta_{13} = 0.11$ (0.14) for the normal (inverted) hierarchy, $\sin^2 2\theta_{23} = 1.00$ and $\delta = 0$. The 68% and 90% C.L. allowed ranges are shown for different values of δ in figure 5.8.

6

Global Three-neutrino Fits

As noted in section 3.1, in the limit $\theta_{13} \to 0$, $\nu_e \to \nu_\mu$, ν_τ oscillations of solar neutrinos and $\nu_\mu \to \nu_\tau$ oscillations of atmospheric neutrinos decouple, i.e., they are governed by separate parameters. However, for $\theta_{13} \neq 0$, solar ν_e and KamLAND $\bar{\nu}_e$ will have a further suppression due to θ_{13} via oscillations at the δm^2_{31} scale, and there will be some $\nu_\mu \to \nu_e$ oscillations for atmospheric neutrinos and in long-baseline experiments. Since the θ_{13} parameter now enters into oscillations in all experiments, a global fit is necessary. Fits to some subsets of the data can also give constraints on θ_{13}.

A three-neutrino oscillation fit to the combined data from Super-K atmospheric neutrinos, the 2008 ν_μ survival data from MINOS, and the CHOOZ constraint showed a slight preference for nonzero θ_{13}; a value $\sin^2\theta_{13} = 0.012 \pm 0.013$ was found in [276] and 0.019 (with 1σ upper bound 0.12) in [277], although $\sin^2\theta_{13} = 0$ was preferred in [278, 279]. As noted, the MINOS and T2K $\nu_\mu \to \nu_e$ searches had a slight preference for nonzero θ_{13}, after which all analyses had a best-fit value for $\sin^2 2\theta_{13}$ above zero [278, 280–282].

It has also been noted that the best-fit values for θ_{12} in solar and KamLAND data disagreed slightly, which could be resolved by nonzero θ_{13} [283, 284]. This tension can be seen in the left panel of figure 6.1, where the solar and KamLAND experiments prefer different values of θ_{12}. The combined solar/KamLAND data then show a slight preference for $\theta_{13} \neq 0$ (see the center panel of figure 6.1), with fitted value [281] $\sin^2\theta_{13} = 0.021 \pm 0.017$. A global fit to all data (right panel of figure 6.1) yielded $\sin^2\theta_{13} = 0.020 \pm 0.010$ [285], or $\theta_{13} = (8 \pm 4)^\circ$, an approximately 2σ effect. An updated fit including newer KamLAND data finds [286]

$$\sin^2\theta_{13} = 0.009^{+0.013}_{-0.007}, \tag{6.1}$$

or $\theta_{13} = (5.4^{+3.1}_{-2.8})^\circ$. A combined fit of the first Double Chooz data with T2K and MINOS gives the range $0.003 < \sin^2 2\theta_{13} < 0.219$ at the 3σ level [80]. A discussion of future experiments that will attempt to measure θ_{13} is presented in chapter 8.

Two large mixing angles and two independent mass-squared differences are now well-determined by solar, reactor, atmospheric, and accelerator neutrino data, while the third mixing angle is small. A comparison of three global fits to these parameters

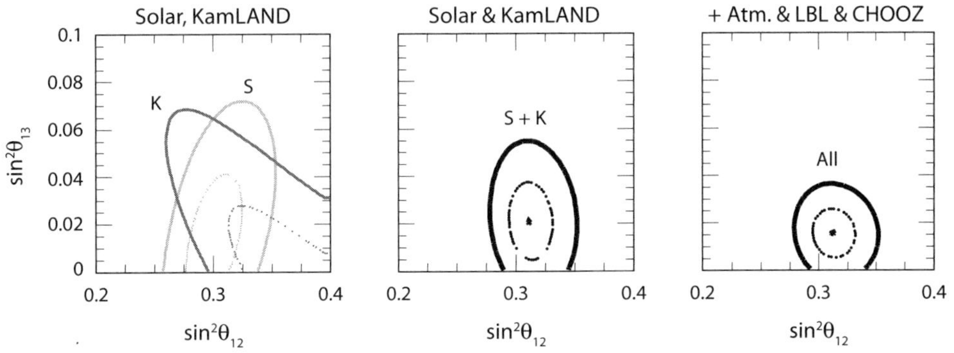

Figure 6.1. The 68% and 95% C.L. allowed regions in the $\sin^2\theta_{12}$-$\sin^2\theta_{13}$ plane for: solar (S) and KamLAND (K) data separately (left panel), the combined solar/KamLAND data (S + K, center panel), and all data including atmospheric, long-baseline, and CHOOZ experiments (right panel). Adapted from [281].

TABLE 6.1
Global three-neutrino fits to the neutrino oscillation parameters.

Parameter	[278]	[280]	[282]
δm^2_{21} (eV^2)	$(7.59^{+0.23}_{-0.18}) \times 10^{-5}$	$(7.59 \pm 0.20) \times 10^{-5}$	$(7.58^{+0.22}_{-0.26}) \times 10^{-5}$
$\sin^2 2\theta_{12}$	$0.868^{+0.026}_{-0.025}$	$0.869^{+0.023}_{-0.024}$	$0.849^{+0.027}_{-0.024}$
$\sin^2\theta_{12}$	$0.318^{+0.019}_{-0.017}$	$0.319^{+0.017}_{-0.016}$	$0.306^{+0.018}_{-0.015}$
θ_{12} (°)	$34.3^{+1.2}_{-1.0}$	34.4 ± 1.0	$33.6^{+1.1}_{-1.0}$
δm^2_{31} (eV^2)	$(2.40^{+0.12}_{-0.11}) \times 10^{-3}$	$(-2.36 \pm 0.11) \times 10^{-3}$	$(2.35^{+0.12}_{-0.09}) \times 10^{-3}$
		$(+2.46 \pm 0.12) \times 10^{-3}$	
$\sin^2\theta_{23}$	$0.500^{+0.070}_{-0.063}$	$0.462^{+0.082}_{-0.051}$	$0.42^{+0.08}_{-0.03}$
θ_{23} (°)	$45.0^{+4.0}_{-3.6}$	$42.8^{+4.7}_{-2.9}$	$40.4^{+4.6}_{-1.8}$
$\sin^2 2\theta_{13}$	$0.051^{+0.051}_{-0.035}$	$0.038^{+0.049}_{-0.028}$	$0.082^{+0.027}_{-0.031}$
$\sin^2\theta_{13}$	$0.013^{+0.013}_{-0.009}$	$0.010^{+0.012}_{-0.007}$	$0.021^{+0.007}_{-0.008}$
θ_{13} (°)	$6.5^{+2.8}_{-2.9}$	$5.6^{+3.0}_{-2.7}$	$8.3^{+1.3}_{-1.8}$

Note: For δm^2_{31}, the fits of [278] and [282] are for the magnitude, while the fit of [280] was done separately for $\delta m^2_{31} > 0$ and $\delta m^2_{31} < 0$. The fit of [282] includes the T2K $\nu_\mu \rightarrow \nu_e$ appearance data of [275]. Adapted from [278], [280] and [282].

is shown in table 6.1; there is a very strong agreement between them, and all show a preference for nonzero θ_{13}.[1]

[1] The Daya Bay experiment has recently measured $\sin^2 2\theta_{13} = 0.092 \pm 0.016\,(\text{stat.}) \pm 0.005\,(\text{syst.})$, ruling out a nonzero value of θ_{13} at 5.2σ; see arXiv:1203.1669 [hep-ex]. The RENO experiment has confirmed this result, measuring $\sin^2 2\theta_{13} = 0.113 \pm 0.013\,(\text{stat.}) \pm 0.019\,(\text{syst.})$, 4.9σ from zero; see arXiv:1204.0626v2 [hep-ex].

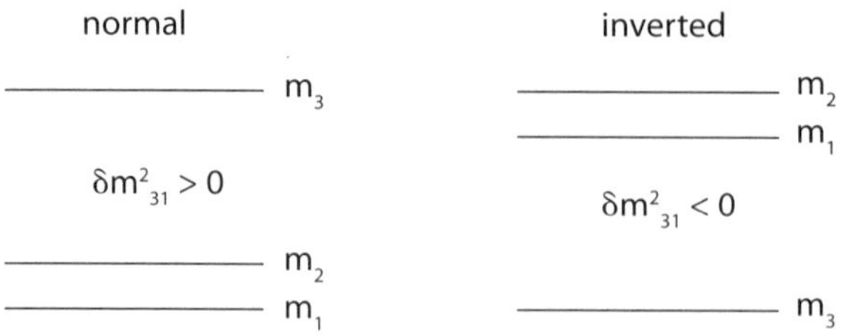

Figure 6.2. The patterns of relative mass differences in normal (left) and inverted (right) neutrino mass hierarchies. From [182].

We note that the sign of δm_{21}^2 has been determined, but the sign of δm_{31}^2 has not. Since the sign of δm_{31}^2 is unknown, there are two possible mass orderings, or hierarchies, as illustrated in figure 6.2. The mass hierarchy is an important discriminant of neutrino mass models. If the scale of the lightest mass is much larger than 0.05 eV, then the neutrino masses are approximately degenerate. Also, the data is mostly insensitive to the quadrant of θ_{23}, although subleading effects in the three-neutrino fit make the uncertainties in θ_{23} slightly asymmetric. Finally, there is currently no constraint on the CP phase δ.

The measurement of the mixing angle θ_{13}, the Dirac CP phase δ, and the sign of δm_{31}^2 will be the main goal of future long-baseline neutrino experiments. In chapter 8 we discuss the the next-generation long-baseline experiments that are being considered after MINOS, ICARUS, and OPERA.

7

Absolute Neutrino Mass

7.1 Beta Decay

Neutrino oscillations tell us nothing about the absolute scale of neutrino masses, except that the heaviest eigenstate has mass above $\sqrt{|\delta m_{31}^2|} \simeq 0.05$ eV. The standard technique for probing the absolute mass is to study the endpoint region of the electron spectrum in tritium beta-decay,

$$^3\mathrm{H} \rightarrow {}^3\mathrm{He}^+ + \mathrm{e}^- + \bar{\nu}_\mathrm{e}. \tag{7.1}$$

The electron energy spectrum is given by

$$\frac{dN}{dE} = \frac{G_F^2 m_e^5}{2\pi^3} \cos^2\theta_c |M|^2 F(Z,E) pE(E_0-E) \sum_i |V_{ei}|^2 [(E_0-E)^2 - m_i^2]^{\frac{1}{2}} \Theta(E_0-E-m_i), \tag{7.2}$$

where E and p are the energy and momentum of the electron, E_0 is the endpoint of the spectrum, θ_c is the Cabibbo angle, M is the nuclear matrix element, and $F(Z, E)$ is the Fermi function. The step function ensures that ν_i is produced only if enough energy is available.

Distinct virtues of tritium as a β emitter are (i) that its decay is a superallowed transition so that nuclear matrix element M is completely known, and (ii) since the fraction of beta decays in the endpoint region $\sim E_0^{-3}$, the low endpoint energy of 18.6 keV maximizes the fraction of the beta decays in this region.

In the foreseeable future, experiments will be sensitive only to the quasidegenerate mass spectrum. In this case, the effect of nonzero neutrino masses is to suppress and cut off the spectrum at the electron energy $E_0 - m_\beta$, where the effective neutrino mass is [287]

$$m_\beta^2 = \sum_i |V_{ei}|^2 m_i^2. \tag{7.3}$$

Note that the elements of the mixing matrix enter as the squares of their absolute values, so that no cancellation can occur in the sum. The present limit from the

Troitsk [15] and Mainz [16] experiments is $m_\beta \leq 2.2\,\text{eV}$ at 2σ. Future sensitivity down to $m_\beta = 0.35\,\text{eV}$ is expected in the KATRIN experiment [288], which will begin collecting data in 2012.

7.2 Cosmological Limits

The sum of neutrino masses $\Sigma \equiv \sum m_\nu$ can be probed in cosmology. Neutrinos of eV masses are relativistic when they decouple, and so their final number density is independent of their mass, $n_\nu = 3/11 n_\gamma$. Since $\langle E_\gamma \rangle = 2.7 T_\gamma$, and $\Omega_\gamma h^2$ is essentially the energy density of the CMB with $T_\gamma = 2.725$ K, n_γ is known from the Planck black-body distribution, and

$$\Omega_\nu h^2 = \frac{n_\nu \Sigma}{\rho_c} \simeq \frac{\Sigma}{94.1\,\text{eV}}, \tag{7.4}$$

where ρ_c is the critical energy density of the universe.

Neutrinos free-stream on scales smaller than their Jeans length scale, which is known as the free-streaming scale. While neutrinos free-stream, their density perturbations are damped, and simultaneously the perturbations of cold dark matter and baryons grow more slowly because of the missing gravitational contribution from neutrinos. The free-streaming scale of relativistic neutrinos grows with the horizon. When the neutrinos become nonrelativistic, their free-streaming scale shrinks, they fall back into the potential wells, and the neutrino density perturbation resumes to trace those of the other species. Free-streaming suppresses the power spectrum on scales smaller than the horizon when the neutrinos become nonrelativistic. (For eV neutrinos, this is the horizon at matter-radiation equality.) Lighter neutrinos free-stream out of larger scales and cause the power spectrum suppression to begin at smaller wavenumbers [289],

$$k_{nr} \simeq 0.026 \left(\frac{m_\nu \omega_M}{1\,\text{eV}} \right)^{1/2} \text{Mpc}^{-1}, \tag{7.5}$$

assuming almost degenerate neutrinos. Here, $\omega_M \equiv \Omega_M h^2$ is the total matter density (which is comprised of baryons, cold dark matter and massive neutrinos). On the other hand, heavier neutrinos constitute a larger fraction of the matter budget and suppress power on smaller scales more strongly than lighter neutrinos [101]:

$$\frac{\Delta P_m}{P_m} \approx -8 f_\nu \simeq -0.8 \left(\frac{\Sigma}{1\,\text{eV}} \right) \left(\frac{0.1}{\omega_M} \right), \tag{7.6}$$

where $f_\nu \equiv \Omega_\nu / \Omega_M$ is the fractional contribution of neutrinos to the total matter density.

Analyses of CMB data are not very sensitive to neutrino masses because at the epoch of last scattering, eV mass neutrinos behave essentially like cold dark matter. However, an important role of CMB data is to constrain other parameters that are degenerate with Σ. Sensitivity to neutrino masses results from the complementarity of galaxy surveys and CMB experiments.

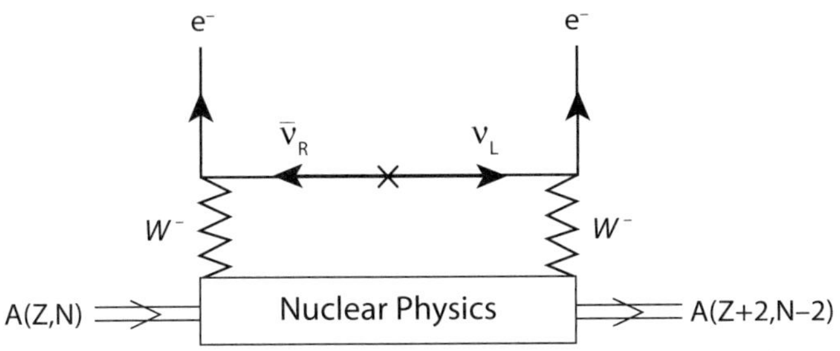

Figure 7.1. Neutrinoless double-beta decay mediated by Majorana neutrinos. From [300].

WMAP data provide a robust 95% C.L. upper limit $\Sigma < 1.3\,\text{eV}$ that is slightly relaxed to 1.5 eV if the equation of state of dark energy is allowed to deviate from −1 [102]. Combining the halo power spectrum obtained from the SDSS Luminous Red Galaxy sample with WMAP5 data results in a bound of 0.62 eV [290]. Much stronger, but less reliable constraints that depend on data whose interpretation requires nonlinear modeling exist in the literature. For example, a joint analysis of Lyman alpha forest data with CMB, galaxy clustering, and supernova data yields $\Sigma < 0.17\,\text{eV}$ [291]. Overall, it is safe to say that $\Sigma \lesssim 1\,\text{eV}$. It is interesting that an argument relying on anthropic selection concluded that $\Sigma \sim 1\,\text{eV}$ so that neutrinos cause a small but non-negligible suppression of galaxy formation [292]. In the future, lensing measurements of galaxies and the CMB by large scale structure are expected to probe a hierarchical neutrino mass spectrum with $\Sigma \approx 0.05\,\text{eV}$ [293].

All neutrino masses are linked to the lightest mass by the values of δm^2_{31} and δm^2_{21} determined by neutrino oscillation studies [294]. If the scale of the lightest mass is small, then the heaviest mass is approximately $\sqrt{|\delta m^2_{31}|} \simeq 0.05\,\text{eV}$ and a neutrino mass hierarchy exists.

7.3 Neutrinoless Double-beta Decay

Two-neutrino double-beta decay conserves lepton number, while neutrinoless double-beta decay ($0\nu\beta\beta$) violates lepton number by 2 units. The two decay modes have very different phase spaces of the outgoing particles and can be distinguished by the spectral shapes of the sum of the electron energies. In the Standard Model with massive neutrinos and no other new physics,[1] $0\nu\beta\beta$ probes the absolute mass, provided that neutrinos are Majorana particles; see figure 7.1. Numerous theoretical analyses have been made of what can be learned about the neutrino sector from $0\nu\beta\beta$ [296–300]. The decay rate depends on the ν_e–ν_e element of the neutrino mass matrix [87]:

$$M_{ee} = \left|\sum V^2_{ei} m_i\right| . \tag{7.7}$$

[1] $0\nu\beta\beta$ can be induced by various new physics mechanisms including R-parity violating supersymmetry, and models in which the global B–L symmetry is spontaneously broken. In the latter case, the massless Goldstone boson is called the Majoron. The singlet Majoron model of [295] predicts $0\nu\beta\beta$ with the emission of a Majoron and is not ruled out by the invisible width of the Z boson since the Majoron is extremely weakly coupled to neutrinos.

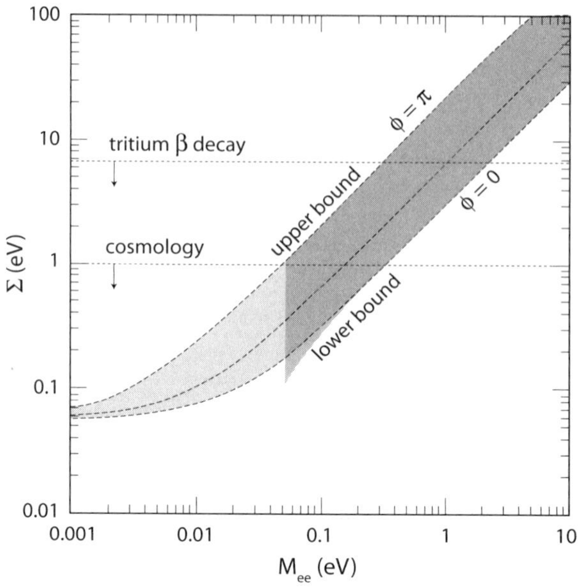

Figure 7.2. Σ vs. M_{ee} for the normal (light shading) and inverted (dark shading) hierarchies. For the inverted hierarchy, $M_{ee} \geq \sqrt{|\delta m_{31}^2|}$. (Here, $|\delta m_{31}^2|$ was taken to be $3 \times 10^{-3}\,\text{eV}^2$.) The 95% C.L. bounds from tritium beta decay and cosmology are shown. Adapted from [297].

In contrast to equation 7.3, the dependence on the unknown Majorana phases permits cancellations in the sum. The prediction is insensitive to θ_{13} and δm_{21}^2 because they are small. Setting $\theta_{13} = 0 = \delta m_{21}^2$, the following relation between M_{ee} and Σ is obtained for both hierarchies [297]:

$$M_{ee} = \left(2\Sigma - \sqrt{\Sigma^2 + 3\delta m_{31}^2}\right) \left|c_{12}^2 + s_{12}^2 e^{i\phi}\right| /3, \tag{7.8}$$

where ϕ is a Majorana phase. For a given measured value of M_{ee} both upper (since $\theta_{12} \neq \pi/4$) and lower bounds are implied for Σ. These bounds are displayed in figure 7.2. The present upper limit on M_{ee} is 0.35 eV at the 90% C.L. [301], with an overall factor of 3 uncertainty associated with the $0\nu\beta\beta$ nuclear matrix elements [302, 303]. A detection of neutrinoless double beta decay, corresponding to $M_{ee} = 0.39\,\text{eV}$, has been reported [304], but this experimental result is highly controversial [305].

Extensive analyses of past, ongoing and future $0\nu\beta\beta$ experiments have been made in [303]. Future experiments include (the decaying nucleus used is shown in parentheses) CUORE (^{130}Te) [306], EXO (^{136}Xe) [307], XMASS (^{136}Xe) [230], GENIUS (^{76}Ge) [308], Majorana (^{76}Ge) [309], MOON (^{100}Mo) [228], and SuperNEMO (^{82}Se) [310]. The upcoming experiments are expected to have sensitivity better than 50 meV, which is the critical mass scale of $\sqrt{|\delta m_{31}^2|}$.

There has been speculation about detecting CP violation using $0\nu\beta\beta$ [298]. However, this will not be feasible until the uncertainties in the nuclear matrix elements can be reliably estimated [299]. Moreover, the further the solar amplitude is constrained away from unity, the more stringent the precision requirement will be on the matrix elements for such a detection to be made even in principle [299]. Even under extremely optimistic assumptions, at best it may be possible to determine whether ϕ is closer to 0 or to π, corresponding to CP conservation.

8

Long-baseline Neutrino Oscillations

Based on our current knowledge and future goals, a future neutrino program will probably include the following objectives:

- Complete the measurement of the neutrino mixing angles;
- Determine the sign of δm_{31}^2;
- Measure δ to determine if CP is violated;
- Search for exotic effects in neutrino oscillations.

Of these future neutrino physics goals, the search for and study of CP violation is of primary importance for several reasons, which we briefly address.

CP violation has so far only been observed in the quark sector of the Standard Model. Its discovery in the neutrino sector should shed additional light on the role of CP violation in nature. Unveiling neutrino CP violation is particularly important because of its potential connection with the observed matter–antimatter asymmetry of our universe, a fundamental problem at the heart of our existence. The leading explanation is currently a leptogenesis scenario in which decays of very heavy right-hand neutrinos created in the early universe give rise to a lepton number asymmetry that later becomes a baryon–antibaryon asymmetry via the B–L conserving 't Hooft mechanism of the Standard Model at weak scale temperatures.

Leptogenesis offers an elegant, natural explanation for the matter–antimatter asymmetry; but it requires some experimental confirmation of its various components before it can be accepted. Those include the existence of very heavy right-handed neutrinos as well as lepton number and CP violation in their decays, but such neutrinos may be well beyond the reach of accelerators.

A number of neutrino mass models have been proposed and precise knowledge of neutrino parameters is essential to test them. Specifically, the value of the mixing angle θ_{13} and whether the mass hierarchy is normal or inverted will help distinguish between models based on lepton flavor symmetries, models with sequential right-handed neutrino dominance, and more ambitious models based on Grand Unified Theory (GUT) symmetries. GUT models naturally yield a normal hierarchy and a relatively large θ_{13} (although in a few unified models, an inverted hierarchy can be obtained with finetuning).

Long-baseline neutrino experiments offer the only way to establish a nonzero θ_{13}, to determine the mass hierarchy *and* to detect neutrino *CP* violation.

8.1 Conventional Neutrino Beams

The K2K [158] and MINOS [71] experiments have both observed a depletion of ν_μ, confirming the value of $\sin^2 2\theta_{23}$ and δm^2_{31} measured in atmospheric neutrinos. Ultimately MINOS will improve the accuracy of both $\sin^2 2\theta_{23}$ and $|\delta m^2_{31}|$; the MINOS measurement of $|\delta m^2_{31}|$ is already more precise than that from atmospheric neutrinos. The CNGS experiments, ICARUS [170] and OPERA [74], at a distance $L = 730\,\text{km}$ but with higher neutrino energy, have begun to take data. Because of the higher beam energy, the appearance of ν_τ should be observed in the CNGS experiments, which would confirm that the primary oscillation of atmospheric neutrinos is $\nu_\mu \to \nu_\tau$; OPERA has observed a ν_τ event [75].

The two parameters that are not determined are $\text{sgn}(\delta m^2_{31})$, which fixes the hierarchy of neutrino masses, and the CP-violating phase δ. The appearance of ν_e in $\nu_\mu \to \nu_e$ oscillations is the most critical measurement, since the probability is proportional to $\sin^2 2\theta_{13}$ in the leading oscillation, for which there is currently only an upper bound (0.16 for $\delta m^2_{31} = 2.4 \times 10^{-3}\ \text{eV}^2$ at the 90% C.L., from the CHOOZ reactor experiment [72]).

The study of $\nu_\mu \to \nu_e$ oscillations also allows one to test for CP violation in the lepton sector [78]. Intrinsic CP violation in the Standard Model requires both $\delta \neq 0, \pi$ *and* $\theta_{13} \neq 0$. In vacuum, the CP asymmetry in the $\nu_\mu \to \nu_e$ channel, to leading order in the mass-squared differences, is

$$\frac{P(\nu_\mu \to \nu_e) - P(\bar{\nu}_\mu \to \bar{\nu}_e)}{P(\nu_\mu \to \nu_e) + P(\bar{\nu}_\mu \to \bar{\nu}_e)} \simeq -\left(\frac{\sin 2\theta_{12} \sin 2\theta_{23}}{2\sin^2\theta_{23}}\right)\left(\frac{\sin 2\Delta_{21}}{\sin 2\theta_{13}}\right)\sin\delta, \qquad (8.1)$$

where $\Delta_{jk} \equiv \delta m^2_{jk} L/4E$. For large-angle solar and atmospheric neutrino mixing the first factor on the right-hand side of equation 8.1 is of order unity. The existence of CP violation therefore requires that the contribution of the sub-leading scale, Δ_{21}, is nonnegligible, so large L/E_ν values are essential. In practice, the CP conserving and CP violating contributions may have similar size [311], depending on the values of L/E_ν and θ_{13}. Furthermore, earth-matter effects can induce fake CP violation, which must be folded into any measurement of δ; on the other hand, matter effects are essential in determining the sign of δm^2_{31}.

The standard proposed method for measuring CP violation is to compare event rates in two charge conjugate oscillations channels, such as $\nu_\mu \to \nu_e$ and $\bar{\nu}_\mu \to \bar{\nu}_e$, for a given L and E_ν. However, there are three two-fold parameter degeneracies that are present when two such measurements are made, which may result in an overall eight-fold degeneracy (a parameter degeneracy occurs when two or more parameter sets are consistent with the same data):

(i) The (δ, θ_{13}) ambiguity [85, 192, 312–314], in which two different parameter pairs, (δ, θ_{13}) and (δ', θ'_{13}), lead to the same values for $P(\nu_\mu \to \nu_e)$ and $P(\bar{\nu}_\mu \to \bar{\nu}_e)$.

(ii) The sgn(δm^2_{31}) ambiguity [85, 312, 315, 316], where (δ, θ_{13}) for one sign of δm^2_{31} gives the same values for the oscillation probabilities as (δ', θ'_{13}) with the opposite sign of δm^2_{31}.

(iii) The $(\theta_{23}, \frac{\pi}{2} - \theta_{23})$, or θ_{23} octant, ambiguity [85, 317], where $(\theta_{23}, \delta, \theta_{13})$ gives the same values for the oscillation probabilities as $(\frac{\pi}{2} - \theta_{23}, \delta', \theta'_{13})$. This ambiguity exists because the channel used to determine θ_{23}, ν_μ survival, only measures $\sin^2 2\theta_{23}$. The ambiguity vanishes at the experimentally preferred value of θ_{23} $(= \frac{\pi}{4})$.

We emphasize that these degeneracies are exact, i.e., there are different sets of parameters that give *identical* predictions. Thus they are present even in the limit of no experimental uncertainties. In each case a duplicity in inferred values of δ and θ_{13} is possible; thus each of these degeneracies may confuse CP-violating parameter sets with CP-conserving ones, and *vice versa*. An overview of these parameter degeneracies can be found in [85, 318].

In many cases these degeneracies persist for all experimentally allowed values in the (δ, θ_{13}) plane. In fact, they often occur for measurements of *any* two neutrino and/or antineutrino appearance probabilities, even if they are not at the same L and E_ν [319], as long as each measurement is made at fixed L and E. Making a third appearance measurement resolves the (δ, θ_{13}) ambiguity and reduces the regions where the remaining degeneracies occur to lines in (δ, θ_{13}) space. Making a fourth appearance measurement reduces the occurence of these degeneracies to isolated points in the (δ, θ_{13}) plane. A fifth measurement then in principle removes all remaining degeneracies.

Different measurements may be obtained by (i) using a different oscillation channel (i.e., antineutrino versus neutrino), or (ii) using a different beam energy and/or baseline. If a neutrino beam with a range of energies is used (a so-called wide-band beam [93–95]), and there is sufficient energy resolution for the detected neutrino, then in principle multiple measurements are possible with a single detector and all degeneracies can be resolved if there are a sufficient number of signal events above background.

There are two special baselines that are valuable to resolve some of these parameter degeneracies:

(i) The detector is located at a distance that corresponds to the first peak of the leading oscillation $(\Delta_{31} = \frac{\pi}{2})$:

$$L \simeq 500 \text{ km} \left(\frac{E}{1 \text{ GeV}}\right) \left(\frac{2.5 \times 10^{-3} \text{ eV}^2}{\delta m^2_{31}}\right). \tag{8.2}$$

Then the $\nu_\mu \to \nu_e$ probability depends only on $\sin\delta$ and not $\cos\delta$ (see equation 3.31), the (δ, θ_{13}) degeneracy is broken, and θ_{13} is uniquely determined for a given sgn(δm^2_{31}) and θ_{23} [85, 312]. There is a residual $(\delta, \pi - \delta)$ degeneracy, but this degeneracy does not mix CP-violating and CP-conserving solutions. Figure 8.1 shows the remaining degeneracies when L/E_ν is chosen to be at the first peak of the oscillation. Furthermore, if L is taken to be very long, large matter effects will break the sgn(δm^2_{31}) ambiguity.

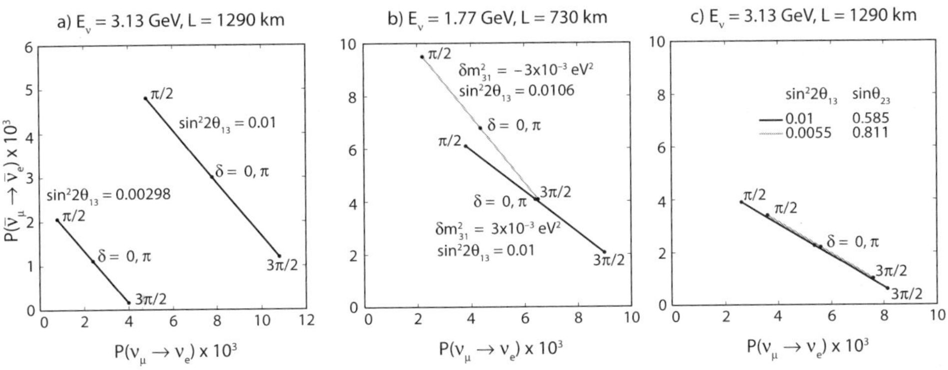

Figure 8.1. Remaining degeneracies when $\Delta_{31} = \frac{\pi}{2}$ for (a) the (δ, θ_{13}) ambiguity, (b) the $\text{sgn}(\delta m^2_{31})$ ambiguity, and (c) the $(\theta_{23}, \frac{\pi}{2} - \theta_{23})$ ambiguity. In (a), each value of $\sin^2 2\theta_{13}$ describes a distinct line in probability space. In (b), the ambiguity in $\sin^2 2\theta_{13}$ is small, but in the overlap region there is still an ambiguity in $\text{sgn}(\delta m^2_{31})$ and a corresponding large uncertainty in δ. In (c), the ambiguity in δ is small, but there may be a large uncertainty in $\sin^2 2\theta_{13}$ when $\theta_{23} \neq \frac{\pi}{4}$. In all cases there remains a $(\delta, \pi - \delta)$ ambiguity since only $\sin\delta$ is being measured. Adapted from [85].

The minimum distance needed depends on the size of θ_{13} and δm^2_{21}, but generally $L \geq 1000\,\text{km}$ is required.

(ii) The detector is located at a distance such that $\hat{A}\Delta_{31} = G_F N_e L/\sqrt{2} \simeq \pi$, which for the earth's density profile implies $L \simeq 7600\,\text{km}$. Then only the leading oscillation term survives in equations 3.31, 3.32, 3.38, and 3.39, and the oscillation probabilities for ν_e appearance are independent of δ *and* δm^2_{21} [85]. This allows an unambiguous measurement of θ_{13} [320, 321] (modulo the $(\theta_{23}, \frac{\pi}{2} - \theta_{23})$ ambiguity). This distance is known by the fanciful name "magic baseline" [320].

For each of these special baselines, additional measurements at different L and/or E_ν values would be necessary to break the remaining degeneracies and determine the precise values of δ and θ_{23}.

There are two neutrino experiments that will use off-axis beams to try to make precision measurements of δ and θ_{13} via the $\nu_\mu \rightarrow \nu_e$ channel: the T2K experiment [83], which uses a neutrino beam from the Japan Proton Accelerator Research Complex (J-PARC) at Tokai and Super-K as the far detector, and the NOνA experiment [84], which uses a beam from NuMI [90, 322, 323] at Fermilab and a 15 kiloton, liquid scintillator far detector near Ash River, Minnesota. NOνA also plans to run in antineutrino mode and search for $\bar{\nu}_\mu \rightarrow \bar{\nu}_e$. As discussed in section 2.6, off-axis beams have a much narrower energy spectrum and a suppression of the high-energy tail compared to on-axis beams, which results in lower backgrounds to ν_e events in the detector. The expected sensitivity of these experiments to $\sin^2 2\theta_{13}$, $|\delta m^2_{31}|$ and $\sin^2\theta_{23}$ are shown in table 8.1.

The determination of the mass hierarchy or discovery of CP violation will be difficult in T2K and NOνA due to the (δ, θ_{13}) and mass hierarchy degeneracies. An independent measurement of θ_{13} helps resolves these degeneracies; this has been done in reactor neutrino experiments, discussed in the next section.

TABLE 8.1
90% C.L. sensitivities of T2K and NOνA for discovering a nonzero $\sin^2 2\theta_{13}$, and for measuring $|\delta m^2_{31}|$ and $\sin^2\theta_{23}$.

Experiment	$\sin^2 2\theta_{13}$	$\|\delta m^2_{31}\|$	$\|\sin^2\theta_{23} - 0.5\|$
T2K	0.004–0.027	$^{+2.0\%}_{-1.9\%}$	0.055
NOνA	0.005–0.014	$^{+2.5\%}_{-2.0\%}$	0.065

Note: From [324]. The range of values for $\sin^2 2\theta_{13}$ indicates the sensitivity depends on δ.

8.2 Reactor Experiments

A different approach to determining θ_{13} without the complication of parameter degeneracies is to measure $P(\bar{\nu}_e \to \bar{\nu}_e)$ at a reactor experiment using a larger version of the CHOOZ experiment, with a far detector of order a few kilometers from the source and a near detector to monitor the flux [325–330]. Ignoring terms cubic or higher in the small parameters θ_{13} and Δ_{21}, the oscillation probability is given approximately by

$$P(\bar{\nu}_e \to \bar{\nu}_e) = 1 - \sin^2 2\theta_{13}\sin^2\Delta_{31} - c^4_{13}\sin^2 2\theta_{12}\sin^2\Delta_{21}, \tag{8.3}$$

and is therefore independent of δ, θ_{23}, and $\text{sgn}(\delta m^2_{31})$. If the baseline for the far detector is chosen such that $\Delta_{31} \sim \pi/2$ (i.e., at the first oscillation maximum), then the last term on the right-hand side of equation 8.3 may also be ignored. The amplitude of the oscillation then provides a direct measurement of θ_{13}. Reactor neutrino experiments are also relatively quick to perform; the canonical running time for these experiments is three years.

Reactor antineutrinos are detected by interactions with free protons via the inverse beta decay reaction $\bar{\nu}_e p \to e^+ n$, which requires a minimum neutrino energy of 1.8 MeV. This requires a detector material with high hydrogen content (usually organic liquid scintillator). The event rate is the product of the flux, which decreases with energy, and the cross section, which increases with energy (see figure 8.2). The peak of the spectrum occurs at $E_\nu \simeq 3.8$ MeV, which means that the first oscillation maximum occurs at around $L \simeq 1.8$ km for $\delta m^2_{31} \simeq 2.5 \times 10^{-3}\,\text{eV}^2$.

Since inverse beta decay has only two particles in the final state, the incident antineutrino energy is directly related to the outgoing positron energy E_e and direction θ_e with respect to the incident antineutrino:

$$E_\nu = \frac{1}{2}\frac{2m_p E_e + m_n^2 - m_p^2 - m_e^2}{m_p - E_e + \sqrt{E_e^2 - m_e^2}\cos\theta_e}, \tag{8.4}$$

where the recoil effect of the neutron has been included. For positron energies much smaller than the nucleon mass

$$E_\nu \simeq E_e + m_n - m_p \simeq E_e + 1.8\,\text{MeV}, \tag{8.5}$$

so that the antineutrino energy may be inferred from a measurement of the positron energy.

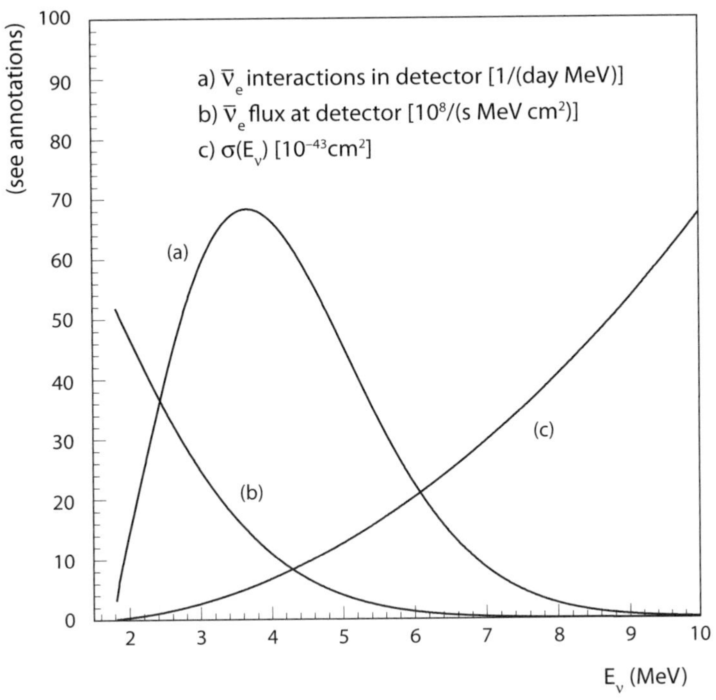

Figure 8.2. (a) Event rate, (b) flux, and (c) cross section for electron antineutrinos from a reactor. From [331].

The neutrino reaction is identified by observation of a prompt positron, followed by a delayed photon from neutron capture. Neutrons may be captured by free protons to form deuterium, with characteristic emission of a 2.2 MeV gamma ray and capture time of 180 μs. The neutron detection efficiency may be increased significantly by loading the detector material with Gadolinium (at about the 0.1% level), which has a much larger neutron capture cross section than a free proton (by more than five orders of magnitude); the subsequent emission of an 8 MeV gamma ray with capture time of about 30 μs provides a distinctive signature for the neutron.

Background reduction is also important, especially at the far detector where the event rate is suppressed by the $1/L^2$ fall-off of the flux. Shielding against natural radiation and spallation neutrons from outside the detector is required. Backgrounds due to cosmic ray muons may be reduced by placing detectors underground; many future experiments will be operated at a reactor site adjacent to hilly terrain, which provides a sizeable overburden (up to several hundred meters) at the far detector position. A muon detection system can help identify any cosmogenic events not eliminated by the shielding.

Reactor experiments measure neutrino survival, which means that a positive signal is a deviation from unity in the oscillation probability. Since θ_{13} is small, uncertainties must be made as small as possible to achieve a precise measurement of θ_{13}; improvements over the CHOOZ experiment can be made by using (i) larger detectors to increase statistics, (ii) near and far detector sites to minimize systematic uncertainties related to the reactor antineutrino flux, (iii) detectors of similar design

TABLE 8.2
Reactor neutrino experiments for measuring θ_{13}.

Experiment	Thermal Power (GW)	Distance (m) Near/Far	Size (tons) Near/Far	$\sin^2 2\theta_{13}$ Sensitivity*
Double Chooz [80]	8.5	275/1050	10/10	0.028
Daya Bay† [81]	11.6	360, 500/1740	40, 40/80	0.008
RENO†† [82]	16.4	500/1450	20/20	0.013
Angra [333]	5.6	300/1500	50/500	0.006

*The $\sin^2 2\theta_{13}$ sensitivity is the 90% C.L. upper bound if no signal is seen, assuming $\delta m^2_{31} = 2.5 \times 10^{-3}$ eV^2.
†The Daya Bay experiment has two values listed for the near detector since there are two reactor sites, and each will have a near detector; the distance to the far detector is the weighted average.
††For RENO the near and far distances are the average values for six reactor sites.

to reduce systematic uncertainties between the near and far detectors, and (iv) the relative spectrum shape (which will be distorted by oscillations) to detect the presence of oscillations. Another key component to the size of the systematic error is how well the Gd fraction and number of free protons is known, both of which affect the normalization of the event rate. While the largest deviation from the unoscillated rate occurs at the first oscillation maximum, the statistical error in measuring the spectrum shape decreases at shorter distances where the event rates are higher. Optimization of detector positions is a complicated function of local overburden variation (which affects the background rate) and the relative size of systematic and statistical uncertainties.

Table 8.2 lists specifics for four different reactor neutrino experiments designed to measure θ_{13}, three of which are taking data or under construction (Double Chooz, Daya Bay, and RENO) and one that is currently in the proposal stage (Angra). Sensitivities as low as $\sin^2 2\theta_{13} = 0.006$ are possible. This value of $\sin^2 2\theta_{13}$ is particularly critical for measuring CP violation, if it exists [332]. If $\sin^2 2\theta_{13}$ is higher, then a combination of beams (such as NOνA and T2K) and reactor measurements may be sufficient to determine θ_{13} and δ.

Double Chooz

The Double Chooz reactor neutrino experiment [80] has begun taking data using identical 10-ton near and far detectors, each doped with 0.1% gadolinium, with the far detector at the underground site of the original CHOOZ experiment, 1.05 km from the reactor cores. The expected overall systematic uncertainty in Double Chooz is 0.6%, compared to 2.7% in CHOOZ. For $\delta m^2_{31} = 2.5 \times 10^{-3}$ eV^2, a running time of three years can place a 90% C.L. upper bound on $\sin^2 2\theta_{13}$ of 0.028, and will give a 3σ discovery reach of $\sin^2 2\theta_{13} = 0.052$. First results from Double Chooz give a measurement of $\sin^2 2\theta_{13} = 0.086 \pm 0.041$ (stat) $\pm$ 0.030 (syst), which is nonzero at the 90% C.L. [80].

The key advantage of the Double Chooz experiment is that a five-fold improvement in the CHOOZ limit may be obtained relatively quickly and cheaply due to the

Figure 8.3. Schematic design of the Daya Bay reactor neutrino experiment. Each reactor site (Daya Bay, Ling Ao, and Ling Ao II) has two closely-spaced reactor cores. Each near detector site has two 20-ton detector modules and the far detector site has four 20-ton modules. Access tunnels are shown by the black lines. From [81].

existing infrastructure at the CHOOZ site. The improvement of Double Chooz over CHOOZ comes from several factors: (i) the use of a near detector to monitor the neutrino flux, (ii) the use of two 10-ton detectors instead of a single 5-ton detector, (iii) a longer running time, (iv) the reactors will be running at full power and (v) significantly improved systematic errors. The overall exposure of Double Chooz will be 250 ton-GW-years, about 20 times that of CHOOZ.

Daya Bay

The Daya Bay reactor neutrino experiment [81] is being constructed near the Daya Bay and Ling Ao nuclear reactors in China, each with a thermal power of 5.8 GW. The two reactor complexes are about 1.1 km apart; therefore two near detectors are needed to reduce systematic errors due to the reactor antineutrino fluxes, one close to Daya Bay and the other close to Ling Ao. There will also be one far detector site. To reduce systematic uncertainties, the Daya Bay experiment will have identical 20-ton detector modules, with four modules at the far site and two modules at each of its near sites (see figure 8.3).

A series of tunnels in the Daya Bay experiment will allow the swapping of detector modules between near and far sites, which can also reduce systematic uncertainties.

TABLE 8.3
Detector to core distances.

Core site*	DB	LA	Far
Daya Bay	363	1347	1985
Ling Ao	857	481	1618
Ling Ao II	1307	526	1613

*Distances (in m) from the Daya Bay near (DB), Ling Ao near (LA), and far (Far) detectors to the midpoints of the Daya Bay, Ling Ao, and Ling Ao II cores [81].

This may be simply illustrated in the case of one reactor and two detectors, one near to and one far from the reactor. If the two detectors have event rates R_1 and R_2 at the near site, then a measurement with detector #1 at the near site and detector #2 at the far site gives a near/far ratio of

$$\frac{N_1}{F_2} = \frac{R_1}{R_2}\frac{L_F^2}{L_N^2}, \tag{8.6}$$

where L_N and L_F are the distances from the reactor to the near and far sites, respectively. After the two detectors are swapped, a second measurement gives a near/far ratio of

$$\frac{N_2}{F_1} = \frac{R_2}{R_1}\frac{L_F^2}{L_N^2}. \tag{8.7}$$

If the rates are related by $R_2 = R_1(1+\delta)$, where δ represents a small fractional difference in detector normalization (due to different numbers of free protons, Gd fractions, efficiencies, etc.), then the average near/far ratio is

$$\frac{1}{2}\left(\frac{N_1}{F_2} + \frac{N_2}{F_1}\right) = \frac{L_F^2}{L_N^2}\left(1 + \frac{\delta^2}{2(1+\delta)}\right) \simeq \frac{L_F^2}{L_N^2}\left(1 + \frac{\delta^2}{2}\right). \tag{8.8}$$

Thus a difference in the detector normalizations as large as 5% leads to a fractional uncertainty in N/F that is only of order 10^{-3}.

After the start of the Daya Bay experiment, a second pair of reactor cores (Ling Ao II) will be added approximately 0.4 km (1.5 km) from the original Ling Ao (Daya Bay) site, which will increase the total neutrino output by 50%. The Ling Ao near detector site will be situated approximately equidistant from Ling Ao and Ling Ao II to help reduce the systematic uncertainty in the reactor fluxes (see figure 8.3). The distances from the three core sites to the three detector sites are listed in table 8.3. With three years of running time, the Daya Bay experiment can place a 90% C.L. upper bound on $\sin^2 2\theta_{13}$ of about 0.008–0.010, depending on the value of δm_{31}^2. Daya Bay has a 3σ discovery reach of $\sin^2 2\theta_{13} = 0.015$.[1]

[1] A rate-only analysis of 55 days of data taken by the Daya Bay experiment gives $\sin^2 2\theta_{13} = 0.092 \pm 0.016\,(\text{stat.}) \pm 0.005\,(\text{syst.})$, a discovery of nonzero θ_{13} at the 5.2σ C.L., see arXiv:1203.1669 [hep-ex].

TABLE 8.4
Distances from the near and far RENO detectors to the reactor cores [82].

Reactor #	Near Detector (m)	Far Detector (m)
1	668	1557
2	452	1456
3	305	1396
4	336	1381
5	514	1414
6	739	1490

RENO

The RENO reactor neutrino experiment [82] is being constructed at the Yonggwang nuclear power plant in South Korea, the second largest in the world. Six reactor cores lie equally spaced along a line, with the near and far detectors an average distance of 500 and 1450 m, respectively, from the reactor cores. Table 8.4 shows the distance from each core to the near and far detectors. Both detectors will be 20 tons. With three years of running time a 90% C.L. upper bound on $\sin^2 2\theta_{13}$ of about 0.013 (0.026) is expected, assuming 0.5% (1.0%) total systematic error.[2]

Angra

The Angra reactor neutrino experiment [333] is planned at the site of a nuclear power reactor located near Angra dos Reis in Brazil. Although the reactor power is somewhat less than that of other experiments (only 5.6 GW), a much larger far detector (500 tons) compensates for this, leading to a very competitive 90% C.L. upper bound on $\sin^2 2\theta_{13}$ of 0.006. This experiment also has the largest overburden at the far detector site, about 700 m.

Combining Results from Accelerators and Reactors

If a precise measurement of $\sin^2 2\theta_{13}$ is made with reactors, it removes the (δ, θ_{13}) degeneracy. Therefore, neutrino beams from accelerators and reactors are very complementary experiments. However, actual sensitivities still depend on the value of δ. For $\sin^2 2\theta_{13} = 0.1$, the mass hierarchy (i.e., the sign of δm^2_{31}) may be determined and CP violation may be discovered for approximately half of the values of δ when the results from T2K, NOνA, and reactor experiments are combined. However, the sensitivity to the mass hierarchy (CP violation) disappears completely for $\sin^2 2\theta_{13} < 0.04(0.02)$ [324]. If $\sin^2 2\theta_{13}$ is below these values, upgrades of the accelerator experiments will be needed, as discussed in the next section.

[2] A rate-only analysis of 229 days of data taken by the RENO experiment gives $\sin^2 2\theta_{13} = 0.113 \pm 0.013\,(\text{stat.}) \pm 0.019\,(\text{syst.})$, which is 4.9σ from zero; see arXiv:1204.0626v2 [hep-ex].

8.3 Superbeams

In order to reduce correlations, especially between θ_{13} and δ, different strategies are possible for superbeam upgrades [92,334], with neutrino fluxes increased by a factor of 2 or more [83, 335, 336], which will allow smaller values of θ_{13} to be probed. Off-axis beams have a narrow beam energy, permitting a counting experiment at an oscillation maximum with low background. For experiments using the off-axis technology, a second detector at a different location can provide complementary information for a different L and/or E. Alternatively, wide-band beams have a higher flux and allow an experiment that utilizes spectral energy information, but requires detectors with relatively good energy resolution and neutral-current rejection to reduce backgrounds. Here we describe some superbeam scenarios:

(i) Having detectors at different off-axis angles allows one to modify both the baseline *and* neutrino energy. Multiple detectors utilizing an off-axis beam such as the one at Fermilab can eliminate degeneracies for larger values of $\sin^2 2\theta_{13}$ [319]. Alternatively, having detectors at two different on-axis distances from the same superbeam can provide multiple measurements that help to remove parameter degeneracies. One such possibility, called T2KK [337], is to combine a moderate distance such as Tokai to Super-K ($L = 295$ km), with a longer distance such as Tokai to Korea ($L \simeq 1050$ km) using a ν_μ beam with average energy around 600 MeV.

(ii) Another approach is to use a wide-band superbeam, for example from Fermilab to the Deep Underground Science and Engineering Laboratory (DUSEL) [94], with $L \simeq 1300$ km. This is the goal of the Long Baseline Neutrino Experiment (LBNE) [95]. The measurement of quasielastic events allows a determination of the neutrino energy with reasonable precision. The lower-energy events are more sensitive to the δ terms in the oscillation probability, while the higher-energy events are more sensitive to the sign of δm^2_{31}. Again, binning the quasielastic events according to energy is roughly equivalent to running many narrow-band beams simultaneously, which can help to resolve neutrino parameter degeneracies. In principle only a neutrino beam is required, but event rates could be low for $\delta m^2_{31} < 0$. Also running with an antineutrino beam would provide essential confirmation (especially of CP violation) and probe lower in θ_{13} if $\delta m^2_{31} < 0$. A short-baseline beam using $\bar{\nu}_\mu$ from pion and muon decay-at-rest could also be a complement to a conventional ν_μ beam; combining the results of the two experiments can give an enhanced sensitivity to CP violation [338].

(iii) For shorter source to detector distances a lower neutrino energy is required to sit near the peak of the first oscillation. One possibility is to direct 250–350 MeV ν_μ and $\bar{\nu}_\mu$ beams from the CERN Super Proton Linac (SPL) to the MEMPHYS detector at Frejus ($L = 130$ km) [339–341].

(iv) Two superbeam experiments do significantly better than one in parameter determinations [342, 343]. For example, the combination of data from an upgraded T2K experiment ($\theta_{\text{off-axis}} = 2$ deg, $E_\nu \simeq 0.6$ GeV, $L = 295$ km, 22.5 kt water Cherenkov) and an upgraded NuMI to southwestern Ontario experiment (1 deg, $E_\nu = 1.8$ GeV, $L \simeq 900$ km, 20 kt low-Z calorimeter), both with 2 years ν_μ running and 6 years $\bar{\nu}_\mu$ running, would be sensitive to the sign of δm^2_{31} and to CP violation for $\sin^2 2\theta_{13} \gtrsim 0.03$ [342]. Running with only ν_μ in both experiments may allow one to determine $\text{sgn}(\delta m^2_{31})$ if θ_{13} is not too small [344], although θ_{13} and δ will not be well-measured without $\bar{\nu}_\mu$ data.

In any long-baseline neutrino experiment there is a trade-off between detector size and the ratio of the ν_e CC signal events to background: generally speaking, detector technologies that allow a bigger reduction in background cannot be built as large [92]. Four types of detectors that have been studied for $\nu_\mu \to \nu_e$ detection are: (i) water Cherenkov (backgrounds of order 10^{-2} of the number of unoscillated CC events, maximum fiducial volume of order 500 kt) [83, 96]; (ii) iron scintillator (3×10^{-3}, 50 kt) [323,345,346]; (iii) liquid argon (3×10^{-3}, 50 kt) [97], and (iv) low-Z calorimetric [347]. There is also an additional background of order 3×10^{-3} due to ν_e contamination in the beam. The larger water Cherenkov detectors generally do better when the neutrino flux is less (such as for a conventional beam before superbeam upgrade or for baselines $\geq$ 4000 km where there is a large $1/L^2$ fall-off of the flux), whereas the smaller detectors that can measure $e^\pm$ positions on a finer scale generally do better when there is more flux (such as with superbeams or for baselines below 4000 km). One study [348] has shown that for the same neutrino luminosity, a liquid Argon detector gives roughly the same sensitivity to $\sin^2 2\theta_{13}$, the mass hierarchy, and CP violation as does a water Cherenkov detector six times larger.

8.4 Neutrino Factories

A neutrino factory (NuFact) [99] is perhaps the ultimate technology for neutrino oscillation studies. Muons are injected into a storage ring, and their decays will give neutrino beams in the directions of the straight sections of the ring. Stored muon energies in the range 5–50 GeV have been considered in design studies.

The neutrino energy spectrum and flux from a neutrino factory are discussed in section 2.6. An entry-level NuFact produces a time integrated number of decaying muons in the straight section of the ring $n_0 \sim 10^{20}$ and a high-performance NuFact has $n_0 \sim 10^{21}$. Early studies of the capabilities of a NuFact can be found in [100, 349–352]. See also the more recent report in [353].

Golden Channel: $\nu_e \to \nu_\mu$

Since the sign of the detected lepton is critical for determining the oscillation channel being observed, most studies have focused on the $\nu_e \to \nu_\mu$ and $\bar{\nu}_e \to \bar{\nu}_\mu$ oscillation channels with final state muon detection. Employing a magnetized iron detector allows the determination of the sign of the detected charged lepton, and the backgrounds are quite small, of order 3×10^{-5}. This small background compared to ν_e detection in a superbeam, plus the fact that NuFact neutrino fluxes can be one or two orders of magnitude greater than that of a superbeam, are the two main reasons a NuFact is superior.

There have been many studies of the physics capabilities of a NuFact using muon appearance [192, 316, 354–357] (for a discussion of the physics that can be done with electron appearance, see [358]). By comparing $\nu_e \to \nu_\mu$ and $\bar{\nu}_e \to \bar{\nu}_\mu$ event rates, factoring in earth-matter effects, very precise determinations of the

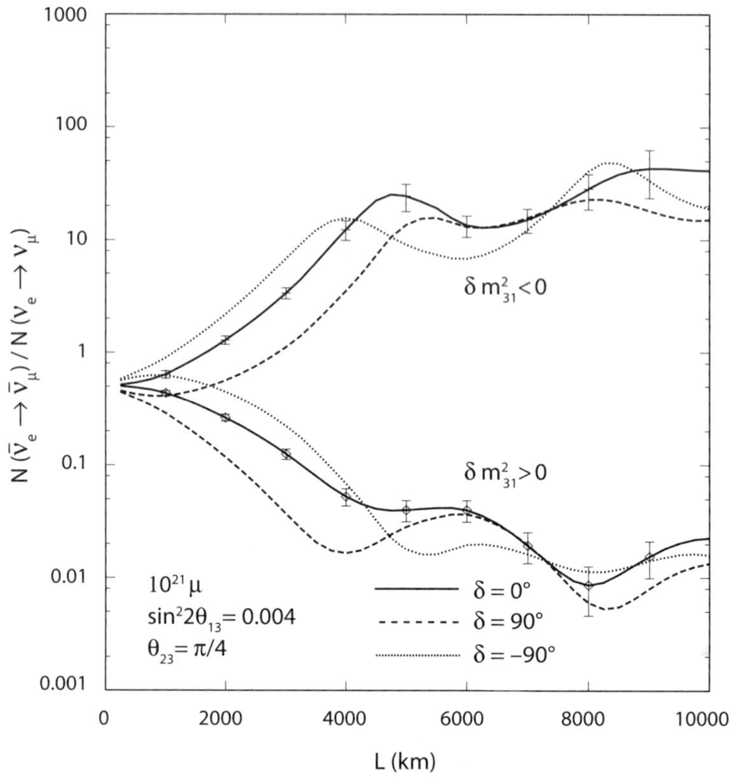

Figure 8.4. Ratio of antineutrino to neutrino appearance events versus baseline in a neutrino factory for $\sin^2 2\theta_{13} = 0.004$ and several values of δ. Both $\delta m^2_{31} > 0$ and $\delta m^2_{31} < 0$ cases are shown. Adapted from [355].

oscillation parameters θ_{13}, δ, and $\mathrm{sgn}(\delta m^2_{31})$ can be made. Figure 8.4 shows the ratio of antineutrino to neutrino muon appearance events versus baseline for $\delta m^2_{31} > 0$ and $\delta m^2_{31} < 0$ for $\sin^2 2\theta_{13} = 0.004$ and several values of δ. The different signs of δm^2_{31} are clearly distinguishable when $L \geq 2000\,\mathrm{km}$, and these measurements are especially sensitive to the amount of CP violation when $L \sim 3000\,\mathrm{km}$. In practice, if a NuFact is run with μ^- about twice as long as with μ^+, then the total number of CC events in the neutrino and antineutrino channels will be about the same since the $\bar{\nu}_e$ cross section is about half that of the ν_e cross section.

One obstacle to making precise measurements of δ and θ_{13} and determining the sign of δm^2_{31} is that the other neutrino mass and mixing parameters, i.e., θ_{23}, θ_{12}, δm^2_{21}, and δm^2_{31}, may not be precisely known. Measurements of ν_μ survival in a superbeam or NuFact will reduce current uncertainties in $\sin^2 2\theta_{23}$ and δm^2_{31}.

As with superbeams, there is the possibility of having an eight-fold parameter degeneracy using neutrino and antineutrino event rates in a NuFact. The $\mathrm{sgn}(\delta m^2_{31})$ ambiguity can be resolved by choosing a baseline $\geq 2000\,\mathrm{km}$. The ambiguity is easier to resolve for large θ_{13} (due to the larger matter effect) and small δm^2_{21} (due to the smaller size of the CP violating term in the oscillation probability). There have been a number of different proposals for resolving the (δ, θ_{13}) ambiguity in the golden channel:

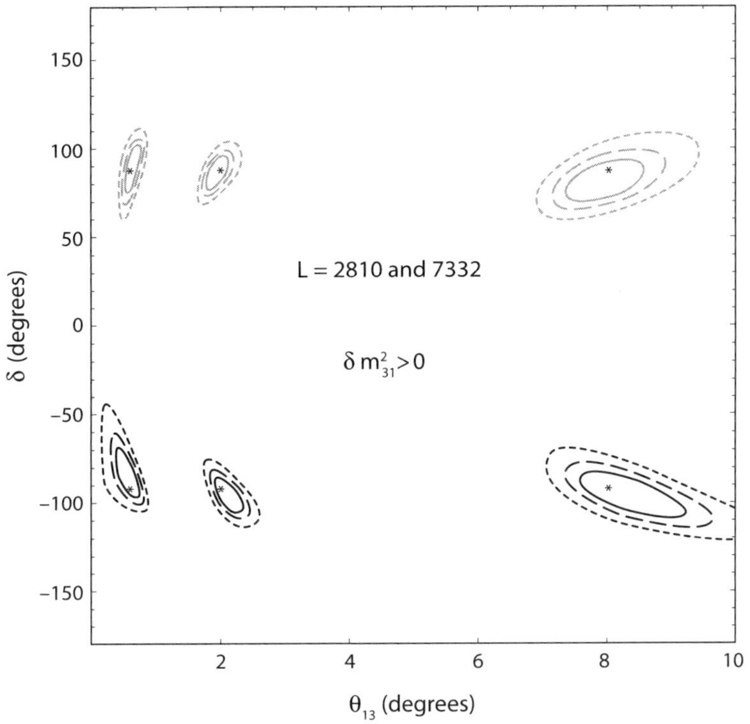

Figure 8.5. Fits to δ and θ_{13} at a neutrino factory using hypothetical results from two baselines, 2810 km and 7332 km, for several input values of δ and θ_{13} (and $\theta_{23} = \pi/4$). The three curves in each case represent the 90%, 95%, and 99% C.L. ranges of allowed parameters. Expected uncertainties in the oscillation parameters have been included, in addition to a 1% uncertainty in the matter density. From [192].

(i) Since a NuFact has a broad spectrum of neutrino energies, measuring the energy of the detected muon gives information about the modulation of the oscillation probability with E_ν. A 10% muon energy resolution is sufficient to remove the (δ, θ_{13}) ambiguity [356]. A combined fit with the ν_μ survival channel can also improve the measurement of δ, θ_{13}, and $\mathrm{sgn}(\delta m^2_{31})$. An understanding of parameter correlations and degeneracies is essential for extracting meaningful constraints from the data [314, 356].

(ii) Another idea is to combine $\nu_e \to \nu_\mu$ and $\bar{\nu}_e \to \bar{\nu}_\mu$ measurements from neutrino factory experiments at two baselines. Having one detector at $L \simeq$ 3000 km and another at $L \simeq$ 7300 km (e.g., the Fermilab to Gran Sasso distance) would provide good discrimination between degenerate solutions for a wide range of δ and θ_{13} [192] (see figure 8.5). Measurements at 7300 km, near the magic baseline, where the δ and δm^2_{21} dependence is minimal [85], would provide an unambiguous measurement of the leading oscillation amplitude $\sin^2\theta_{23}\sin^2 2\theta_{13}$ [320]. With a race-track design (see figure 2.11) this scenario would require two separate runs for both stored μ^+ and μ^-, whereas a triangular configuration could allow data to be taken simultaneously at two baselines.

TABLE 8.5
Reach* in $\sin^2 2\theta_{13}$.

Machine	Discovery	$\mathrm{sgn}(\delta m^2_{31})$	CP violation
High-energy NuFact	$2\times10^{-5} - 7\times10^{-5}$	$3\times10^{-5} - 1\times10^{-4}$	1×10^{-4}
Low-energy NuFact	$4\times10^{-5} - 3\times10^{-4}$	$6\times10^{-4} - 2\times10^{-2}$	2×10^{-4}

*Approximate 3σ reaches in $\sin^2 2\theta_{13}$ for measuring θ_{13}, determining the $\mathrm{sgn}(\delta m^2_{31})$, and detecting CP violation in high-energy and low-energy neutrino factories. For θ_{13} and $\mathrm{sgn}(\delta m^2_{31})$ the upper end of the range is possible for all values of δ, while the lower range is achievable only for the optimal value of δ. For CP violation the reach assumes that a positive signal can be achieved for 50% of δ values. From [362].

(iii) Since a superbeam will most certainly be a precursor to a NuFact, it is quite natural to combine data from a superbeam and a NuFact to help resolve the (δ, θ_{13}) ambiguity. Studies show that it is possible to remove this ambiguity for $\sin^2 2\theta_{13} \geq 0.0005$ [359]. In this sense, superbeams and neutrino factories are complementary.

It should be noted that even if θ_{13} were zero, subleading terms in the oscillation probability associated with δm^2_{21} can lead to observable effects in appearance experiments [355].

Since the luminosity of the neutrino beams from a neutrino factory increases as the square of the parent muon energy (see equation 2.49), generally muon energies from 20 to 50 GeV are considered. Since θ_{13} is not below about 2°, then a low-energy neutrino factory [360, 361] with muon energy less than 5 GeV also has good sensitivity to the parameters for baselines less than about 2000 km.

A summary of typical capabilities of future neutrino factory experiments is given in table 8.5 for both high-energy and low-energy options, assuming only that the ν_e and $\bar{\nu}_e$ appearance *and* the ν_μ and $\bar{\nu}_\mu$ survival channels are used; a summary of many possible baseline options is given in [362]. A high-energy neutrino factory would have sensitivity down to 10^{-4} and possibly lower in $\sin^2 2\theta_{13}$ for both determining $\mathrm{sgn}(\delta m^2_{31})$ and detecting CP-violation at the 3σ level. A low-energy neutrino factory would be two to three times less sensitive in detecting CP violation and one to two orders of magnitude less sensitive in determining $\mathrm{sgn}(\delta m^2_{31})$; the latter difference is primarily due to the shorter baseline required at lower energies, for which matter effects are smaller. A 1% determination of δm^2_{31} should also be possible at a NuFact. Thus, it appears that precision reconstruction of the neutrino mixing matrix would be possible with a neutrino factory.

There are many factors to consider when determining the sensitivity of a given long-baseline experiment, e.g., the neutrino energy spectrum, the level of the background and the energy resolution of detected particles. The software package General LOng Baseline Experiment Simulator (GLoBES) [363] is a good resource for calculating sensitivities in long-baseline experiments.

Silver Channel: $\nu_e \to \nu_\tau$

Another advantage of a NuFact over a superbeam is that it has available the oscillation channel $\nu_e \to \nu_\tau$. The theoretical oscillation probabilities for $\nu_e \to \nu_\tau$ and $\bar{\nu}_e \to \bar{\nu}_\tau$ can be obtained from those for $\nu_e \to \nu_\mu$ and $\bar{\nu}_e \to \bar{\nu}_\mu$ by the transformations $\sin\theta_{23} \leftrightarrow \cos\theta_{23}$ and $\delta \to -\delta$. Comparing the muon and tau channels in a NuFact therefore allows one to (i) help resolve the $(\theta_{23}, \frac{\pi}{2} - \theta_{23})$ ambiguity since the leading term in the oscillation probability is proportional to $\sin^2\theta_{23}$ for $\nu_e \to \nu_\mu$ and $\cos^2\theta_{23}$ for $\nu_e \to \nu_\tau$ [85], and (ii) help resolve the (δ, θ_{13}) ambiguity since the δ dependence is different in the two channels [364].

8.5 Beta Beams

Beta beams [98, 365] provide a pure source of either ν_e or $\bar{\nu}_e$, depending on whether the parent nucleus is an e^+ or e^- emitter, respectively. For lifetimes much less than 1 s, decay losses during acceleration are too large, while if the lifetimes are too long there are not enough decays to produce an intense beam; most nuclei being considered have a lifetime of order 1 s. The most common isotope discussed for e^+ (e^-) emission is ^{18}Ne (^{6}He), with an average ν_e ($\bar{\nu}_e$) energy of 1.86 MeV (1.94 MeV) in the nuclear rest frame. In the standard scenario [366], the nuclei are completely stripped of electrons and accelerated to $\gamma \approx 60$ for ^{6}He and 100 for ^{18}Ne, which leads to average neutrino energies of 230 MeV and 370 MeV in the lab frame for $\bar{\nu}_e$ and ν_e, respectively (see equation 2.45). The ratio $\gamma_{\rm He}/\gamma_{\rm Ne} = 3/5$ allows one to simultaneously circulate ^{6}He and ^{18}Ne, since these will have the same radius of curvature, $r = mv\gamma/(Bq)$, for a given magnetic field.

The peak of the leading oscillation occurs for $L/E_\nu = 2\pi/\delta m^2_{31} \approx 500$ km/GeV for $\delta m^2_{31} = 2.5 \times 10^{-3}\,\text{eV}^2$. Since the cross section $\sigma \propto E_\nu$, $E_\nu \propto \gamma$ (see equation 2.45), and the neutrino flux $\Phi \propto \gamma^2/L^2$ (see equation 2.46), the event rate $\sigma\Phi \propto \gamma$ when sitting at the peak of the oscillation.

For this standard scenario, the peak of the oscillation occurs at a baseline of $L \approx 130$ km, which is approximately the distance from CERN to Frejus. A large water Cherenkov detector is ideally suited to the relatively low neutrino energy. Using the appearance channels $\nu_e \to \nu_\mu$ and $\bar{\nu}_e \to \bar{\nu}_\mu$ with 4400 kiloton-year exposure, the 90% C.L. sensitivity to $\sin^2 2\theta_{13}$ lies in the 3×10^{-4} to 3×10^{-3} range, depending on the value of δ [366]. The (θ_{13}, δ) ambiguity can be resolved by measuring the energy dependence of the signal. The sgn(δm^2_{31}) ambiguity is not resolved because the matter effects are small at this baseline. A subsequent study has shown that increasing γ to more than 100 can improve the sensitivity to θ_{13} and δ even for $L \sim 130$ km; a γ ratio of 150/250 would have an optimal baseline near 300 km [367].

A higher-γ option [368] with $\gamma_{\rm He}/\gamma_{\rm Ne} = 350/580$ (average ν_e ($\bar{\nu}_e$) energy 2160 MeV (1360) MeV) has also been discussed, which may be possible at CERN with a Super Proton Synchrotron (SPS) accelerator upgrade. In this case the optimal baseline is near 730 km, the CERN to Gran Sasso distance. Sensitivity to θ_{13} is improved and the larger matter effects also allow a sgn(δm^2_{31}) determination for $\sin^2 2\theta_{13} \geq 0.02$ at 99% C.L. for 4000 kt-yr exposure. Using the LHC, $\gamma_{\rm He}/\gamma_{\rm Ne} = 1500/2500$ would be possible, with an optimal baseline near 3000 km. In this case, average neutrino energies are 6–9 GeV, and the water Cherenkov detector is no longer suitable. Due to large matter effects, a 40 kt iron calorimeter at 3000 km with γ ratio 1500/2500 has increased sensitivity to sgn(δm^2_{31}), with some loss in *CP* sensitivity [368].

Another beta beam option is to use nuclear decays with higher average neutrino energy in the nuclear rest frame [369]. In this case ^{8}B (^{8}Li), with average neutrino energy 7.37 MeV (6.72 MeV) in the nuclear rest frame, could be used to produce a ν_e ($\bar{\nu}_e$) beam. Simultaneous ν_e and $\bar{\nu}_e$ beams are possible for $\gamma_{\rm B}/\gamma_{\rm Li} = 5/3$.

For $\gamma = 500$, the average neutrino energy of around 7 GeV is close to the resonance energy when the baseline has the "magic" value of ≈ 7600 km (where the appearance probabilities depend on θ_{13}, but not δ or δm^2_{21}). The increased oscillation probabilities on resonance help compensate for the reduced flux at the longer distance. For 250 kt-yr exposures in both the $\nu_e \to \nu_\mu$ and $\bar{\nu}_e \to \bar{\nu}_\mu$ channels, a 90% C.L. sensitivity in $\sin^2 2\theta_{13}$ of about 6×10^{-4} is possible [370].

For a monoenergetic neutrino beam using electron capture, the neutrino energy is precisely known. Measurements of both the survival ($\nu_e \to \nu_e$) and appearance ($\nu_e \to \nu_\mu$) channels made at two different boost factors are sufficient to disentangle the (δ, θ_{13}) ambiguity. Electron capture in ^{150}Dy gives a final state neutrino with $E = 1.4$ MeV in the nuclear rest frame. With baseline $L = 650$ km and 2200 kt-yr exposures using a water Cherenkov detector at the boost factors $\gamma = 195, 400$, $\sin^2 2\theta_{13}$ sensitivity of order 10^{-3} is possible [371]. Combining superbeam and beta beam results can also improve sensitivity [372].

8.6 Comparing Long-baseline Experiments

Designing for *CP* violation studies in next generation neutrino programs has important benefits. First, the degree of difficulty to establish *CP* violation is high but achievable. It requires an intense proton beam of about 1–2 MW and a very large detector, 100–500 kt Water Cherenkov (WC) or a liquid argon (LArTPC) detector of size ~ 100 kt, which could be equivalent in sensitivity due to its better performance [373]. Water Cherenkov is an established technology, while liquid argon, which promises superior particle identification and control over backgrounds, is still under development. An ambitious infrastructure will allow very precise measurements of all neutrino oscillation parameters as well as the mass hierarchy via $\nu_\mu \to \nu_\mu$ disappearance and $\nu_\mu \to \nu_e$ appearance studies.

The design characteristics of some future long-baseline experiments (superbeams, neutrino factories and beta beams) are listed in table 8.6. The superbeam experiments are WBB-WC (wide-band beam with a water Cherenkov detector), T2KK (Tokai to Kamioka and Korea) and SPL (CERN to Frejus). The neutrino factory experiments include both a high-energy (International Design Study Neutrino Factory, IDS-NF) and low-energy (LENF) option, while the beta beam experiments have one detector and two sources at a single baseline (BB2, with ^{6}He and ^{18}Ne sources) or two detectors each with two sources at a different baseline (BB4, with ^{6}He and ^{18}Ne at a shorter baseline and ^{8}B and ^{8}Li at a longer baseline). For the beta beam experiments, where neutrinos are not created by colliding a proton beam on a target, the proton power is not applicable; the number of useful ion decays per year are 3×10^{19} for BB2 and 9×10^{18} for BB4. A summary of typical capabilities for these experiments for discovering a nonzero θ_{13}, determining $\text{sgn}(\delta m^2_{31})$, and measuring *CP* violation is given in figure 8.6.

For θ_{13} discovery, a high-energy neutrino factory has the best reach, followed by low-energy neutrino factories and beta beams; superbeams are not sensitive to $\sin^2 2\theta_{13} < 10^{-3}$. For $\text{sgn}(\delta m^2_{31})$, high-energy neutrino factories provide the best sensitivity; the wide-band superbeam is competitive in some cases with a beta beam or a low-energy neutrino factory. For larger values of $\sin^2 2\theta_{13}$, close to 0.10 or higher, superbeams are competitive with the other technologies in measuring *CP* violation.

TABLE 8.6
Design Characteristics of some future long-baseline experiments.*

Expt	$t_\nu + t_{\bar{\nu}}$ [yr]	P_{Target} [MW]	L [km]	Detector technology	m_{Det} [kt]
WBB-WC	$5+5$	$1(\nu)+2(\bar{\nu})$	1290	WC	300
T2KK	$4+4$	4	$295+1050$	WC	$270+270$
SPL	$5+5$	4	130	WC	440
BB2	$5+5$	n/a	650	WC	500
BB4	$2.5+2.5$	n/a	$650+7000$	WC + MIC	$500+30$
LENF	$5+5$	4	1480	TAS	30
IDS-NF	$5+5$	4	$3000+7500$	MIC + MIC	$50+50$

*Neutrino t_ν and antineutrino $t_{\bar{\nu}}$ running times, corresponding target power P_{Target}, baseline L, detector technology (WC = water Cherenkov, MIC = Magnetized Iron Calorimeter and TAS = Totally Active Scintillator), and detector mass m_{Det} for the experiments considered in figure 8.6. For experiments with two baselines, each baseline operates with the running times shown.

Different assumptions are often made in the sensitivity calculations for long-baseline experiments, and therefore one must be careful when attempting to determine which experiment is better. To make an unbiased comparison of the physics potentials of the experimental setups the sensitivities as functions of *exposure* may be compared, where exposure is defined to be $\mathcal{L}$ = detector mass [Mt] $\times$ target power [MW] $\times$ running time [10^7 s]. The assumptions for the three superbeam scenarios, a beta beam experiment, and neutrino factory (NuFact) experiment, are listed in table 8.7. In figure 8.7 we show the discovery reaches for $\sin^2 2\theta_{13}$, *CP* violation, and normal mass hierarchy versus the exposure for a fraction of δ of 0.5 (see figure caption). The experiments we consider are a future narrow-band beam experiment from Fermilab to Ash River (F2AR) with average neutrino energy $E_\nu = 2.6$ GeV, a wide-band beam experiment from Fermilab to DUSEL, and a narrow-band beam experiment T2KK with $E_\nu \simeq 0.8$ GeV.

Since θ_{13} is not too small, it may be possible to mount experiments that will permit us to determine the ordering of the states in the neutrino mass spectrum and to measure *CP* violation in the neutrino sector.

For small $\sin^2 2\theta_{13} \ll 10^{-2}$, it is well-known that a neutrino factory complex has the optimal physics potential for all of the considered performance indicators. This is a consequence of the high neutrino energies and high event rates. However, the oscillation maximum sits at relatively low energies, where the backgrounds from event misidentification are large, and the event rates are comparatively moderate. Therefore, for large $\sin^2 2\theta_{13}$, a beta beam or superbeam experiment tuned to the oscillation maximum may have the better performance. Since the effort for a beta beam may be larger than for a superbeam upgrade, and the technology needs further exploration, it is an interesting question if the superbeam upgrades can compete with a neutrino factory or beta beam for large $\sin^2 2\theta_{13}$. We use the neutrino factory and beta beam setups from table 8.7 for this comparison.

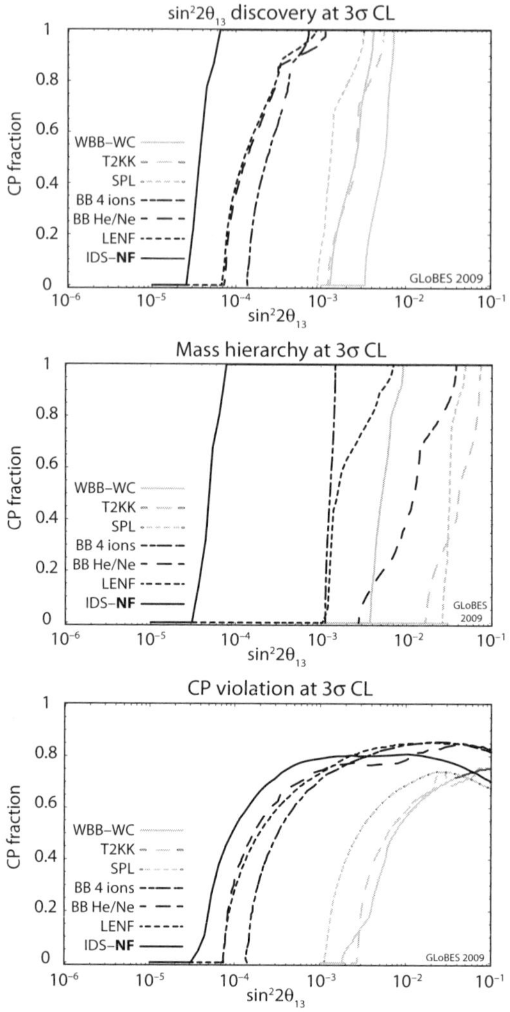

Figure 8.6. Summary of 3σ sensitivities for discovering a nonzero θ_{13}, determining $\mathrm{sgn}(\delta m^2_{31})$ and measuring CP violation. In each case, the fraction of δ values for which a 3σ measurement can be made is shown. Curves are taken from [373] (WBB-WC and T2KK), [374] (SPL), [375] (BB2 and BB4), [353] (IDS-NF) and [361] (LENF). Adapted from [376].

All the experimental facilities under consideration have good sensitivity to nonzero $\sin^2 2\theta_{13}$ and the mass hierarchy for large $\sin^2 2\theta_{13}$. We therefore do not discuss the $\sin^2 2\theta_{13}$ and mass hierarchy sensitivities and focus on the *CP* violation measurement.

In figure 8.8 we show the *CP* fraction for the 3σ discovery of *CP* violation as a function of exposure. The different panels correspond to different true values of $\sin^2 2\theta_{13}$. The shaded region marks the potential between a beta beam (solid line) and neutrino factory (dashed line) as given in table 8.7. For $\sin^2 2\theta_{13} = 0.1$, the superbeam upgrades perform at least as well as the neutrino factory, and a

TABLE 8.7
Experimental setups* considered in figure 8.7.

Expt	$t_\nu + t_{\bar{\nu}}$ [yr]	P_{Target} [MW]	L [km]	Detector	m_{Det} [kt]	$\mathcal{L}$
F2AR	$3+3$	$1.13(\nu/\bar{\nu})$	810	LArTPC	100	1.15
WBB	$5+5$	$1(\nu)+2(\bar{\nu})$	1290	LArTPC	100	2.55
T2KK	$4+4$	4	$295+1050$	WC	$270+270$	17.28
β beam	$4+4$	n/a	730	WC	500	n/a
NuFact	$4+4$	4	$3000+7500$	MIC	$50+50$	n/a

*With LArTPC = Liquid Argon Time Projection Chamber and exposure $\mathcal{L}$ [Mt MW 10^7 s]; other notation is the same as in table 8.6.

moderate increase of exposure can make their physics potential optimal. Note that, for instance, the neutrino factory requires a target power of 4 MW. The Fermilab-based experiments therefore still have space for an increase of target power. For $\sin^2 2\theta_{13} = 0.01$, the situation is already very different. In this case, WBB can compete with the neutrino factory with a factor of two or three increase in the exposure. This upgrade basically corresponds to an upgraded proton source and a somewhat longer running time. The increase in exposure necessary for F2AR and T2KK to be competitive would be unrealistic.

This discussion indicates that superbeams may be the technology of choice for large $\sin^2 2\theta_{13}$.

To summarize,

- The scientific goals of a program of long-baseline neutrino oscillation experiments are to measure the mixing parameter $\sin^2 2\theta_{13}$, to determine the order of the states of the neutrino mass spectrum, and to determine whether there is *CP* violation in the neutrino sector. Measurement of these quantities is an important goal of elementary particle physics.

- Determination of the ordering of the neutrino mass spectrum, searching for *CP* violation, and resolution of parameter degeneracies with sensitivity down to $\sin^2 2\theta_{13} \simeq 0.01$ will require a new generation of experiments with detectors with masses of 100 kilotons or more. This represents an increase in sensitivity of more than one order of magnitude over the experiments that will begin to acquire data in the next few years.

- The wide-band beam approach to neutrino oscillation physics can, in principle, utilize either a liquid argon detector or a water Cherenkov detector. If located more than 1000 km from Fermilab, there is good sensitivity for determining the mass hierarchy and measuring the amount of *CP* violation. The optimal baseline for a wide-band beam experiment is between 1200 and 1500 km.

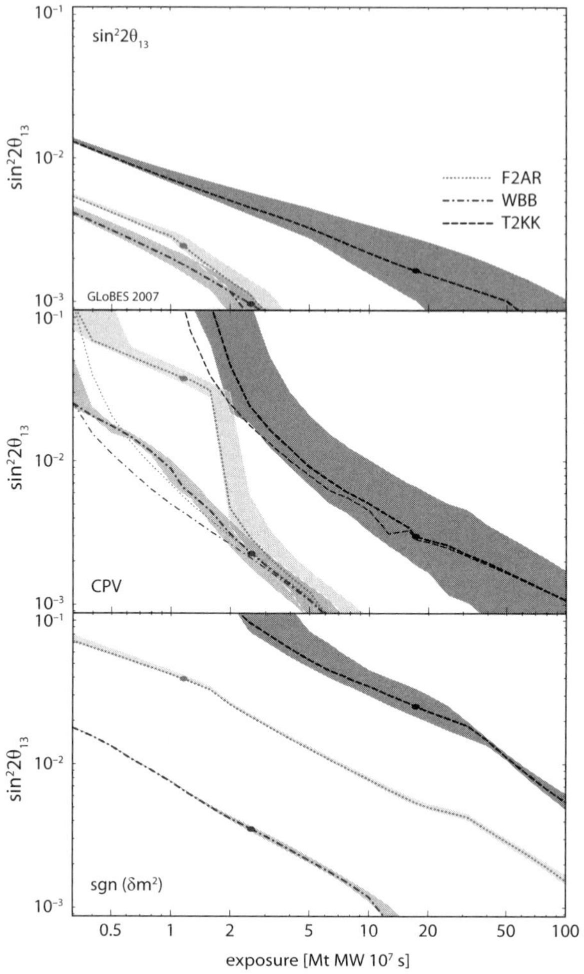

Figure 8.7. The $\sin^2 2\theta_{13}$ reach at 3σ for the discovery of nonzero $\sin^2 2\theta_{13}$, *CP* violation, and the normal hierarchy as a function of exposure. The curves are for a fraction of the *CP* phase δ of 0.5, which means that the performance will be better for 50% of all values of δ, and worse for the other 50%. The light curves in the *CPV* panel are made under the assumption that the mass hierarchy is known to be normal. The dots mark the exposures of the setups as defined in table 8.7. The shaded regions result by varying the systematic uncertainties from 2% (lower edge) to 10% (upper edge). Adapted from [373].

- Among experiments with superbeams, wide-band beam experiments have the most robust performance and the best mass hierarchy performance. The sensitivity of experiments with narrow-band beams is significantly affected by the true value of $|\delta m^2_{31}|$. Overall, wide-band beam experiments are the best experimental concept.
- A wide-band beam is a competitive experimental concept compared to a neutrino factory or beta beam if $\sin^2 2\theta_{13} \gtrsim 0.01$.

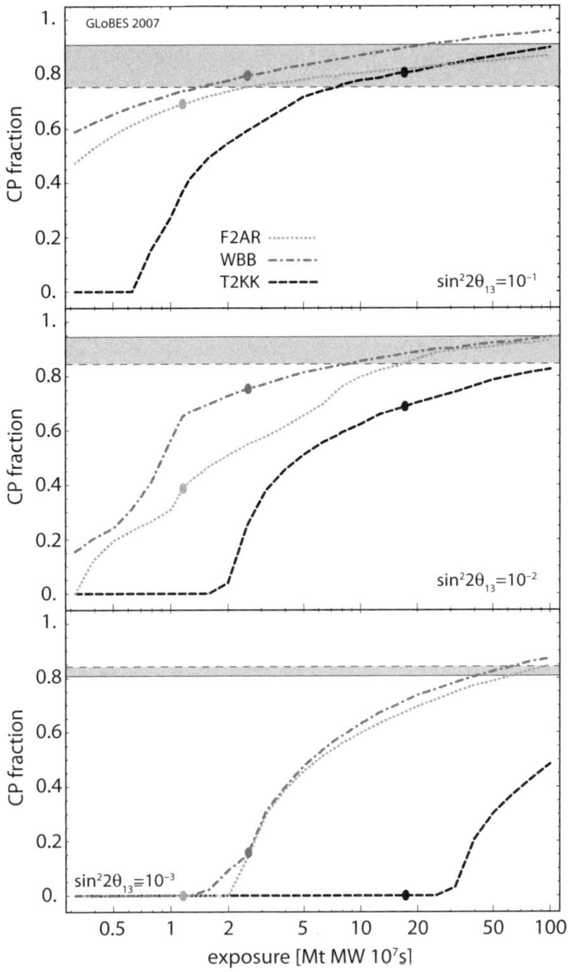

Figure 8.8. *CP* fraction for a 3σ discovery of *CP* violation as a function of exposure. The different panels correspond to $\sin^2 2\theta_{13} = 0.1$, $\sin^2 2\theta_{13} = 0.01$, and $\sin^2 2\theta_{13} = 0.001$. The shaded region marks the potential reach between a beta beam (solid line) and neutrino factory (dashed line) as defined in table 8.7. Adapted from [373].

8.7 *T* and *CPT* Symmetries

The *CPT* symmetry (the combination of Charge conjugation *C*, Parity *P*, and Time reversal *T*) is conserved in a local quantum field theory that is Lorentz invariant.[3] Then *CP* violation implies *T* violation, which can be measured, e.g., by comparing $\nu_e \to \nu_\mu$ to $\nu_\mu \to \nu_e$. There have been many phenomenological studies of *T* violation in the literature [318, 357, 380]. Unlike *CP* violation, matter does not induce *T*

[3] *CPT* violation requires Lorentz invariance violation, although not all theories with Lorentz invariance violation have *CPT* violation [377]. The authors of [378] claim to have found a class of non-local theories in which *CPT* invariance is violated while Lorentz invariance is preserved, but time-ordered products are not covariant in those theories [379].

violation in a long-baseline neutrino oscillation experiment, due to the symmetric matter distribution (i.e., the matter distribution is the same from the detector to the source as for the source to the detector), although matter can modify the amount of T violation.

If CPT is not conserved, then $P(\nu_\alpha \to \nu_\beta)$ is not necessarily equal to $P(\bar{\nu}_\beta \to \bar{\nu}_\alpha)$ in vacuum; matter can induce fake CPT violation (for a study of matter-induced CPT violation see [381]). In a NuFact, CPT violation can be tested down to very low levels by comparing the survival channels $\nu_\mu \to \nu_\mu$ and $\bar{\nu}_\mu \to \bar{\nu}_\mu$. There have been a number of studies of possible tests of CPT violation in the neutrino sector [382].

9

Model Building

9.1 The Seesaw Mechanism

The renormalizable SM Lagrangian [383] does not allow neutrino mass terms because there are no right-handed neutrino fields. Consequently, beyond the SM physics is mandated in the neutrino sector. A simple scheme for neutrino mass generation is to use the SM fields to construct a non-renormalizable addition to the Lagrangian. The unique dimension-5 lepton-number violating operator that conserves SM symmetries is schematically [384]

$$\left(\frac{\kappa}{\Lambda}\right) L_i L_j H H, \tag{9.1}$$

where $L_i = (\nu_{iL}, \ell_{iL})$ and $H = (\phi^+, \phi^0)$ are $SU(2)_L$ lepton and Higgs doublets, respectively, i and j are generation indices, κ is a dimensionless coupling and Λ is the energy scale associated with the generation of this effective operator.[1] Λ can be interpreted as the scale at which lepton number is violated.

When the neutral Higgs field develops its vacuum expectation value (VEV) $v_0 = 246$ GeV to break the electroweak symmetry, a neutrino mass term is generated with

$$m_\nu = \kappa v_0^2/\Lambda\,. \tag{9.2}$$

The smallness of $m_\nu \lesssim 1$ eV is explained by having $\Lambda \gg v_0$. This is the spirit of the "seesaw" mechanism explanation for why neutrino masses are small. For κ of order unity, $m_\nu \lesssim 1$ eV implies that $\Lambda \gtrsim 10^{14-15}$ GeV. If the Yukawa coupling is fine-tuned to be very small, the heavy neutrino mass can be at the TeV scale and accessible to discovery at colliders. The neutrino is a Majorana (self-conjugate) field in this mechanism and the mass-term is said to be of Majorana type. In specific beyond the SM extensions, the effective low-energy limit of the high-energy theory will relate the effective parameters κ and Λ to the fundamental parameters of the ultraviolet theory.

[1] Higher dimension effective operators are discussed in [385].

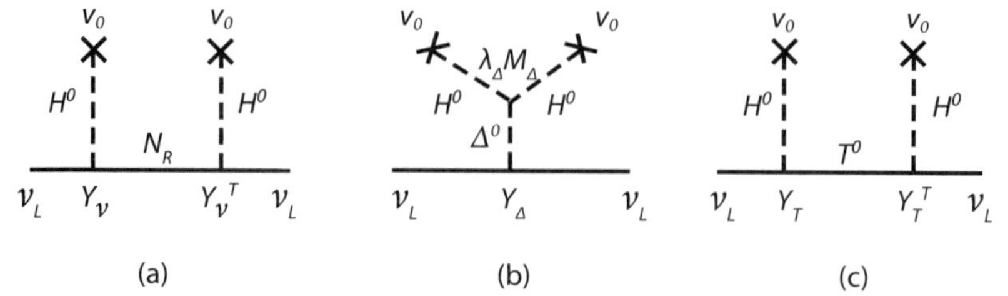

Figure 9.1. Three seesaw mechanisms for generating light Majorana neutrino masses: (a) Type I, (b) Type II, and (c) Type III. Adapted from [389].

The effective operator in equation 9.1 may be realized at tree-level in only three ways [386]: (I) L and H form a fermion $SU(2)_L$ singlet, (II) L_i and L_j form a scalar triplet, and (III) L and H form a fermion triplet. (The scalar singlet combination of L_i and L_j does not generate a neutrino mass term.) In each case one or more fields must be added to the SM. Then Majorana neutrino masses may be generated in a renormalizable extension of the SM as follows:

- *Type I seesaw mechanism* [387, 388]:
Neutral right-handed lepton singlet fields N_i are added to the SM fields, and they combine with the $L_j H$ fermion singlets to form a Dirac mass term. A large Majorana mass term for the N_i is also allowed at tree-level. Upon spontaneous symmetry breaking the neutrino mass matrix for a single generation in the (ν_L, N_R^c) basis is then

$$\begin{pmatrix} 0 & m_D \\ m_D & m_R \end{pmatrix}, \tag{9.3}$$

where m_D is a Dirac mass and m_R is the heavy right-handed neutrino mass. The eigenmasses are then approximately $-m_D^2/m_R$ and m_R (the negative value of the lighter state can be made positive by a redefinition of the phases of the neutrino fields). If Y_ν is the Yukawa coupling that gives the Dirac mass, then $m_D = Y_\nu v_0$ and the resulting light neutrino mass is

$$m_\nu = Y_\nu^2 v_0^2 / m_R, \tag{9.4}$$

where m_R takes on the role of the effective scale Λ. This mass generation mechanism is shown diagrammatically in figure 9.1a.
- *Type II seesaw mechanism* [390]:
The Higgs sector is extended by adding a Higgs triplet Δ in the adjoint representation of $SU(2)_L$, which then couples to the scalar $SU(2)_L$ triplet combination of L_i and L_j to form a Majorana mass for the left-handed neutrino. The light neutrino mass is $m_\nu = Y_\Delta v_\Delta$, where v_Δ is the VEV of the neutral component of the triplet Δ^0 and Y_Δ is the Yukawa coupling. If $\lambda_\Delta m_\Delta$ is the coefficient of the $HH\Delta$ coupling, it can be shown [391] that

upon minimization of the Higgs potential $v_\Delta \sim (\lambda_\Delta m_\Delta)v_0^2/m_\Delta^2 = \lambda_\Delta v_0^2/m_\Delta$, where m_Δ is the mass of Δ^0. Then the light neutrino mass is

$$m_\nu = \lambda_\Delta Y_\Delta v_0^2/m_\Delta \,. \tag{9.5}$$

In this case the effective scale Λ is m_Δ; see figure 9.1b for a diagrammatic representation.

- *Type III seesaw mechanism* [392]:

  Right-handed lepton triplets T_i in the adjoint representation of $SU(2)_L$ are added to the SM fields. They couple to the fermion $SU(2)_L$ triplet combination of LH to form a Dirac mass term. The neutral triplet leptons T^0 obtain a heavy Majorana mass m_T. The neutrino mass matrix is the same as equation 9.3 with Y_ν and m_R replaced by Y_T and m_T, respectively, and the light neutrino mass is

$$m_\nu = Y_T^2 v_0^2/m_T \,, \tag{9.6}$$

  where Y_T is the Yukawa coupling of the SM neutrinos to the neutral triplet leptons and SM Higgs. Here the high scale Λ is the mass of the extra neutral leptons; see figure 9.1c.

In certain extensions of the SM, including some grand unified models, neutrino mass generation can be a hybrid of these three basic mechanisms. For example, in a Left-Right symmetric model with a pair of Higgs triplets, Δ_L and Δ_R, a light Majorana neutrino mass is generated through a combination of the Type I and Type II seesaws [388]. A combination of the Type I and Type III seesaws [386] may be realized if SM lepton singlets N_i and neutral lepton fields in the adjoint representation of $SU(2)_L$ are added [393].

To generalize to three generations, the mass matrix for the light neutrinos in the Type I and Type III seesaws takes the form

$$M_\nu = -M_D(M_R)^{-1}M_D^T \,, \tag{9.7}$$

where M_D is now a 3×3 matrix describing the Dirac neutrino masses (possibly related by symmetries to the charged-lepton mass matrix) and M_R is the 3×3 matrix describing the right-handed Majorana neutrino masses, with its six complex parameters. The a priori lack of information about M_R is a problem for predictivity. In special models, such as the minimal $SO(10)$ models, the right-handed mass matrix can be related to SM masses and mixings. If M_D has a hierarchical form (similar to the charged-lepton mass spectrum), then the neutrino mixing angles in V_{MNS} tend to be small, contrary to what is inferred from data, unless there is an unnatural conspiracy between M_D and M_R [394, 395]. The choice of particular forms for M_D and M_R can avoid this problem [395,396]. A more detailed discussion of such models and relevant references can be found in [397].

For Type II seesaws, the left-handed Majorana neutrino mass matrix for three generations is

$$M_L = \lambda_\Delta v_0^2 Y_\Delta/m_\Delta \,, \tag{9.8}$$

where Y_Δ is the 3×3 matrix describing the $L_i L_j \Delta$ Yukawa couplings. For the combined Type I and II, the most general 6×6 Majorana neutrino mass matrix is

$$M = \begin{pmatrix} M_L & M_D \\ M_D^T & M_R \end{pmatrix} ; \tag{9.9}$$

assuming $M_L \ll M_D \ll M_R$, the 3×3 left-handed Majorana neutrino mass matrix is $M_\nu \simeq M_L - M_D M_R^{-1} M_D^T$ and the 3×3 right-handed Majorana neutrino mass matrix is $M_N \simeq M_R + \frac{1}{2}(M_R^{-1} M_D^T M_D + M_D^T M_D M_R^{-1})$.[2] The 6×6 matrix that block diagonalizes M is

$$U \simeq \begin{pmatrix} 1 - \frac{1}{2} M_D (M_R^{-1})^2 M_D^T & M_D M_R^{-1} \\ -M_R^{-1} M_D^T & 1 - \frac{1}{2} M_R^{-1} M_D^T M_D M_R^{-1} \end{pmatrix}, \tag{9.10}$$

where "1" in equation 9.10 refers to a 3×3 identity matrix. Since $M_D \ll M_R$, the mixing between the left-handed and right-handed Majorana neutrinos is small, as expected.

Although the seesaw mechanism leads to Majorana masses for neutrinos, it is also possible to generate small neutrino masses in other ways (see section 9.4), some of which may give Dirac masses to neutrinos. In the next section we discuss in more detail the possible structure of the 3×3 mass matrix for the light neutrinos.

9.2 Patterns of Neutrino Masses and Mixings

One of the most important challenges in particle physics is to understand the spectrum of fermion masses [182]. The mixing matrix in the quark sector, V_{CKM}, is given by the product $V_u^\dagger V_d$, where V_u and V_d are the unitary transformations applied to the left-handed up and down quarks to diagonalize the up and down quark mass matrices. The mixing matrix that enters into neutrino oscillations is $V_{MNS} = V_L^\dagger V_\nu$. The charged leptons have Dirac masses and their mass matrix M_D may be diagonalized by a bi-unitary transformation $\tilde{M}_D = V_R^\dagger M_D V_L$, where $\tilde{M}_D$ is diagonal and V_L and V_R are unitary transformations acting on the left- and right-handed charged leptons. Then by taking the product $\tilde{M}_D^\dagger \tilde{M}_D = (V_L^\dagger M_D^\dagger V_R)(V_R^\dagger M_D V_L) = V_L^\dagger M_D^\dagger M_D V_L$, we see that V_L is the matrix that diagonalizes $M_D^\dagger M_D$. There are two possibilities for the neutrino mass matrix M_ν: (i) the neutrinos have Dirac masses, in which case V_ν diagonalizes $M_\nu^\dagger M_\nu$, in analogy with the charged-lepton case, or (ii) the neutrinos have Majorana masses, in which case M_ν is complex symmetric and V_ν diagonalizes M_ν via the transformation $V_\nu^T M_\nu V_\nu$. In the latter siutation, M_ν is often part of a larger 6×6 neutrino mass matrix that includes both Dirac and Majorana mass terms, as discussed in the previous section.

In the quark sector, all mixing angles in V_{CKM} are small and there is a mass hierarchy among the generations, whereas in the lepton sector a mass hierarchy exists with two large mixing angles and one small mixing angle in V_{MNS} (although

[2] Note that $\mathrm{Tr}(M_\nu) + \mathrm{Tr}(M_N) = \mathrm{Tr}(M_L) + \mathrm{Tr}(M_R) = \mathrm{Tr}(M)$, as it should.

not necessarily in the neutrino sector). A remarkable property of neutrino masses is that they are so much lighter than the charged leptons. Any theory of fermion mass must reconcile the extreme differences between quark and lepton masses and mixings. Diverse ideas have been advanced to explain neutrino masses and mixings; for comprehensive reviews and more references, see [397–399]. We will discuss a variety of the interesting proposals.

Since absolute neutrino masses are not yet known, there are three possible mass patterns for neutrinos: (i) normal hierarchy ($m_1 \ll m_2 \ll m_3$), (ii) inverted hierarchy ($m_2 \gtrsim m_1 \gg m_3$), and (iii) quasidegenerate ($m_1 \simeq m_2 \simeq m_3$). Because V_{MNS} is a product of the mixing matrices for the charged leptons and neutrinos, the observed mixing in neutrino oscillations can originate from V_L, V_ν, or a combination of the two. Viable models exist with different combinations of mass pattern and origins of the mixing angles.

In models where where the charged-lepton mixing matrix is approximately diagonal, V_{MNS} derives directly from V_ν. If there are three Majorana neutrinos, then there are nine independent parameters in the mass matrix: three absolute masses and six mixing matrix parameters (see equation 3.2). Six of these may be measured in neutrino oscillations (three mixing angles, the Dirac phase, and two mass-squared differences). The absolute mass scale may be determined by measuring tritium beta decay or the cosmological matter power spectrum. The magnitude of the $\nu_e - \nu_e$ mass matrix element (which depends on the three mixing angles, absolute masses, and two Majorana phases) may be determined from $0\nu\beta\beta$-decay experiments (although the value of the associated nuclear matrix elements makes this measurement less than precise). Therefore the complete 3×3 mass matrix for Majorana neutrinos cannot be fully determined by experiment in the near future, and the Majorana phases may never be measurable.

Since mixing in the lepton sector is not small, the simplest assumption is to have the same magnitude for all off-diagonal elements in the charged-lepton mass matrix. This can be done by postulating a charged-lepton mass matrix that is invariant under cyclic permutations of the three flavors, which gives

$$M_D^\dagger M_D = \begin{pmatrix} A & B & B^* \\ B^* & A & B \\ B & B^* & A \end{pmatrix}. \tag{9.11}$$

This matrix is diagonalized by

$$V_L = \frac{1}{\sqrt{3}} \begin{pmatrix} 1 & 1 & 1 \\ 1 & \omega & \omega^* \\ 1 & \omega^* & \omega \end{pmatrix}, \tag{9.12}$$

where $\omega = e^{2i\pi/3}$ and $\omega^* = e^{-2i\pi/3}$ are the complex roots of unity. This so-called trimaximal mixing matrix [400] is equivalent to $\theta_{12} = \theta_{23} = \pi/4$, $\tan\theta_{13} = 1/\sqrt{2}$, and $\delta = \pi/2$ after phase redefinitions. However, it does not provide a good description of the data.

Soon after the initial Super-K discovery that atmospheric neutrinos oscillate with maximal or nearly maximal amplitude, it was noted that the neutrino sector might exhibit bimaximal mixing [401], i.e., maximal or nearly maximal mixing of solar *and* atmospheric neutrinos. Then the mixing matrix has the unique form (up to state redefinitions)

$$V_{MNS} = \begin{pmatrix} \frac{1}{\sqrt{2}} & \frac{1}{\sqrt{2}} & 0 \\ -\frac{1}{2} & \frac{1}{2} & \frac{1}{\sqrt{2}} \\ \frac{1}{2} & -\frac{1}{2} & \frac{1}{\sqrt{2}} \end{pmatrix}, \tag{9.13}$$

where $\theta_{12} = \theta_{23} = \pi/4$ and $\theta_{13} = 0$. Note that there is no CP violation because $V_{e3} = 0$. This mixing can be achieved in different ways depending on whether the mixing occurs entirely in the neutrino sector, or as a combination of charged-lepton mixing and neutrino mixing. In the latter case, mass matrices of the form

$$M_D = m_D \begin{pmatrix} 1 & 0 & 0 \\ 0 & \tilde{\alpha} & \tilde{\beta} \\ 0 & \tilde{\beta}^* & \tilde{\alpha} \end{pmatrix}, \qquad M_\nu = m \begin{pmatrix} \alpha & \beta & 0 \\ \beta' & \alpha & 0 \\ 0 & 0 & 1 \end{pmatrix}, \tag{9.14}$$

lead to bimaximal mixing, where $\beta' = \beta^*$ for Dirac neutrinos and $\beta' = \beta$ for Majorana neutrinos. Normal hierarchy in the neutrino sector requires $|\alpha|, |\beta| \ll 1$, inverted hierarchy requires $|\alpha \pm |\beta|| \gg 1$ for Dirac neutrinos and $|\alpha \pm \beta| \gg 1$ for Majorana neutrinos, while the quasidegenerate case must have $|\beta| \ll |\alpha| \simeq 1$.

If instead the charged-lepton mass matrix is diagonal, then the most general form for the neutrino mass matrix that leads to bimaximal mixing is

$$M_\nu = m \left[\begin{pmatrix} 0 & 0 & 0 \\ 0 & 1 & 1 \\ 0 & 1 & 1 \end{pmatrix} + \alpha \begin{pmatrix} 2 & 0 & 0 \\ 0 & 1 & -1 \\ 0 & -1 & 1 \end{pmatrix} + \beta \begin{pmatrix} 0 & -1 & 1 \\ -1 & 0 & 0 \\ 1 & 0 & 0 \end{pmatrix} \right], \tag{9.15}$$

where $m = m_3/2$, $\alpha = (m_1 + m_2)/4m$, and $\beta = \sqrt{2}(m_1 - m_2)/4m$. For normal hierarchy $|\beta|, \alpha \ll 1$, for inverted hierarchy $\alpha \gg 1$, while for the quasidegenerate case $|\beta| \ll \alpha \simeq 1$.

Perturbations to this basic form can yield mixing that is not quite maximal, and can make V_{e3} $(= s_{13}e^{-i\delta})$ nonzero. The combined solar and KamLAND data disfavor maximal mixing ($\sin^2 2\theta_{12} \leq 0.95$ at the 3σ level), and the emphasis is now on finding models that can give bilarge mixing, although maximal mixing of atmospheric neutrinos is still favored. One class of models has maximal mixing of atmospheric neutrinos ($\theta_{23} = \pi/4$), but θ_{12} can vary [402].

Now that the solar neutrino mixing angle is established to be nonmaximal, a popular mixing scheme is tribimaximal mixing [103], where the charged leptons have the trimaximal form of mixing in equation 9.12 and neutrinos have maximal mixing in the $\nu_1 - \nu_3$ sector. This gives

$$V_{MNS} = \begin{pmatrix} \frac{2}{\sqrt{6}} & \frac{1}{\sqrt{3}} & 0 \\ -\frac{1}{\sqrt{6}} & \frac{1}{\sqrt{3}} & \frac{1}{\sqrt{2}} \\ \frac{1}{\sqrt{6}} & -\frac{1}{\sqrt{3}} & \frac{1}{\sqrt{2}} \end{pmatrix}, \tag{9.16}$$

or, equivalently, $\theta_{23} = \pi/4$, $\tan\theta_{12} = 1/\sqrt{2}$, and $\theta_{13} = 0$. As in the bimaximal case, there would be no CP violation if V_{e3} were exactly zero. The predicted value of θ_{12} is 35.3°, in good agreement with the measured value, $\theta_{12} = (34.1^{+1.2}_{-0.8})^\circ$. Tribimaximal mixing can also be achieved entirely from mixing in the neutrino sector; then the most general form for the neutrino mass matrix is

$$M_\nu = m\left[\begin{pmatrix} 0 & 0 & 0 \\ 0 & 1 & 1 \\ 0 & 1 & 1 \end{pmatrix} + \alpha\begin{pmatrix} 2 & 0 & 0 \\ 0 & 1 & -1 \\ 0 & -1 & 1 \end{pmatrix} + \beta\begin{pmatrix} 1 & -1 & 1 \\ -1 & 0 & 0 \\ 1 & 0 & 0 \end{pmatrix}\right], \tag{9.17}$$

where $m = m_3/2$, $\alpha = (m_1 + 2m_2)/6m$ and $\beta = (m_1 - m_2)/3m$. As was the case with bimaximal mixing, $|\beta|, \alpha \ll 1$ for normal hierarchy $\alpha \gg 1$ for inverted hierarchy, while $|\beta| \ll \alpha \simeq 1$ for the quasidegenerate case; small deviations from exact tribimaximal mixing can be understood as perturbations, allowing $V_{e3} \neq 0$ [403].

On the other hand, such perturbations may not be so small. The quark mixing matrix V_{CKM} can be thought of as the identity matrix modified by perturbations of order $\lambda \simeq \sin\theta_C$, where θ_C is the Cabibbo angle. This is the rationale behind the Wolfenstein parametrization [404]. In the lepton sector, phenomenologically we know that two of the mixing angles are large. However, it is not unrealistic to expect Cabibbo-sized shifts in the leptonic mixing angles from their initial values [405]; one might expect additional mixing of order m_2/m_3, which in a hierarchical structure would be $\sqrt{\delta m_{21}^2/\delta m_{31}^2} \simeq \sqrt{1/30} \approx \theta_C$. Then the bare solar mixing could be maximal, and the bare atmospheric mixing would not need to be. The size of the CP violation parameter J can be as large as λ or as small as λ^4, but in all cases will still be larger than in the quark sector [405]. This opens up many other possibilities for the underlying symmetries that drive the form of the neutrino mass matrices [406]. Perturbations from tribimaximal mixing that also have a relatively large V_{e3} will lead to constraints on the Majorana phases affecting $0\nu\beta\beta$-decay [407].

9.3 GUT Models

In the simplest grand unified model, down-type quarks and charged leptons are both in the **5** multiplet of $SU(5)$, leading to the mass relations between quarks and leptons [408, 409]. There is no similar relation for neutrino masses, which for left-handed neutrinos are phenomenologically much smaller than other fermion masses.

An attractive way to understand small masses for the left-handed neutrinos is the seesaw mechanism (see section 9.1). Heavy right-handed neutrinos exist in most grand unified models, either as an $SU(5)$ singlet or as part of the **16** representation of $SO(10)$, which decomposes into a $\mathbf{10} + \mathbf{5} + \mathbf{1}$ in $SU(5)$. Therefore

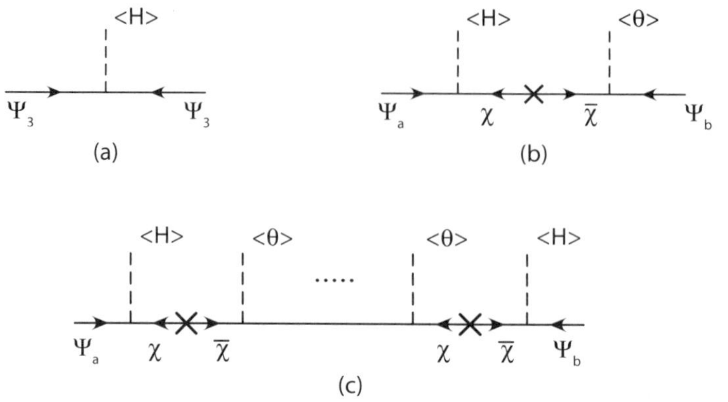

Figure 9.2. Froggatt-Nielsen diagrams. Here a and b are family-indices and $(\chi, \bar{\chi})$ are vector-like fields of mass M and $\langle\theta\rangle$ is the VEV of the flavor Higgses or flavons. The tree-level diagram (a) generates the mass of the third family and the lighter masses are obtained by $\mathcal{O}(\langle\theta\rangle/M)^n$ suppressions from diagrams (b) and (c). From [413].

a GUT/seesaw model is very attractive. If the heaviest light neutrino has mass of order $\sqrt{|\delta m^2_{31}|} \simeq 0.05\,\text{eV}$, and the Dirac mass m_D is the τ lepton mass, then the required Majorana masses for the right-handed neutrinos are $m_R \sim 10^{11}$ GeV. Other interesting possibilities for m_D are the electroweak VEV or the top quark mass, which would imply $m_R = 10^{15}$ GeV (close to the GUT scale).

Because GUT models relate quarks and leptons, the fact that lepton mixing angles are large and quark mixing angles are small is potentially a problem for these models. In fact, many GUT models predicted small solar neutrino mixing [410], now ruled out by data from SNO and KamLAND. However, there is a way around this difficulty, due to the fact that GUT theories relate leptons and quarks of opposite chiralities, e.g., right-handed down quarks are in the same fermion multiplets as left-handed charged leptons, and vice versa. Therefore the mixing we observe in the lepton sector is connected to the right-handed quark mixing, which is unknown and not constrained. So-called lopsided models [411] take advantage of this fact. The three-generation Dirac mass matrices for the charged leptons and down quarks have the forms

$$M_\ell \propto \begin{pmatrix} x & x & x \\ x & 0 & \epsilon \\ x & \sigma & 1 \end{pmatrix}, \qquad M_d \propto \begin{pmatrix} x & x & x \\ x & 0 & \sigma \\ x & \epsilon & 1 \end{pmatrix}, \tag{9.18}$$

where $\epsilon \ll \sigma \sim 1$. The entries involving only the second and third generations come from specific Higgs Yukawa interactions, with no contribution to the middle diagonal. The entries with an "x" involve the first generation and are very small, and in many (although not all) models are generated by Froggatt-Nielsen diagrams mediated by exotic vector-like matter fields [412]; see figure 9.2. The up-quark and neutrino mass matrices are approximately diagonal. Then the quark mixing element V_{ub} is small, but the leptonic mixing element $V_{\mu 3}$, which relates to the mixing of atmospheric neutrinos, is large. In lopsided models, the large atmospheric neutrino mixing comes from the diagonalization of the charged-lepton mass matrix; the solar neutrino mixing angle arises from the structure of the three-generation right-handed Majorana neutrino mass matrix, also determined in some models by Froggatt-Nielsen diagrams.

A Monte Carlo study suggests that lopsided textures are favored by the data [414]. Lopsided models can yield either large or small solar neutrino mixing [415]; for a discussion of which models can naturally yield the LMA solution, see [416]. Most lopsided models are embedded in a GUT [417], but some are not [418].

There are many GUT models that can yield the LMA solution for solar neutrinos, and which are consistent with current data [419]. Most known GUT models that satisfy all experimental data favor a normal hierarchy [420], i.e., $m_3 > m_1, m_2$ in the neutrino sector, and therefore $\delta m^2_{31} > 0$. However, there are exceptions [421, 422]. Predictions for V_{e3} vary; $\sin^2 2\theta_{13}$ can be as small as a few $\times 10^{-4}$ to near the experimental upper bound [420].

In any GUT framework, proper comparison with data can only be made after allowing for the renormalization group running of the neutrino mass terms in the Lagrangian [423]; for detailed discussions, see [424]. Of particular importance are how zeroes in the mass matrices behave under renormalization and the stability of the large mixing angles [425]. In many cases, horizontal (also called family or flavor) symmetries determine the textures of the mass matrices. For a more complete discussion and further references on GUT models, see [397].

Another interesting possibility is to assume that the unification group is replicated at the Planck scale, i.e., there is one copy for each generation. This leads to family-dependent $U(1)$ symmetries as the theory breaks down to the Standard Model at low energies. Some consequences of such models are discussed in [426]. Models that utilize family symmetries can have either (i) both large mixing angles in the neutrino sector deriving from the neutrino mass matrix [427], or (ii) the large atmospheric neutrino mixing angle deriving from the charged-lepton mass matrix and the large solar neutrino mixing angle deriving from the neutrino mass matrix [428]. Supersymmetric (SUSY) GUT models with R-parity violation are discussed in [429].

9.4 Non-GUT-specific Models

There are many alternatives to explicit GUT models (although it is often assumed that they could emerge from an unspecified grand unified theory). Although many models lead to Majorana masses, Dirac masses are also possible in some cases. Some popular possibilities include (by no means mutually exclusive): (i) the Zee model and models with $L_e - L_\mu - L_\tau$ symmetry; (ii) models with other horizontal symmetries; (iii) models with specific textures for the neutrino mass matrix or special relationships involving the entries in the mass matrix; (iv) models with low-energy new physics in which neutrino masses are generated at one- or two-loop level, such as supersymmetry with R-parity violation or models in which neutrinos have interactions with colored fields; (v) models with a $U(1)'$ gauge symmetry; (vi) models in which neutrino masses derive from higher-dimension operators; (vii) models with triplet Higgs bosons; (viii) models with dynamical electroweak symmetry breaking; (ix) models in which neutrino masses have explicit string theory origin; (x) models

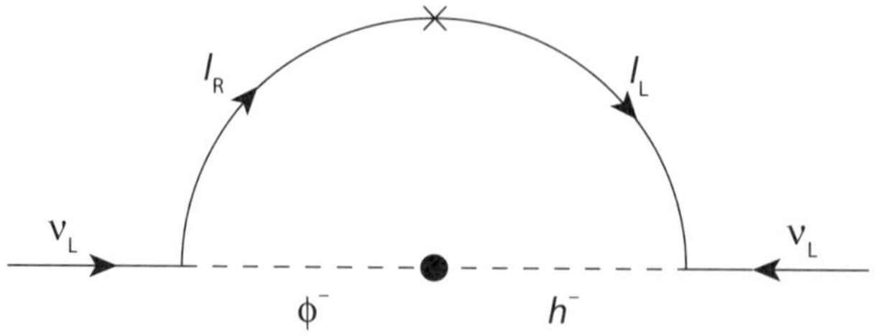

Figure 9.3. Neutrino mass is generated at one loop in the Zee model.

with large extra dimensions; and (xi) models where the right-handed Majorana neutrino mass matrix has a hierarchical structure. In non-GUT models, M_ν is unrelated to the charged-fermion mass matrices, and hence a priori there are few constraints on its structure. Models with horizontal symmetries or other restrictions on the entries in the mass matrix are then desirable to produce a phenomenologically compelling model. It is also possible that neutrino masses derive from an essentially random mass matrix, a scenario called neutrino anarchy.

Zee Model

In the Zee model [430], which invokes radiative neutrino masses via a charged $SU(2)_L$ singlet h and a Higgs field ϕ in the loop (see figure 9.3), the neutrino mass matrix has the approximate form

$$M_\nu = \begin{pmatrix} 0 & A & B \\ A & 0 & 0 \\ B & 0 & 0 \end{pmatrix}, \tag{9.19}$$

where $A \sim B$. In Zee-type models, the diagonal elements of M_ν are zero, and the remaining off-diagonal elements may be nonzero (but small compared to A and B). Such a texture can also result from an approximate $L_e - L_\mu - L_\tau$ symmetry [431]; for some other possibilities see [432]. The mass matrix in equation 9.19 yields large mixing for both solar and atmospheric neutrinos (although some specific models yield vacuum solar neutrino oscillations that are now excluded). The mass hierarchy for the Zee-type mass matrix is inverted, i.e., $m_3 \ll m_1, m_2$ and $\delta m^2_{31} < 0$.

Models with the exact form of equation 9.19 for the neutrino mass matrix predict that the solar neutrino mixing is nearly maximal mixing and are now excluded by data. Models consistent with the LMA solar solution are also possible if there are deviations from that form, e.g., in $L_e - L_\mu - L_\tau$-symmetric models when the symmetry is strongly broken in the neutrino sector or broken in the charged-lepton sector [433].

Horizontal Symmetries

Horizontal symmetries place restrictions (or conditions) on the neutrino mass matrix elements, which in some cases can lead to a predictive model that is also consistent with neutrino data. Often such a symmetry leads to tribimaximal or bimaximal mixing at leading order, which allows an acceptable phenomenology.

Perhaps the simplest models are those that are μ–τ (or 2-3) symmetric [434]. The neutrino mass matrix then has the form

$$M_\nu = \begin{pmatrix} A & B & -B \\ B & C & D \\ -B & D & C \end{pmatrix}, \tag{9.20}$$

and naturally gives maximal mixing in the $\mu - \tau$ sector, or nearly maximal mixing if the symmetry is softly broken. The neutrino mass matrix that leads to bimaximal mixing (equation 9.15) is explicitly 2-3 symmetric, and the neutrino mass matrix that leads to tribimaximal mixing (equation 9.17) can be seen to be 2-3 symmetric after application of a phase transformation to the third row and column. A special case of the 2-3 symmetric models are those that have an $L_\mu - L_\tau$ symmetry [435], which makes $B = C = 0$ in equation 9.20 before any symmetry breaking terms are included. Both normal and inverted hierarchies are allowed in 2-3 symmetric models.

The continuous groups $SO(3)$ [436] and $SU(3)$ [437] have been considered as possible horizontal symmetries. Discrete groups have many distinct low-dimensional representations, and therefore not many exotics, if any, need be present in a viable model, and many different textures are possible for the mass matrices. Discrete groups also avoid the potential difficulties of additional interactions (if gauged) or Goldstone bosons (if not) that can occur if the horizontal symmetry is continuous.

The permutation groups of three or more objects have been suggested as possible family symmetries. The group S_3 has a long history of being studied for both quark and lepton masses [438]. In the neutrino sector these models [439] tend to have small $\sin^2 2\theta_{13}$ and can give either a normal or inverted hierarchy. The group S_4 is another possible horizontal symmetry [440], and generally leads to a normal hierarchy [441], although it can give quasidegenerate light neutrinos [442]. The S_4 group has the distinction of being the smallest discrete subgroup of $SU(3)$ that gives tribimaximal mixing *without* parameter tuning [443].

The set of even permutations of four objects, A_4, is a subgroup of S_4 and has received much attention recently [104, 444]. It can also accommodate tribimaximal mixing. Predictions for $\sin^2 2\theta_{13}$ tend to be small [445], although in some cases they can be somewhat close to the experimental upper bound [446]. Both normal and inverted hierarchies are possible.

Some more exotic discrete groups have also been studied. The dihedral groups $\Delta(3n^2)$ are an infinite set of discrete subgroups of $SU(3)$ [447]. The group $\Delta(12)$ is just A_4, as discussed, and the group $\Delta(27)$, for example, can also lead to tribimaximal mixing [448]. The binary tetrahedral group T', the covering group of A_4 and also a discrete subgroup of $SU(3)$, can give tribimaximal mixing of neutrinos as well as an accurate prediction for the Cabibbo angle [449]. Yet another possibility that can give tribimaximal mixing is the Frobenius group, $Z_7 \times Z_3$ [450], where Z_n is the set of cyclic permutations of n objects. Going beyond discrete subgroups of $SU(3)$, the group $\Sigma(81)$, a subgroup of $U(3)$, has been proposed to avoid additional constraints to obtain tribimaximal mixing [451].

Finally, there are other ways to arrive at the same approximate form as equation 9.16, such as models where the solar mixing angle is related to the golden ratio [428,452], which give $\sin^2 2\theta_{12} = 0.80$ and may be realized with an icosahedral (A_5) symmetry [453].

Textures

Since it will not be possible to experimentally determine all nine parameters of the neutrino mass matrix, many studies have examined simpler structures with fewer independent parameters (with the assumption that the charged-lepton mass matrix is diagonal). Some recent examples:

(i) It has been shown that 3×3 Majorana mass matrices with three or more independent zero entries are excluded by current neutrino data but there are seven distinct textures with exactly two independent zeroes that are acceptable [454]. Note that since a Majorana mass matrix is symmetric, a reflected off-diagonal zero is not counted as independent. Two of these textures lead to a normal mass hierarchy and the other five to a quasidegenerate mass spectrum. In fact, it is possible to fully determine the neutrino mass spectra corresponding to these textures [455]. Several aspects of these seven matrices have been studied in [456]. Other examples are given in [457]. Some of these textures can be realized in a seesaw model with or without extra $U(1)$ flavor symmetries [458]. These textures can also be obtained in models with three Higgs triplets and a sufficiently massive triplet Majoron [459].

(ii) The weak-basis independent condition $\det(M_\nu) = 0$ (which would be approximately true if the lightest neutrino is nearly massless) can also lead to a complete determination of the neutrino mass matrix [460].

(iii) Another possible condition that the neutrino mass matrix might obey is form invariance, $UMU^T = M$, where U is a specific unitary matrix such that U^N represents a well-defined discrete symmetry in the neutrino flavor basis [461]. This condition leads to a variety of possible mass matrices, including all three of the allowed neutrino mass patterns [462]. If the discrete symmetry is the non-Abelian group A_4, the mass pattern is quasidegenerate [463].

(iv) If the sum of the neutrino masses is zero [464], which can occur in models whose neutrino mass matrix can be expressed as the commutator of two matrices, only the inverted hierarchy and quasidegenerate mass patterns are allowed by current neutrino data [421,465].

If the charged-lepton mass matrix is not assumed to be diagonal, then there are more independent parameters. A simplifying ansatz may be used to reduce the number of parameters. For example, the Fritzsch ansatz [466] assumes $M_{11} = M_{22} = M_{13} = M_{31} = 0$ for both the charged leptons and neutrinos, which can lead to acceptable phenomenology. The large mixing of atmospheric neutrinos can come from V_L [428].

In models with flavor democracy [467], the quark and charged-lepton mass matrices have the form

$$M \propto \begin{pmatrix} 1 & 1 & 1 \\ 1 & 1 & 1 \\ 1 & 1 & 1 \end{pmatrix}, \tag{9.21}$$

while the neutrino mass matrix is approximately diagonal. This scenario also leads to large mixing for solar and atmospheric neutrinos and a normal mass hierarchy. The democratic model (and many other non-seesaw models) predicts that the solar neutrino mixing angle is maximal and the atmospheric neutrino mixing angle is

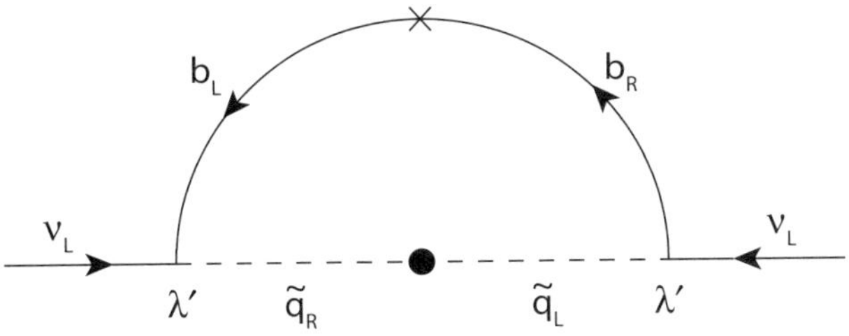

Figure 9.4. The dominant one-loop diagram that generates Majorana neutrino masses for left-handed neutrinos in R-parity violating models. The coupling λ' violates lepton number as well as R-parity. From [474].

large but not necessarily maximal, whereas the data indicate the opposite. However, if there are Cabibbo-sized deviations from the initial structure, this is not a problem.

New Physics at Low Energy

In many models with new low-energy physics, neutrinos are coupled to a heavy fermion in the theory. Mass terms for the light neutrinos are generated by loop diagrams involving the neutrino and the heavy fermion. If the heavy fermion coupling to the second and third generation neutrinos is larger than to the first generation, a normal mass hierarchy and large mixing for atmospheric neutrinos results. In the minimal supersymmetric extension of the Standard Model (MSSM) radiative neutrino mass generation is a direct consequence of R-parity violation [468] (see figure 9.4). For specific realizations to explain the neutrino anomalies see [469]. R-parity violating SUSY models that reproduce the neutrino mass and mixing parameters can have specific signatures in future collider experiments, such as lepton-number violating final states [470], neutralino decay within the detector [471, 472], neutralino decay branching ratios [472, 473], multi-b-jet events with an isolated charged lepton [474], and multi-lepton events [475].

Another one-loop mechanism for generating neutrino masses is to introduce extra fields that are in the adjoint representation of color $SU(3)$ [476]. The minimal such model is to have one (two) scalar octet(s) and two (one) fermion octets (octet). The scalars are in an $SU(2)_L$ doublet and the fermions may transform as either a singlet or triplet under $SU(2)_L$. Both normal and inverted hierachies are allowed in these models.

It is also possible that neutrino masses are generated at the two-loop level. In the Zee-Babu model [477] two charged scalars, h^+ and k^{++}, are introduced that are singlets under both color $SU(3)$ and $SU(2)_L$. They have Yukawa couplings to the leptons and can contribute to neutrino masses via the two-loop diagram in figure 9.5. The magnitude of the neutrino mass is

$$m_\nu \sim \frac{f^2 h}{(16\pi^2)^2} \frac{m_\tau^2}{\Lambda}, \tag{9.22}$$

where f and h are dimensionless Yukawa couplings and Λ is the scale of new physics. For $f \sim h \sim 0.1$, the neutrino mass will be of order 0.1 eV for $\Lambda \sim 1$ TeV. One consequence of the Zee-Babu model is potentially observable flavor-changing neutral currents in the lepton sector.

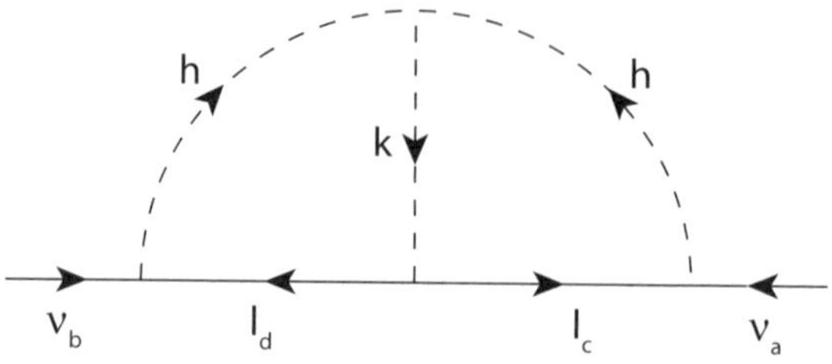

Figure 9.5. Two-loop diagram that contributes to neutrino mass in the Zee-Babu model. From [478].

Higher-dimension Operators

Higher-dimension operators can be an alternative to the standard seesaw operator of equation 9.1 as a source for light neutrino masses [479, 480]. Such models can give either Majorana or Dirac neutrino masses.

Gauged $U(1)'$

If the right-handed neutrinos carry a nonzero $U(1)'$ charge, then they cannot obtain the large Majorana mass needed for the conventional seesaw mechanism. In this case small Dirac neutrino masses are possible if they are suppressed, such as in theories with higher-dimension operators. However, their couplings to the Z' boson can lead to their creation in large numbers in the early universe, which could result in a too-large ^{4}He abundance unless the Z' mass is above $\simeq 2$ TeV [481]. If the right-handed neutrinos are neutral under $U(1)'$, then large Majorana masses are possible and there can be a conventional seesaw mechanism [481–483]. For a discussion of neutrino masses in a theory with a gauged $U(1)'$, see [484].

The simplest supersymmetric model with a gauged $U(1)'$ group has three generations of right-handed neutrinos and the minimal SUSY Higgs sector with a gauged $U(1)_{B-L}$ [485]. The sneutrino achieves a VEV at the TeV scale, which breaks both the $U(1)_{B-L}$ and R-parity, and the scale of $B-L$ breaking is identified with the soft SUSY-breaking mass scale. Left-handed neutrino masses are generated through an extended seesaw mechanism.

Triplet Higgs Bosons

In models with triplet Higgs bosons, horizontal symmetries are employed to constrain the texture of the neutrino mass matrix [459]. An example that uses an $S_2 \times S_2$ permutation symmetry in a four-neutrino theory (with one neutrino becoming heavy) is given in [486].

Dynamical Electroweak Symmetry Breaking

The seesaw mechanism can be realized in models with dynamical electroweak symmetry breaking (extended technicolor, or ETC models). By having a smaller Dirac mass term m_R in equation 9.4, the heavy Majorana scale need not be too high.[3] Dynamical electroweak symmetry breaking due to a neutrino condensate has also been considered [488].

String Theory

Neutrino masses can also be generated nonperturbatively in string theories, i.e., not by the usual Higgs mechanism, but via exponentially suppressed Yukawa couplings deriving from D-brane instantons. The resulting neutrino masses may be either Majorana [489] or Dirac [490].

Extra Dimensions

Theories with large extra dimensions [491, 492] have been postulated to avoid the hierarchy problem. In such theories, there is no very high scale (e.g., the GUT or Planck scale), and so the smallness of the neutrino masses cannot be obtained from a conventional seesaw mechanism; instead, it is a consequence of the suppressed coupling between the active neutrinos on the brane (the usual four-dimensional world) and sterile neutrinos in the bulk (Kaluza-Klein modes) or on other branes, associated with the small overlap of their wavefunctions. According to the particular model and coupling mechanisms, the neutrino masses can be either Dirac or Majorana. Some examples of models of neutrino mass in extra dimensions are given in [493]. However, no evidence of Kaluza-Klein modes, whose effects are like those of sterile neutrinos, has been found in the oscillation data. For a more complete discussion of theories with large extra dimensions, see [399].

Alternatively, extra dimensions can be generated dynamically at low energies from a theory that is four-dimensional and renormalizable at high energies (a process called dimensional deconstruction [494]). Acceptable neutrino phenomenology appears to be possible in such a scenario [495].

Hierarchical Right-handed Neutrinos

An attractive aspect of GUT models is that they often lead to relationships between quark and lepton masses. However, their different mixing characteristics (small for quarks, bilarge for neutrinos) can make it difficult to explain both. It has been shown that if the right-handed Majorana neutrino mass matrix has a hierarchical form (the likely form of the quark mass matrix), after the seesaw mechanism it is possible to have phenomenologically acceptable mixing for neutrinos in GUT models [496], models with a family symmetry [497], or models in extra dimensions [498].

Neutrino Anarchy

Finally, even though it is aesthetically pleasing to think that symmetries in one form or another account for the structure seen in the neutrino masses and mixings, it could be that an essentially random three-neutrino mass matrix can give the appropriate phenomenology [499]. In models with neutrino anarchy, it was originally thought that large mixing angles are quite natural, and the value of θ_{13} could lie just below the current experimental bound. However, a subsequent study suggests that large

[3] Typically the m_D terms can be much smaller than the lightest ETC scale, which can be as low as a few TeV [487].

mixing angles are not preferred if the mass matrix elements are truly random in a basis-independent way [500]. Statistical analyses of nonrandom structures have also been performed [414, 501].

9.5 Leptogenesis

The origin of the baryon asymmetry of our universe is an open question, and many scenarios for *baryogenesis* have been put forward including GUT baryogenesis, electroweak baryogenesis, the Affleck-Dine mechanism, and leptogenesis [502].

The smallness of the neutrino masses lends credibility to the seesaw mechanism. A direct consequence of the seesaw mechanism is *thermal leptogenesis* [503]. A net lepton asymmetry $Y_L \equiv (n_L - n_{\bar{L}})/s$ can be generated in the early universe because all of Sakharov's conditions [12] are met: (i) the heavy right-handed neutrinos N_i decay into a lepton-Higgs pair (lH) and into the CP conjugate pair with different partial widths, thereby violating lepton number; (ii) CP violation results from phases in the Yukawa couplings and neutrino mass matrices; (iii) the cosmological expansion yields the departure from thermal equilibrium.

As the universe cools and the N_i drop out of equilibrium, their decays lead to a CP asymmetry [504],

$$\epsilon_i = \frac{\Gamma(N_i \to lH) - \Gamma(N_i \to \bar{l}H^*)}{\Gamma(N_i \to lH) + \Gamma(N_i \to \bar{l}H^*)}. \tag{9.23}$$

The lepton asymmetry generated is $Y_L \sim \sum \epsilon_i / g_*$, where g_* is the number of degrees of freedom.

Since sphaleron [505] interactions preserve $B - L$ but violate B and L [506], the lepton asymmetry is partially converted to a baryon asymmetry Y_B. In terms of the initial $B - L$ [507],

$$Y_B = a Y_{B-L} = \frac{a}{a-1} Y_L, \tag{9.24}$$

where a depends on the processes in equilibrium. In the seesaw extended SM (MSSM), $a = 28/79$ ($a = 8/23$). Note that equation 9.24 is valid only for temperatures far above the weak scale.

In $SO(10)$-inspired scenarios where the Type-I seesaw mechanism provides the dominant contribution to neutrino masses, thermal leptogenesis typically fails to explain the observed baryon asymmetry of the universe. The reason is that the lightest right-handed neutrino is generally too light to generate enough asymmetry [508]. Including flavor effects [509, 510], the situation improves since the next-to-lightest right-handed neutrino (not accounted for in [508]), can generate a large asymmetry [511]. Nevertheless, the scenario remains tightly constrained [512, 513], which may lend itself to tests because definite predictions for rates of lepton flavor violating processes are possible.

It is worth mentioning that a mixed type-I + type-II seesaw mechanism can be naturally obtained within $SO(10)$, and successful leptogenesis is much more easily accomplished [512, 514].

TABLE 9.1
Present bounds and projected sensitivities for lepton flavor violating processes.

	Present	*Futur*
$\mathrm{BR}(\mu \to e\gamma)$	1.2×10^{-11} [518]	10^{-13} [524]
$\mathrm{BR}(\tau \to \mu\gamma)$	4.5×10^{-8} [519]	10^{-9} [525]
$\mathrm{BR}(\tau \to e\gamma)$	3.3×10^{-8} [520]	10^{-9} [525]
$\mathrm{BR}(\mu \to eee)$	1.0×10^{-12} [521]	10^{-14} [526]
$\mathrm{BR}(\tau \to \mu\mu\mu)$	3.2×10^{-8} [522]	10^{-9} [525]
$\mathrm{BR}(\tau \to eee)$	3.6×10^{-8} [522]	10^{-9} [525]
$\mathrm{CR}(\mu\,\mathrm{Ti} \to e\,\mathrm{Ti})$	4.3×10^{-12} [523]	10^{-18} [527]
$\mathrm{CR}(\mu\,\mathrm{Al} \to e\,\mathrm{Al})$	-	10^{-16} [528]

CR is the ratio of the $\mu \to e$ conversion rate to the muon capture rate.

Although a very appealing idea, leptogenesis is difficult to test; for recent assessments see [515]. The ϵ_i can be expressed independently of the neutrino mixing matrix V of equation 3.2 [516]. Any connection between the CP phase δ and the CP violation required for leptogenesis requires assumption about the texture of the Yukawa matrix, and is therefore model-dependent [517].

The only way to test leptogenesis directly is to constrain the Yukawa matrix and the masses of right-handed neutrinos, which may be possible in supersymmetric seesaw frameworks. This can be achieved by searching for lepton flavor violating decays and the electric dipole moments of the charged leptons to which much improved sensitivity is expected in the near future; see table 9.1. Even so, any determination of the Yukawa matrix will depend on how precisely the Higgs sector is known and on assumptions about the GUT model.

✳ 10 ✳

Supernova Neutrinos

10.1 General Description of a Supernova

Stars more massive than about $8M_{\odot}$ undergo gravitational collapse that leads to the production of a neutron star or a black hole. Neutrinos have a crucial role in the evolution of these core collapse supernovae. The variety of core collapse SN are listed in table 10.1. In stars whose mass is 8–10 $M_{\odot}$, the low mass core, $M_{core} < 1.44M_{\odot}$, undergoes O-Ne-Mg core collapse but the core mass is too small to ignite Ne burning. Stars with mass $\gtrsim 10M_{\odot}$ have iron cores that exceed the Chandrasekar limit of about $1.44M_{\odot}$; they can no longer be supported against gravitational collapse by electron degeneracy pressure and catastrophic collapse ensues.

Once the core of the star becomes constituted primarily of iron, further compression of the core does not ignite nuclear fusion and the star is unable to thermodynamically support its outer envelope. As the surrounding matter falls inward under gravity, the temperature of the core rises and iron dissociates into α particles and nucleons. Electron capture on protons becomes heavily favored and electron neutrinos are produced as the core gets neutronized (a process known as neutronization). When the core reaches densities above 10^{12} g/cm^3, neutrinos become trapped (in so-called neutrinospheres). The collapse continues until 3–4 times nuclear density is reached, after which the inner core rebounds, sending a shock-wave across the outer core and into the mantle. This shock-wave loses energy as it heats the matter it traverses and incites further electron capture on the free protons left in the wake of the shock. During the few milliseconds in which the shock-wave travels from the inner core to the neutrinosphere, electron neutrinos are released in a pulse. This neutronization burst carries away approximately 10^{51} ergs of energy, which is similar to the optical and kinetic energy release. However, 99% of the binding energy $E_b \sim 10^{53}$ ergs of the protoneutron star (which is about 10% of the star's rest mass energy) is released in the following $\sim$ 10 s. The primary processes are beta decay (providing a source of electron antineutrinos), $\nu_e\bar{\nu}_e$ annihilation [529], e^+e^- annihilation, and nucleon bremsstrahlung ($N + N \rightarrow N + N + \nu + \bar{\nu}$, which give all flavors of neutrinos: $\nu_e\bar{\nu}_e$, $\nu_\mu\bar{\nu}_\mu$, and $\nu_\tau\bar{\nu}_\tau$), in addition to electron capture. The neutrino luminosity components of a typical Type II supernova are illustrated in figure 10.1.

TABLE 10.1
Varieties of Type-II supernovae, including the initial mass, core composition, collapse event, and type of remnant.

$M/M_{\odot}$	*Core*	*Event*	*Remnant*
8 – 10	O-Ne-Mg	Low-mass SN	Neutron star
10 – 25	Fe	Normal SN	Neutron star
25 – 50	Fe	Hypernova	Black hole

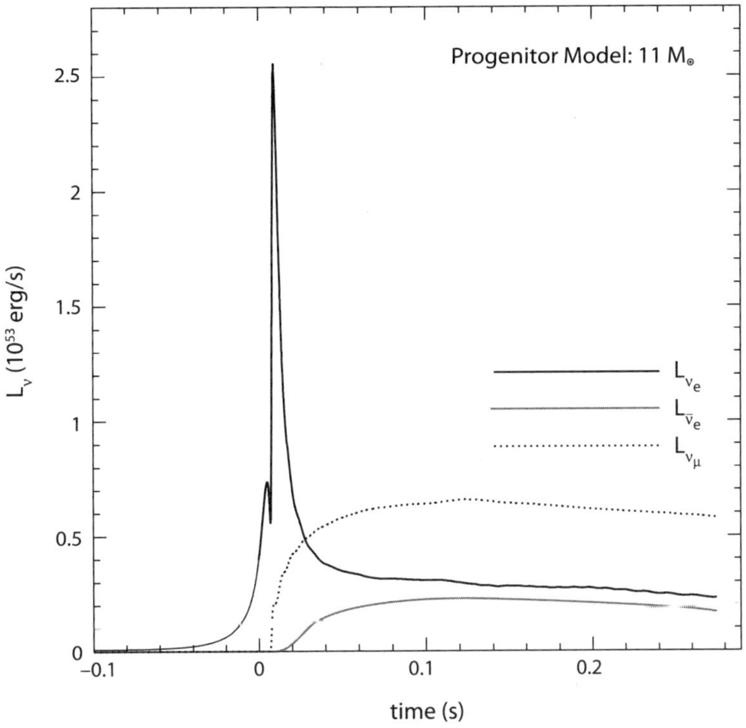

Figure 10.1. Total ν_e, $\bar{\nu}_e$, and ν_μ luminosity versus time for a supernova originating from a star with 11 solar masses. The neutronization burst of ν_e is clearly evident. From [530].

The essential role of neutrinos in supernova explosions is universally agreed upon but a full theoretical understanding of the complexity of the neutrino process involved is still under development [531]. The traditional approach took into account only flavor conversions of neutrinos without self-interactions. Later, it was realized that neutrino-neutrino interactions are significant near the neutrinospheres due to the large neutrino densities there. These give rise to neutrino refraction, which also goes under the label of "neutrino collective effects" [532–534]. These collective oscillations of supernova neutrinos cause the flavor swapping of neutrino spectral distributions over energy intervals bounded by sharp spectral splits. At distances for which collective effects become insignificant, typically a few hundred kilometers, MSW flavor conversions are encountered. In the following sections we describe the

TABLE 10.2
Typical predictions for primary average neutrino energies.

Flux model	$\langle E^0_{\nu_e} \rangle$	$\langle E^0_{\bar{\nu}_e} \rangle$	$\langle E^0_{\nu_x} \rangle$	$\frac{L_{\nu_e}(0)}{L_{\nu_x}(0)}$	$\frac{L_{\bar{\nu}_e}(0)}{L_{\nu_x}(0)}$
M1 [535]	12	15	24	2.0	1.6
M2 [536]	12	15	18	0.85	0.75

Note: Indicated by the superscript 0, in MeV, and flux ratios in two neutrino flux models. From [537].

qualitative aspects of neutrino flavor conversions in iron-core SN, first via collective effects and later by MSW transitions.

10.2 Neutrino Fluxes from the SN Core

The SN core is essentially a neutrino blackbody with fluxes that are flavor dependent. The initial fluxes of ν_μ, $\bar{\nu}_\mu$, ν_τ, and $\bar{\nu}_\tau$, are almost the same, and it is convenient to work in the ν_e, ν_x, ν_y basis, where

$$\nu_x = (\cos\theta_{23})\nu_\mu - (\sin\theta_{23})\nu_\tau\,, \qquad \nu_y = (\sin\theta_{23})\nu_\mu + (\cos\theta_{23})\nu_\tau\,. \tag{10.1}$$

The primary neutrino spectra are parametrized by

$$F_\alpha(E,t) = \frac{L_\alpha(t)\beta_\alpha^{\beta_\alpha}}{\langle E_\alpha\rangle\Gamma(\beta_\alpha)}\left(\frac{E}{\langle E_\alpha\rangle}\right)^{\beta_\alpha-1}\exp\left(-\beta_\alpha\frac{E}{\langle E_\alpha\rangle}\right), \tag{10.2}$$

where $\alpha = \nu_e, \bar{\nu}_e, \nu_x$, and $\langle E_\alpha\rangle$ (β_α) is the average energy (dimensionless shape parameter that quantifies the width of the spectrum) of species α. Typically $\beta_{\nu_e} = 4$, $\beta_{\bar{\nu}_e} = 5$, and $\beta_{\nu_x} = 4$. Here $L_\alpha(t)$ is the neutrino emission rate or luminosity of species α.

For typical model calculations of neutrino fluxes from supernovae, see [535] and [536]. The primary fluxes are highly dependent on the SN model, as illustrated in table 10.2, and the predictions of the neutrino flavor conversion probabilities depend on the primary spectra.

The hierarchy of primary average energies

$$\langle E^0_{\nu_e}\rangle < \langle E^0_{\bar{\nu}_e}\rangle < \langle E^0_{\nu_x}\rangle\,, \tag{10.3}$$

evident in table 10.2, is simply related to the interaction strengths of neutrinos with matter. Since the protoneutron star is opaque to neutrinos, it takes a few tens of seconds for the neutrinos to diffuse out. The ν_e and $\bar{\nu}_e$ interact with nuclear matter via both charged- and neutral-current reactions (with a smaller cross section for $\bar{\nu}_e$), while the ν_x experience only neutral current scattering. Neutrinos that interact more strongly have neutrinospheres (transient regions where neutrinos exist in thermal equilibrium) at higher radii: $R_{\nu_e} > R_{\bar{\nu}_e} > R_{\nu_x}$. Each neutrino species decouples at a temperature characterized by the temperature at the surface of its neutrinosphere.

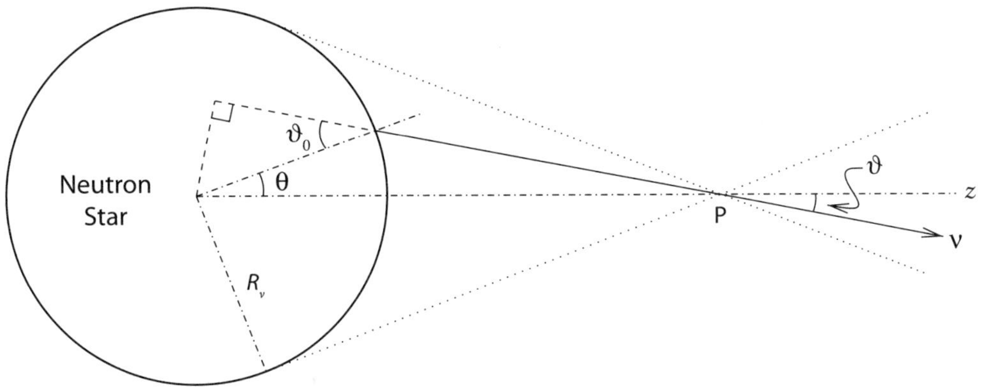

Figure 10.2. Geometric picture of neutrinos originating from the neutrinosphere (with radius R_ν). Because neutrinos are emitted from the neutrinosphere, neutrinos at point P only see neutrinos traveling within the cone delimited by the dotted lines. From [534].

10.3 Flavor Swapping from Collective Effects

Above the protoneutron star, neutrinos free-stream and the coherent evolution of the flavor state of neutrino i, $\psi_{\nu,i}$, is given by

$$i\frac{\partial}{\partial t}\psi_{\nu,i} = (\mathcal{H}_{vac,i} + \mathcal{H}_{e,i} + \mathcal{H}_{\nu\nu,i})\psi_{\nu,i}, \tag{10.4}$$

where t is an affine parameter along neutrino i's world line, and the three terms of the flavor-changing Hamiltonian along this trajectory represent the vacuum Hamiltonian, forward ν_e-electron scattering via W-exchange, and $\nu_i\nu_i$ and $\nu_i\nu_j$ forward scattering via Z-exchange, respectively. The new ingredient is the neutrino self-coupling,

$$\mathcal{H}_{\nu\nu,i} = \sqrt{2}G_F\sum_j\left(1-\hat{k}_i\cdot\hat{k}_j\right)n_{\nu,j}\psi_{\nu,j}\psi^\dagger_{\nu,j} - \sqrt{2}G_F\sum_j\left(1-\hat{k}_i\cdot\hat{k}_j\right)n_{\bar{\nu},j}\psi_{\bar{\nu},j}\psi^\dagger_{\bar{\nu},j}, \tag{10.5}$$

where $\hat{k}_j$ is the unit vector tangent to the trajectory for ν_j or $\bar{\nu}_j$, and $n_{\nu,j}$ is the local number density of neutrinos in state j. This contribution to the Hamiltonian is a source of nonlinearity and anisotropy in the neutrino flavor evolution, and of collective neutrino flavor transformation.

The intersecting neutrino trajectories are illustrated in figure 10.2. Although directional-dependent flavor evolution is in principle possible, it has been found that multi-angle effects are small and a single-angle approximation is usually adequate [538].[1]

The neutrino self-coupling can alter neutrino flavor evolution in SN and cause large neutrino flavor conversions deep in the supernova envelope. Near the

[1] Multi-angle effects suppress collective oscillations close to the neutrinosphere at late times (after the shock is pushed out) due to the interplay between neutrino-neutrino interactions and the vacuum oscillation term [539]. In the case studied in [539], the effect on the final neutrino spectra is qualitatively insignificant. However, there is evidence that this is not a general result [540].

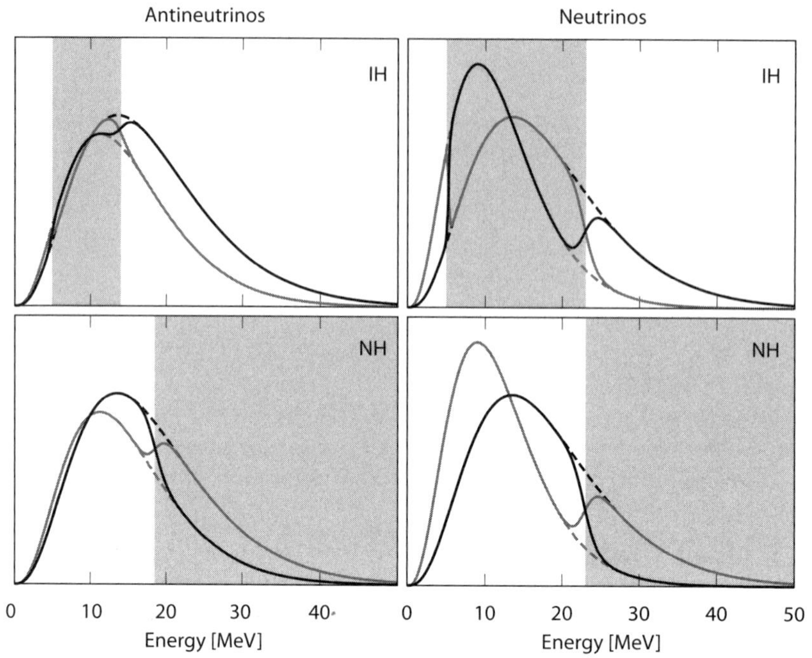

Figure 10.3. Supernova neutrino spectra before (dashed lines) and after (solid) collective oscillations, but before possible MSW conversions. The upper (lower) panels are for inverted (normal) hierarchy. The gray (black) lines show the spectra for ν_e (ν_y). The shaded regions show the energy ranges where spectral swaps occur. From [541].

neutrinospheres, the neutrino and antineutrino densities are so high ($>10^{30}/\text{cm}^3$) that neutrino-neutrino interactions make the time-evolution of the dense relativistic neutrino gas nonlinear up to a few 100 km. Thus the simple idea of neutrinos crossing MSW resonances must be replaced by a more complicated description of neutrino propagation as it leaves the supernova.

For model M2 of table 10.2 the outcome for flavor *swaps* (an exchange of the ν_e ($\bar{\nu}_e$) spectrum with the ν_y ($\bar{\nu}_y$) spectrum in a specific energy range) as the neutrinos emerge from the collective region is illustrated in figure 10.3 for both mass hierarchies. The number and locations of the *spectral splits* (sharp features at the edges of a swap interval) are dependent on the initial spectra and the value of the effective mixing angle. Further, in a three-flavor calculation, some results are qualitatively different from the two-flavor calculation [542]. These splits can be smeared by multi-angle effects of the neutrino-neutrino interactions. Note that the nature of the SN flavor spectra depend on both the accretion phase (example M1) and cooling phase (example M2) of the explosion.

10.4 MSW Conversions in a Supernova

Beyond the collective effects region, at distances of order 1000 km and outward, the neutrinos undergo MSW flavor interchanges. The neutrinos following the neutronization burst are the ones of interest in the following discussion.

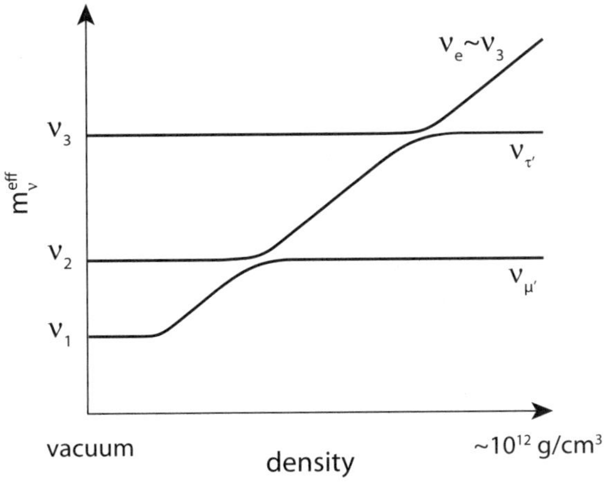

Figure 10.4. Schematic level-crossing diagram for neutrinos emitted by a SN in the case of a normal mass hierarchy. $\nu_{\mu'}$ and $\nu_{\tau'}$ are basis states that diagonalize the (ν_μ, ν_τ) submatrix of the Hamiltonian governing the neutrino evolution. From [182].

Because of the extremely high density of the matter in the region of the neutrinospheres, all flavors of neutrinos start out in pure mass eigenstates. As the neutrinos stream out they undergo MSW conversion and the relative numbers of each species can change. The density profile is represented by $V_0(R/r)^3$ [543], where R is the star radius and V_0 is a constant. Due to the wider range of densities that the neutrinos encounter, both the solar and atmospheric scales contribute to the oscillation dynamics. The hierarchical nature of the two scales ($|\delta m^2_{21}| \ll |\delta m^2_{31}|$) and the smallness of the mixing parameter $\sin^2 2\theta_{13}$ imply that the dynamics can be approximately factored so that oscillations are governed by δm^2_{31} and $\sin^2 2\theta_{13}$ at high densities ($\sim 10^3$ g/cm^3), and by δm^2_{21} and $\sin^2 2\theta_{12}$ at low densities (~ 20 g/cm^3 for the LMA solution) [544]; see figure 10.4. Transitions in the latter region are adiabatic. In the high density region, neutrinos (antineutrinos) pass through a resonance if $\delta m^2_{31} > 0$ ($\delta m^2_{31} < 0$). The jumping probability is the same for both neutrinos and antineutrinos [545] and is of the form $P_H \sim e^{-\sin^2\theta_{13}(|\delta m^2_{31}|/E_\nu)^{2/3} V_0^{1/3}}$ [546]. Note the exponential dependence of P_H on $\sin^2\theta_{13}$. That SN neutrinos could provide a handle on the sign of δm^2_{31} can be seen from table 10.3, which shows predictions for survival probabilities in the "swapped" and "unswapped" energy intervals.

As the shock traverses the resonance, adiabaticity is severely affected, causing oscillations to be temporarily suppressed, as first pointed out in [547]. After the shock moves beyond the resonance, oscillations are restored. Then one expects a dip in the time evolution of the average neutrino energy and the number of events. This modulation is visible in the neutrino (antineutrino) channel for a normal (inverted) mass hierarchy and only if $\sin^2\theta_{13} \gtrsim 10^{-3}$, i.e., only for oscillations that would occur adiabatically for a static density profile. Therefore, the scenarios of table 10.3 can be identified by shock wave effects irrespective of collective effects.

As SN neutrinos travel to a detector, they may pass through the earth, which may change the survival probabilities in the energy ranges where $P \neq 0$ and $\bar{P} \neq 0$; see table 10.3.

TABLE 10.3
Approximate survival probabilities.

	Hierarchy	$\sin^2\theta_{13}$	P *unswapped*	P *swapped*	$\bar{P}$ *unswapped*	$\bar{P}$ *swapped*
A	Normal	$\gtrsim 10^{-3}$	0	$\sin^2\theta_{12}$	$\cos^2\theta_{12}$	0
B	Inverted	$\gtrsim 10^{-3}$	$\sin^2\theta_{12}$	0	0	$\cos^2\theta_{12}$

Note: For neutrinos, P, and antineutrinos, $\bar{P}$, in different mixing scenarios in the "swapped" and "unswapped" energy intervals. From [537].

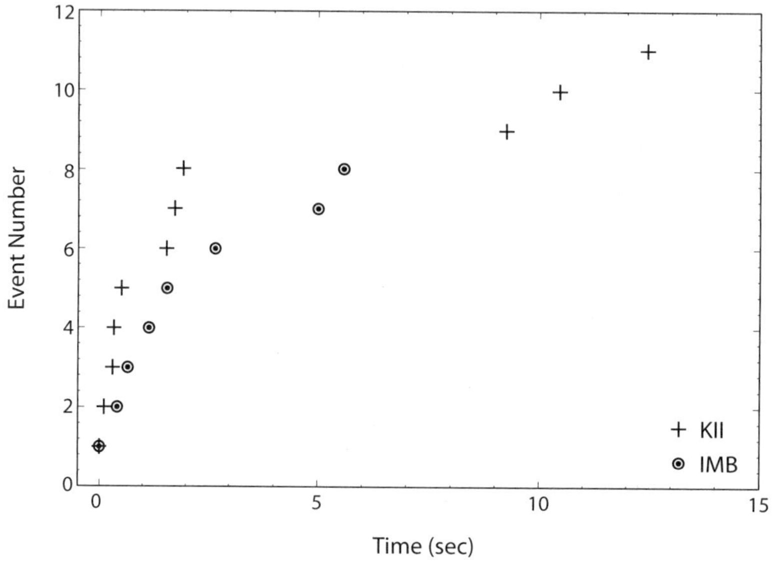

Figure 10.5. Timing of events for the SN1987A signal in the Kamiokande II and IMB detectors. The first event in each case is taken as $t = 0$. From [550].

A complete understanding of the complexities of neutrino oscillation phenomena in a supernova is still a work in progress. Collective, MSW, and 3-flavor effects all need to be taken into account, with a matching of the fluxes in the transition region between the high-density collective region and the lower-density MSW region.

10.5 Detection of Supernova Neutrinos

The detection of neutrinos from SN 1987A was momentous. The 11 events at Kamiokande II [548] and 8 events at the Irvine-Michigan-Brookhaven [549] detectors (see figure 10.5) have lent strong support to the generic model of core collapse supernovae [551]. The significance of these few events provides a tantalizing glimpse into the physics potential offered by a future galactic SN event.

Several neutrino detectors have conducted searches for supernovae occurring in our galaxy [552]. The LVD detector set an upper limit of 0.18 supernovae per year in our galaxy at 90% C.L. [553]. The Super-Kamiokande experiment set a limit of

TABLE 10.4
Expected number of events from a supernova 10 kpc from Earth in current and proposed neutrino detectors.

Detector	# of events	Status
HALO	40	running
Borexino	100	running
MiniBooNE	200	running
KamLAND	300	running
Super-K	7000	running
IceCube	600000	running
MiniCLEAN	30	proposed
OMNIS	1000	proposed
UNO	100000	proposed

less than 0.32 supernovae per year at a distance of 100 kpc, at 90% C.L. [554]. MiniBooNE set a limit on the core-collapse supernova rate out to a distance of 13.5 kpc to be less than 0.69 supernovae per year at 90% C.L. [555].

Despite the fact that only a few galactic SN are expected per century, the potential payoff is so huge that a number of experiments dedicated to SN neutrino detection have been proposed. Furthermore, since SN neutrinos arrive at Earth before the optical signal, the SuperNova Early Warning System (SNEWS) has been established for detecting SN neutrinos [556]. Table 10.4 shows the expected number of events from a supernova 10 kpc from Earth for a number of current and proposed experiments. IceCube in particular is expected to determine the bounce time to within a few milliseconds for a SN within 10 kpc, which can then be compared to the timing of the supernova's gravitational wave signal [557].

Neutrinos from a galactic SN could in principle provide much information on neutrino oscillations. A determination of θ_{13} and the neutrino mass hierarchy from SN neutrinos is special in that degeneracies [85] arising from the unknown CP phase δ and whether θ_{23} is above or below $\pi/4$ do not contaminate it, i.e., the eight-fold parameter degeneracies that are inherent in long-baseline experiments [85] are absent. This cleanness results because (i) nonelectron fluxes do not depend on the CP phase δ [558], and so SN neutrinos directly probe θ_{13}, and (ii) whether θ_{23} is above or below $\pi/4$ is immaterial since this parameter does not affect the oscillation dynamics.

Many investigations of the effect of neutrino oscillations on SN neutrinos have been made in [544, 550, 559, 560]. Whether or not the mass hierarchy can be determined and θ_{13} constrained depends strongly on how much $\langle E_{\nu_x}\rangle/\langle E_{\bar{\nu}_e}\rangle$ is greater than unity [544], and on details described in section 10.4. The higher the value of $\langle E_{\nu_x}\rangle/\langle E_{\bar{\nu}_e}\rangle$, the better the possible determinations.

Simulations of supernovae explosions depend critically on the assumed mechanisms of explosions (neutrino-driven wind/jet, hydrodynamic, acoustic power, etc.). A key feature of all mechanisms that produce an explosion is the breaking of spherical symmetry. The breakout burst of neutrinos predicted from one of these simulations is shown in figure 10.1. The energy and luminosity predictions for the ν_e neutronization burst are reasonably robust [561]. Consequently, the ν_e survival probability can be inferred from the burst signal and thus distinguish a scenario with a small oscillation probability [544].

Detections of shock modulations in either the ν_e spectrum in a kt-class liquid Ar detector or in the $\bar{\nu}_e$ spectrum in a Mt-class water Cherenkov detector can provide a probe of the true neutrino mass hierarchy [562]. The modulations should be observed in one and only one of the two channels (ν_e, $\bar{\nu}_e$). If a reverse shock also develops in the SN envelope, a double-dip signature should be seen in either the ν_e or $\bar{\nu}_e$ spectrum if θ_{13} is not too small. In addition to probing oscillation parameters, the presence of a double dip would confirm that SN simulations correctly depict shock propagation.

Observationally, pulsars have been seen with kicks exceeding 1000 km/s [563]. Possible explanations include the recent discovery of the standing accretion shock instability [564], and initial asymmetries produced by the excitement of unstable g-modes by the ϵ-mechanism [565]; for the shortcomings of these explanations see [111]. Sterile neutrinos with keV masses may explain the pulsar velocities through an anisotropy in their emission from a cooling neutron star born in a supernova explosion [111].

10.6 Supernova Relic Neutrinos

Supernova explosions in our galaxy are rare but about one supernova explosion occurs each second in the universe. The bulk of the energy in these supernova explosions is carried off by neutrinos. The Diffuse Supernovae Neutrino Background (DSNB) flux from all these relic supernovae neutrinos encodes information about the history of star formation and stellar evolution, so the detection of this neutrino background is of considerable interest [566].

The flux of the DSNB neutrinos integrated over all energies is estimated to be a few tens per cm^2 per second [567]. Potentially, the DSNB electron-antineutrino flux could be observed with the Super-Kamiokande (at antineutrino energies > 19 MeV) [568] and KamLAND (at antineutrino energies > 6 MeV) detectors, and DSNB electron-neutrinos could be seen at SNO (with neutrino energies in the range 22.5 to 32.5 MeV) [569].

The differential flux for the DSNB is given by [570]

$$\frac{d\Phi_\nu}{dE} = \int_0^\infty R_{SN}(z) \frac{dN(E(1+z))}{dE}(1+z)\left|\frac{dt}{dz}\right| dz, \tag{10.6}$$

where $R_{SN}(z)$ is the supernova rate as a function of redshift, $E(1+z)$ is the neutrino energy at emission and E is the energy at detection and dN/dE is the neutrino spectrum from an individual Type-II supernova event, modeled by a Fermi-Dirac spectrum with an effective chemical potential

$$\frac{dN}{dE} \propto \frac{E^2}{\exp((E-\mu)/T)+1}, \tag{10.7}$$

normalized to $\sim 10^{53}$ ergs, the typical energy released in a SN. The last factor in the integrand is the inverse of the differential distance

$$\left|\frac{dz}{dt}\right| = H_0(1+z)\sqrt{\Omega_m(1+z)^3+\Omega_\Lambda}\,, \tag{10.8}$$

where $H_0 = 70$ km/s/Mpc, $\Omega_m = 0.3$ and $\Omega_\Lambda = 0.7$. A parametrization of the supernova rate density at redshift z of the form $R_{SN}(z) = R_{SN}(0)(1+z)^\alpha$ is assumed, where α is a parameter that depends on redshift range. The resulting predictions are 3.6 events per year in Super-K and 0.4 events per year KamLAND. However, there is a factor of 2 normalization uncertainty between DSNB density of bright and faint SN that impacts the predicted DSNB. Super-K has so far seen no evidence for DSNB neutrinos and reported an upper bound on the DSNB $\bar{\nu}_e$ flux [571] that is close to the theoretical predictions.

The problem in Super-K is how to differentiate this rare DSNB signal from the much higher solar neutrino flux. To identify the signal as an electron antineutrino event, the scattered neutron must be detected in addition to the Cherenkov light from the relativistic positron. A 0.2% concentration of gadolinium in the water could be used for neutron detection, since the neutron capture cross section on Gd is very large and after the capture Gd emits a cascade of observable gamma rays. This would enable Super-K to detect up to 20 relic neutrinos over 5 years with essentially no background. Super-K has a gadolinium test facility underway to confirm that the addition of Gd would not impact its other neutrino observations [572].

11

High-energy Astrophysical Neutrinos

It is anticipated that neutrino telescopes will map out the sky over the next few decades. The neutrinos will point back to their sources, like gamma rays but unlike cosmic rays that are bent by magnetic fields. Neutrino telescopes should allow probes further back in time and deeper into the sources. The energy range from a TeV to a ZeV may permit the study of astrophysics and particle physics in conjunction.[1]

High-energy neutrino fluxes from cosmologically distant sources are generally expected in association with the production of cosmic-rays (CR), whose energy spectrum extends to 10^{20} eV and is likely dominated above $\sim 3 \times 10^{17}$ eV by protons, neutrons, and nuclei of extragalactic origin. The energy distribution of the CR flux is shown in figure 11.1. Source candidates include galactic sources like supernova remnants and extragalactic sources like Active Galactic Nuclei (AGN) and Gamma Ray Bursts (GRB).

High-energy neutrino production is thought to be associated with the interactions of high-energy protons that produce energetic charged pions by $p\gamma$ or by $p\bar{p}$ interactions. In sources that are optically thin to meson-nucleon interactions, the $\pi^+ \to \mu^+ \nu_\mu$ decays and subsequent $\mu^+ \to e^+ \nu_e \bar{\nu}_\mu$ decays (and corresponding π^- decay chain) lead to high-energy neutrinos. The decays of neutral pions, $\pi^o \to \gamma\gamma$, may be observed as gamma ray signals in experiments such as observations by the Fermi Gamma-ray Space Telescope (FGST). A schematic illustration of the beam dump origin of astrophysical neutrinos is given in figure 11.2.

11.1 Cosmogenic Neutrinos

Ultra high-energy cosmic ray protons (by which one often means protons having energies above an EeV) will undergo inelastic interactions, mainly $p\gamma \to \Delta^+ \to n\pi^+$ and $p\pi^0$, on the cosmic microwave background. The threshold energy for this reaction is the Greisen-Zatsepin-Kuzmin (GZK) energy, $E_{GZK} \sim 4 \times 10^{19}$ eV [575].

[1] TeV = 10^{12} eV, PeV = 10^{15} eV, EeV = 10^{18} eV, and ZeV = 10^{21} eV.

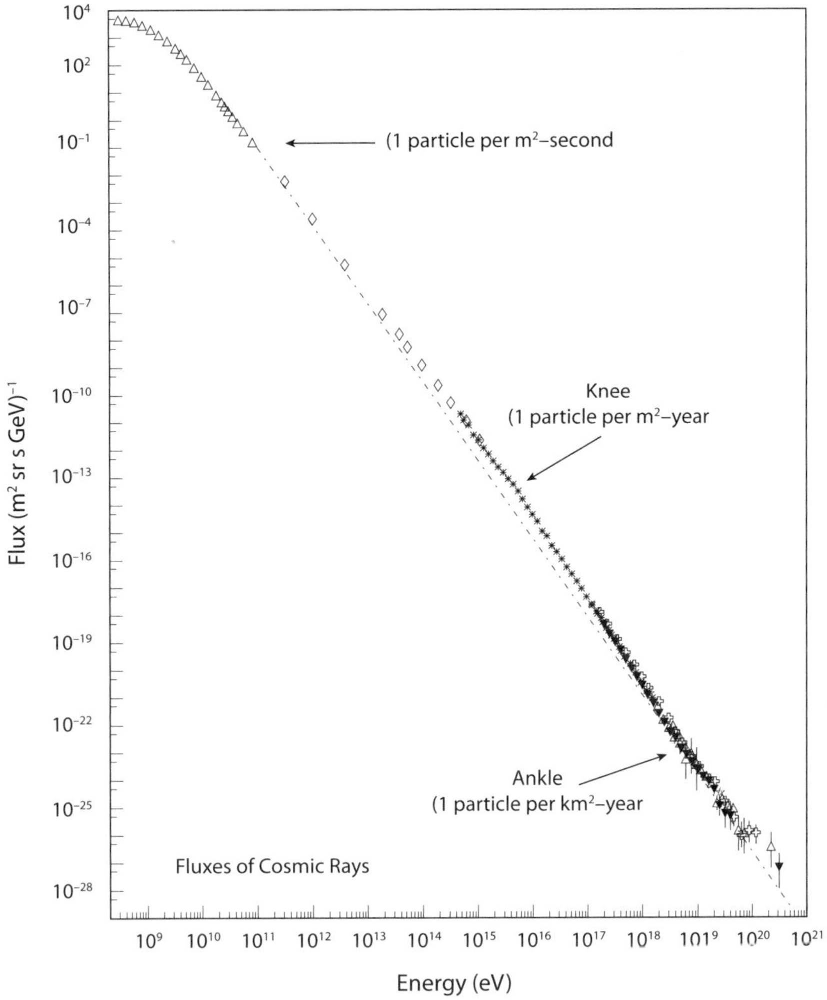

Figure 11.1. Cosmic ray flux versus energy. From [573].

At such high energies, the gyroradius of a proton in the galactic magnetic field is larger than the size of the galaxy and it is therefore expected that these cosmic ray protons are of extragalactic origin. Since the attenuation caused by this reaction has a length scale of about 50 Mpc [576, 577], a strong suppression in the cosmic ray spectrum is expected above E_{GZK}, which has in fact been observed by the HiRes [578] and Auger [579] experiments: see figure 11.3.

The decays of charged pions produced in the GZK process are a source of *cosmogenic* neutrinos as predicted by Berezinsky and Zatsepin (BZ) [581]; neutral pion decays are sources of gamma rays. However, there are two caveats to the BZ prediction of cosmogenic neutrinos. The first is that cosmogenic accelerators may only produce protons up to the energy of the observed cut-off of the cosmic ray spectrum. The second is that Auger finds evidence that the high-energy cosmic rays are a mixture of protons and heavy nuclei [582] and the GZK threshold energy for

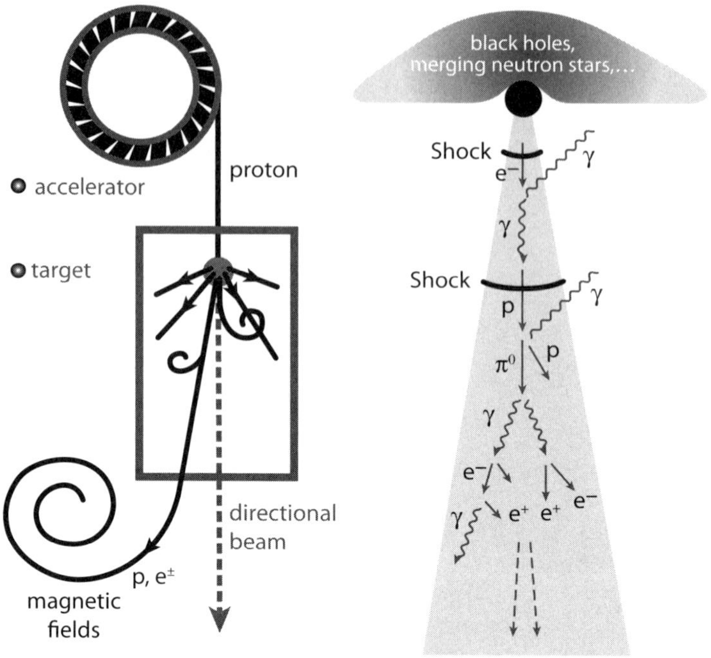

Figure 11.2. Mechanism for production of astrophysical neutrinos (right), compared to a man-made neutrino beam (left). From [574].

heavy nuclei is higher than that for protons. In these circumstances the BZ neutrinos may not be found.

11.2 IceCube

High-energy neutrinos can be detected from secondary particles produced by neutrino interactions in large volumes of transparent ice or water that are instrumented by a lattice of photomultiplier modules. The principle of detection is to record the Cherenkov light from relativistic particles moving faster than the speed of light in medium, which has an index of refraction 1.3. The IceCube experiment [112, 583] is designed to detect neutrinos from astrophysical sources by their interactions in the glacial ice cap at the geographic South Pole. The earth acts as a filter to absorb the intense background flux from cosmic-ray induced muons, thus permitting the identification of upward moving events. The neutrino events from atmospheric cosmic rays are a background to neutrino events from cosmic sources, but this background falls steeply with energy.

The layout of the IceCube detector is shown in figure 11.4. The first-generation AMANDA detector, the precursor/prototype of IceCube, demonstrated the feasibility of the concept in measuring the atmospheric neutrino flux up to 10^5 GeV; see figure 11.5. The fully implemented cubic kilometer IceCube detector consists of 80 kilometer-length strings, each instrumented with 60 10-inch photomultipliers spaced vertically every 17 meters. The digital optical modules are buried between 1450 and 2450 meters in vertical holes melted in the ice. The array covers a surface area of

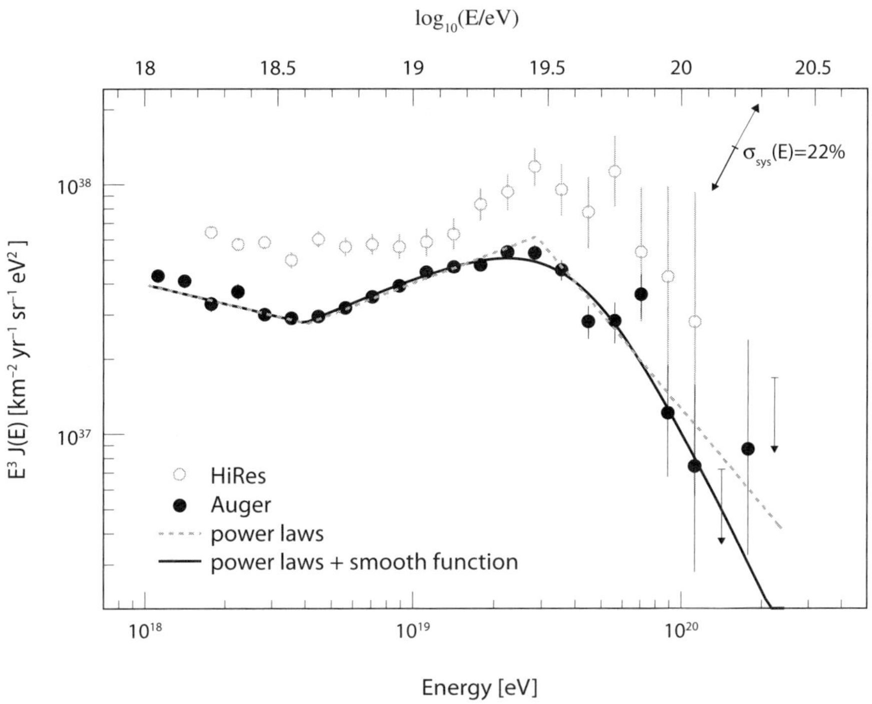

Figure 11.3. Comparison of the cosmic ray flux in the Auger and HiRES experiments to a power law spectrum E^{-3}. The spectrum falls steeply above the GZK cut-off energy of $E \sim 4 \times 10^{19}$ eV. The diagonal arrow shows the size of the energy uncertainty for the Auger data. The curves indicate fits to two functions: three power laws with free breaks between them (dotted), and two power laws in the ankle region ($E \simeq 10^{18.6}$ eV) with a smoothly changing function at higher energies (solid). From [580].

one square kilometer. A surface air shower detector, IceTop, augments the deep ice detector by background rejection and cosmic ray studies.

The DeepCore sub-array with an infill of 6 more closely spaced strings and 7 nearby IceCube strings is deployed in the highly transparent ice in the lower half of the IceCube detector, with the optical modules buried 2000 to 2450 m below the surface. Its instrumented volume is of order 10 Mt. The IceCube detector is optimized for neutrino energies above 10 TeV, but has sensitivity down to 0.1 TeV. The DeepCore detector reduces the neutrino threshold energy to 10 GeV. The IceCube detector acts as a filter to remove neutrino events that occur outside the DeepCore volume. Thus, the contained events in DeepCore allow 4π coverage.

Scientific Objectives of IceCube

IceCube is unique in providing access to atmospheric neutrino beams for energies of $1 - 10^5$ TeV. IceCube is expected to collect a data set of order one million atmospheric neutrinos over ten years.

The DeepCore significantly extends the physics capabilities of IceCube as an atmospheric neutrino detector. Both the disappearance of ν_μ and the appearance

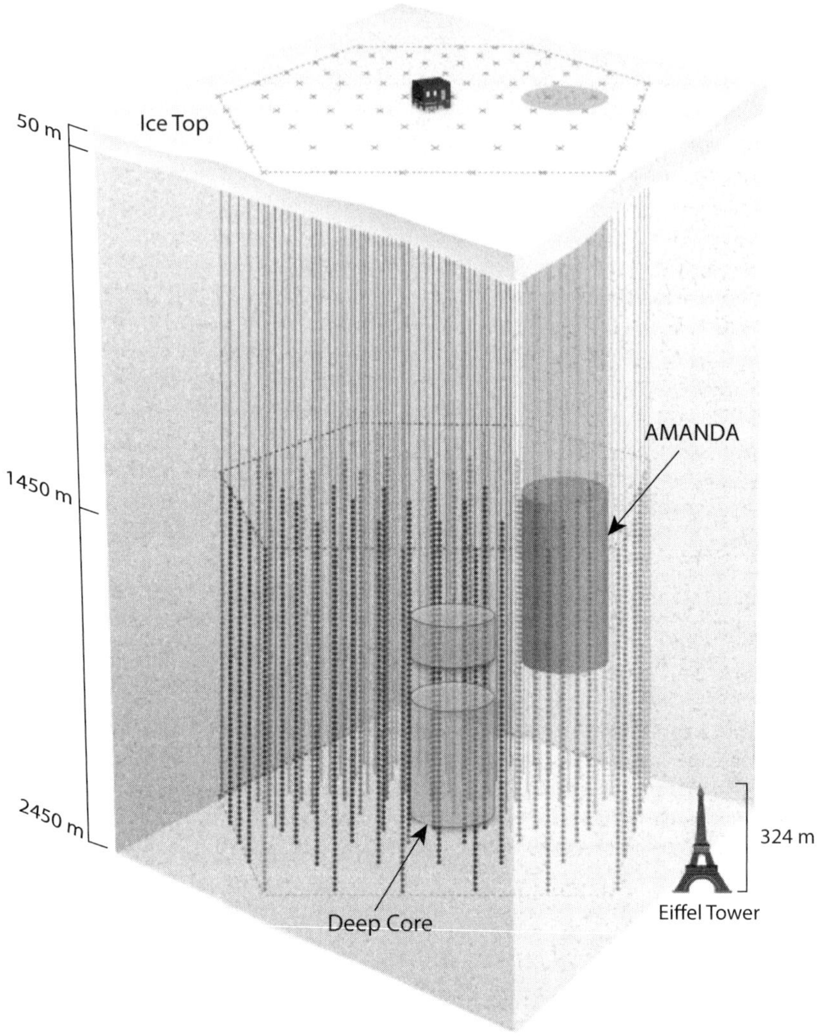

Figure 11.4. Schematic diagram of the IceCube detector. The positions of the AMANDA, DeepCore, and IceTop detectors are also shown. From [584].

of ν_τ and ν_e can be observed. High statistics will be accumulated, which may cover the first oscillation dip of ν_μ near 20 GeV. For values of θ_{13} close to the present bound, it may be possible to determine the mass hierarchy with DeepCore. The key is that the resonance condition is satisfied for neutrino energies of order 15 GeV for the baselines of thousands of kilometers relevant to IceCube/DeepCore.

To study astrophysical neutrinos, one needs to differentiate between astrophysical and atmospheric neutrinos. One means is by way of their energy spectra but this is challenging because the expected astrophysical neutrino flux is low. Another way is to detect several neutrinos from the same direction, especially if the direction is associated with a known source of high-energy gamma-rays. It is even better to see

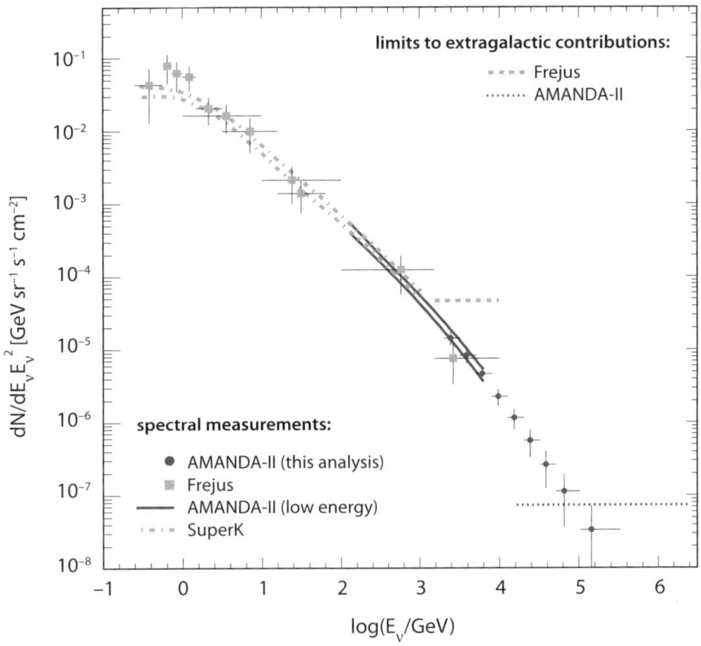

Figure 11.5. Measurement of the high-energy atmospheric neutrino flux in the AMANDA-II detector (circles), a lower-energy measurement from AMANDA [585] (solid curves), and earlier results from Super-K [586] (dash-dotted) and the Frejus detector [587] (squares). Upper bounds on extragalactic contributions to the neutrino flux from Frejus [588] and AMANDA [589] are also shown. From [590].

two or more neutrinos at the same time and from the same direction as a gamma-ray burst (GRB).

Potential neutrino sources for IceCube include extra-galactic AGN and GRBs, both of which are likely to be cosmic particle accelerators. The neutrinos would be produced when accelerated protons (or nuclei) interact with photons or gas near the sources. Galactic cosmic-ray accelerators such as supernova remnants and interacting X-ray binaries, or micro-quasars, may also be viable neutrino sources.

IceCube can search for neutrinos as a signature of dark matter by looking for neutrinos from the sun where weakly interacting massive particles (WIMPs) could accumulate and annihilate each other. The indirect search for WIMPs in the sun depends on the spin-dependent cross section for their interactions with protons in the sun. DeepCore will extend the sensitivity to solar WIMP annihilations (see figure 11.6) by virtue of its lower energy threshold and the capability to isolate solar WIMP signals even when the sun is above the horizon.

An important signature of exotic physics in IceCube is a track propagating at less than the speed of light. Possibilities for such signals are massive magnetic monopoles, Q-balls and massive nuclearites.

A galactic supernova could give a large flux of MeV-energy neutrinos in IceCube over a period of about 10 seconds that would be detected as an excess of the background counting rate in all individual optical modules. Although only a counting experiment, IceCube could measure the time profile of a neutrino burst

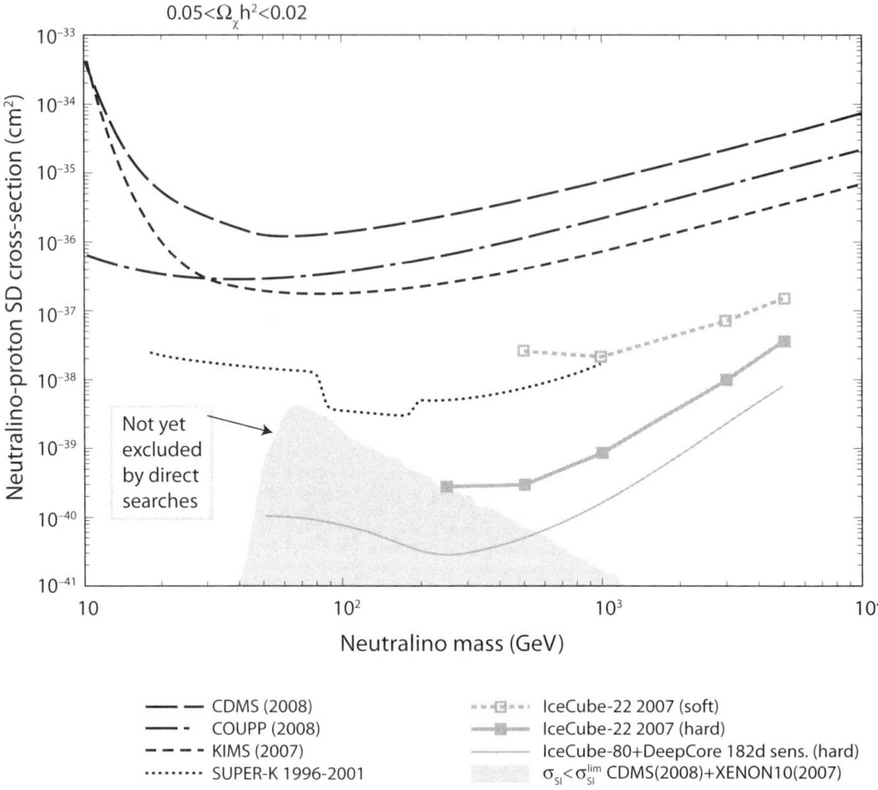

Figure 11.6. Limit on spin-dependent neutralino-proton cross section (solid curve) achievable by the fully-deployed IceCube detector (IceCube-80), shown versus neutralino mass. Current limits [591] are also shown. From [592].

near the center of the galaxy with high statistics of about one million events, equivalent to the sensitivity of a 2 megaton water detector.

Cherenkov radio and acoustic signals have attenuation lengths in ice that can be as high as 1 km, making it practical to build a sparse array that instruments a much larger effective volume than IceCube. The radio array would be co-deployed with the IceCube strings between 80–400 m below the surface.

11.3 Waxman-Bahcall Flux

Assuming that the ultra high-energy cosmic rays are extragalactic protons and lose their energy in interactions producing pions, Waxman and Bahcall (WB) estimated a neutrino flux from astrophysical sources to be [593]: $E_\nu^2 \Phi_\nu \sim 2 \times 10^{-8}\,\mathrm{GeV/cm^2/s/sr}$. This flux assumes an injection spectrum for protons that is typically expected for Fermi acceleration. According to the WB flux, a detector of $(\mathrm{km})^2$ area is necessary to detect 10 to 50 neutrino interactions per year at TeV scale neutrino energies. The IceCube detector at the South Pole and the KM3 detector under construction in the Mediterranean Sea will both have $(\mathrm{km})^2$ area. As noted

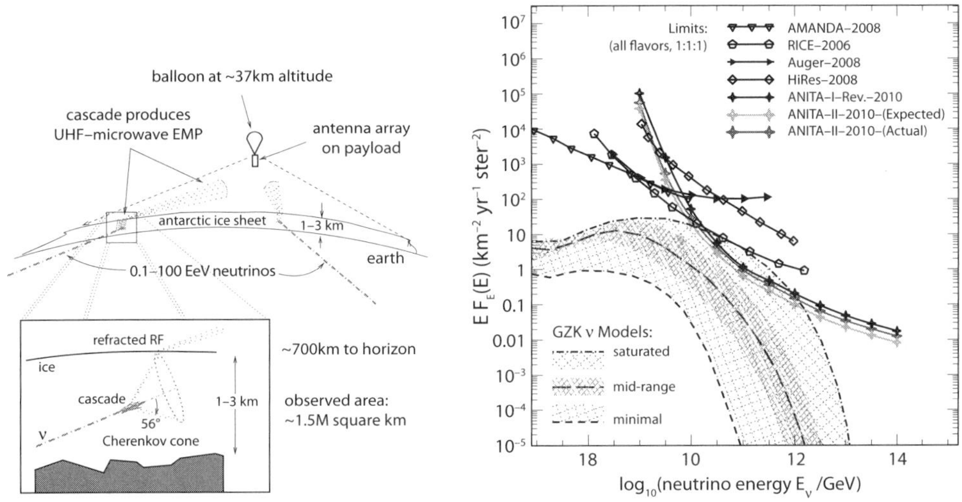

Figure 11.7. Left: Schematic of neutrino detection in ANITA; neutrinos interact in the Antarctic ice sheet, producing an ultra high-frequency EM pulse that is measured by the ANITA detector [595]. Right: Model predictions and experimental bounds on the cosmogenic neutrino flux (from [596]).

before, the AMANDA and IceCube detectors can view only the northern celestial hemisphere (upward events that pass through the earth), because the backgrounds from cosmic-ray muons dominate events from downward neutrino interactions.

11.4 Ultra High-energy Neutrino Cross Sections

Neutrino cross sections at very high energies could be quite different from extrapolations from lower energies, and deviations could signal new physics. Therefore measuring neutrino cross sections in the asymptotic region is essential.

A cosmogenic flux of ultra high-energy neutrinos can be measured in experiments that detect radio signals from neutrino interactions, because of still larger area coverage. The RICE [594] and ANITA [595] experiments at the South Pole have thereby placed upper bounds on the cosmogenic neutrino flux. The ANITA balloon experiment can detect horizontal neutrinos that interact with the full polar ice cap (see figure 11.7); ANITA thus has an effective area $\sim 10^6$ km^2. Current experimental limits on the cosmogenic neutrino flux are shown in figure 11.7.

Theoretical estimates of the cosmogenic neutrino flux are very model-dependent and differ by 2 to 3 orders of magnitude. This uncertainty must be dealt with in order to extract or constrain the high-energy neutrino-nucleon cross section with cosmogenic neutrino flux measurements or upper bounds. There are two ways to reliably calculate the neutrino flux. The first way is to assume a power-law energy injection spectrum for the protons in analyzing the CR data, taking into account possible redshift evolution of the sources with a power law index, $(1 + z)^n$. The second way is a model-independent unfolding procedure in a CR analysis. The results from the two methods are in excellent agreement. The 95% C.L. upper bounds obtained from the RICE data on the neutrino-nucleon cross section relative

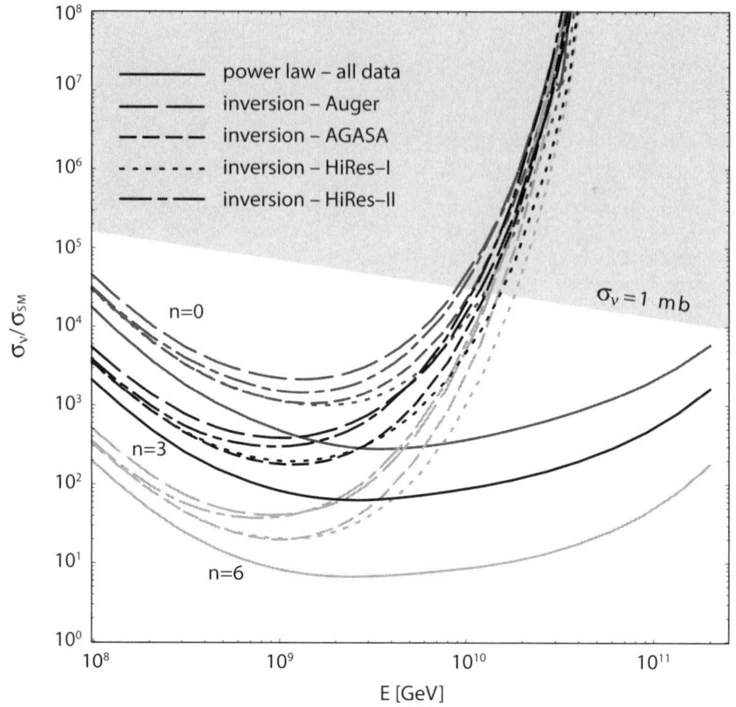

Figure 11.8. 95% C.L. upper bounds on the ratio of the neutrino-nucleon cross section to its SM value using RICE data for power law indices $n = 0$ (gray curves), $n = 3$ (black), and $n = 6$ (light gray). The solid curves assume a power law injection spectrum and the non-solid curves are obtained by using a propagation inversion procedure on the indicated CR spectrum. The bounds are valid only in the unshaded region. Adapted from [597].

to the Standard Model are shown in figure 11.8. New physics contributions are not tightly constrained by the present experimental bounds; see figure 11.9.

11.5 Z-burst Mechanism

The direct detection of the cosmic neutrino background neutrinos from the Big Bang is of fundamental interest. The Z-dip/burst mechanism was proposed by T. Weiler to observe the cosmic neutrino background via resonant annihilation of ultra high-energy cosmic neutrinos through the Z-boson [130]. The $\nu + \bar{\nu} \to Z$ cross section has a several orders of magnitude enhancement at an energy

$$E_{res} = \frac{M_Z^2}{2m_\nu} = 4.2 \times 10^{21} \left(\frac{\text{eV}}{m_\nu}\right) \text{ eV}. \tag{11.1}$$

In principle the annihilation could be detected as absorption dips in the incident neutrino flux or as emission features in the Z-decays to nucleons and photons. The viability of the Z-burst mechanism depends critically on the size of the neutrino flux. It was suggested that the background neutrinos could gravitationally cluster on

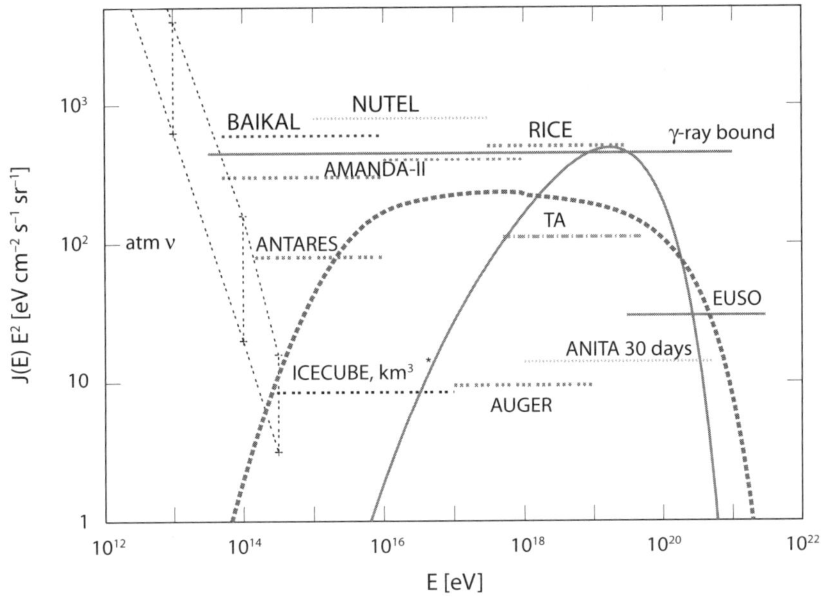

Figure 11.9. Neutrino fluxes for scenarios where ultra high-energy cosmic rays are explained as new hadrons produced as secondaries of accelerated protons (solid and dashed curves). Also shown are expected sensitivities. From [598].

cold dark matter with the neutrino overdensity enhancing the flux; the nearby Virgo cluster is the prime site. Unfortunately, the fact that the absolute neutrino mass is now known to be in the sub-eV range, and the corresponding resonance energy is two orders of magnitude above the GZK cut-off energy, makes it unlikely that the signal of the Z-burst mechanism is observable.

11.6 Astrophysical Neutrino Flavor Content

The neutrino detectors of km^2 or larger area were designed primarily to detect high-energy muons. The muon energy threshold of the IceCube detector is about 100 GeV. The signals of ν_e can be distinguished from ν_μ by characteristic differences in the particle showers produced by the passing charged electrons and muons. In water Cherenkov and ice detectors, electrons produced by the ν_e and the neutral current interactions of the ν_e, ν_μ, and ν_τ give localized electromagnetic and/or hadronic showers. High-energy ν_μ produced by the charged-current interactions are identified via the long tracks of the muons. The production of τ's by high-energy ν_τ ($\sim$PeV) can give characteristic "double bang" events due to the showers from both the initial hadronic interaction and the subsequent τ decay, whose locations can be separated because of the long decay length of the relativistic τ; the separation can be of the order of meters for the relativistic tau leptons. Figure 11.10 shows a simulated representation of these three event types. The ν_τ's can also give "lollipop" events, where the hadronic lobe of the double-bang is not detected – these can be distinguished from CC ν_e and NC events by the lower level of light produced by the τ lepton in flight.

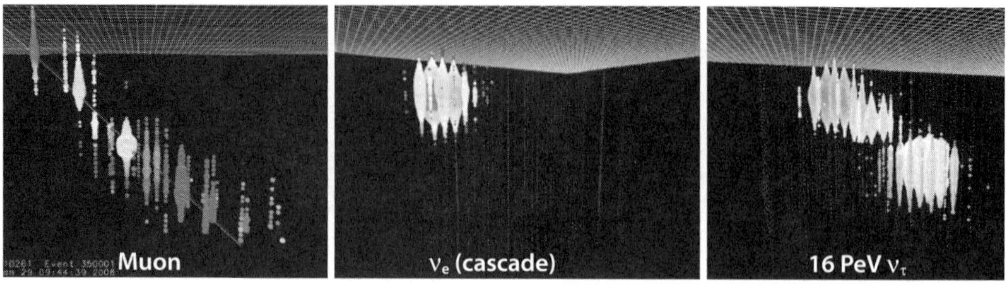

Figure 11.10. Energy deposition patterns in IceCube for ν_μ events (left panel), CC ν_e and NC events (middle), and "double bang" ν_τ events (right). Also possible are "lollipop" ν_τ events where one lobe of the signature is not detected. Adapted from [599].

Using all this information, the $\nu_e : \nu_\mu : \nu_\tau$ ratio may in principle be estimated, though in practice this will be challenging. If electromagnetic and hadronic showers can be distinguished, the determination of the flux ratios will be improved. The energy resolution of the detectors and the presence of the background from atmospheric neutrinos are complicating factors.

If astrophysical sources of neutrinos exist and neutrino detectors are of sufficiently large size to detect significant signals and the three neutrinos flavors (ν_e, ν_μ, ν_τ) can be distinguished by the detectors, then a variety of interesting physics possibilities can be tested [600]. The following is a qualitative discussion of these prospects. (Note that the signals of neutrinos and antineutrinos are the same in detectors without magnetic fields.)

Neutrinos from AGN beam dump sources originate from the decays of pions (and to a lesser extent kaons) and the subsequent decays of muons in flight. If neutrinos are created from purely hadronic processes (e.g., $p + p \to \pi^\pm + X$), approximately equal numbers of neutrinos and antineutrinos are expected at the source with twice as many ν_μ as ν_e and essentially no ν_τ. Thus the neutrino fluxes (neutrino + antineutrino) at the source are roughly in the ratios

$$\phi_{\nu_e} : \phi_{\nu_\mu} : \phi_{\nu_\tau} = 1 : 2 : 0. \tag{11.2}$$

However, the ν_e to ν_μ ratio could have some energy dependence. A small prompt ν_τ component could be present due to the production and semileptonic decay of charmed particles. If the dominant production process is $p + \gamma \to \pi^+ X$, then the flavor mix is the same, but there is no initial $\bar{\nu}_e$ flux (to a first aproximation).

Other possible ratios for the flavor content of the high-energy neutrino flux are $0 : 1 : 0$ if the muons from pion decay lose energy or get absorbed, or $1 : 0 : 0$ if the high-energy sources emit primarily neutrons; however, as noted the ratios of the actual fluxes could differ slightly from these ideal relationships.

Neutrino Oscillation Effects

Since the oscillation arguments $\delta m^2 L/4E$ for solar and atmospheric δm^2 scales are very large ($> 10^7$ even for PeV energies with $L > 10^{10}$ km), the vacuum oscillations en route to the earth average $\sin^2(\delta m^2 L/4E)$ to 1/2. Then the oscillation probabilities are given by

$$P_{\alpha\alpha} = \sum_i |V_{\alpha_i}|^4, \tag{11.3}$$

$$P_{\alpha\beta} = \sum_i |V_{\alpha_i}|^2 |V_{\beta_i}|^2. \tag{11.4}$$

If there is maximal atmospheric neutrino mixing ($\theta_{23} = \pi/4$) and under the approximation that $V_{e3} = 0$, the survival probability matrix is

$$P = \begin{pmatrix} c_{12}^4 + s_{12}^4 & s_{12}^2 c_{12}^2 & s_{12}^2 c_{12}^2 \\ s_{12}^2 c_{12}^2 & \frac{1}{2}(1 - s_{12}^2 c_{12}^2) & \frac{1}{2}(1 - s_{12}^2 c_{12}^2) \\ s_{12}^2 c_{12}^2 & \frac{1}{2}(1 - s_{12}^2 c_{12}^2) & \frac{1}{2}(1 - s_{12}^2 c_{12}^2) \end{pmatrix}, \tag{11.5}$$

where

$$\phi_{\nu_\alpha}(\text{earth}) = \sum_\beta P_{\alpha\beta}\, \phi_{\nu_\beta}(\text{source}). \tag{11.6}$$

The initial flux ratio of equation 11.2 is thereby converted to a universal mixture at the earth of

$$\nu_e : \nu_\mu : \nu_\tau = 1 : 1 : 1. \tag{11.7}$$

The observation of this democratic flavor distribution would provide evidence for the conventional wisdom about the beam dump nature of the production process. However, for maximal atmospheric neutrino mixing and $V_{e3} = 0$ any initial flux combination that has one-third ν_e, including equal fluxes of all flavors at the source, gives the same result. Also, it will not be easy to confirm this prediction, particularly for the ν_τ.

Other possibilities for the neutrino flux mix at the source have been considered [600, 601], leading to predictions for the flux ratios at earth of the generic form

$$\nu_e : \nu_\mu : \nu_\tau = \alpha : 1 : 1. \tag{11.8}$$

For tribimaximal mixing, $\alpha = 4/7$ if there is an initial pure ν_μ beam (with the muons losing energy from interactions) and $\alpha = 5/2$ for a pure $\bar{\nu}_e$ beam (from a neutron decay source). A summary of possible initial neutrino flux ratios and the ratios seen on earth assuming tribimaximal mixing are shown in table 11.1.

The difference between pp and $p\gamma$ production can also be probed according to the predicted ratios seen at the earth (for tribimaximal mixing)

$$pp: \quad \frac{\bar{\nu}_e}{\nu_{\text{tot}}} = \frac{1}{6}, \quad \frac{\bar{\nu}_e}{(\nu_\mu + \bar{\nu}_\mu)} = \frac{1}{2} \tag{11.9}$$

TABLE 11.1
Predicted $\nu_e : \nu_\mu : \nu_\tau$ ratios for various production mechanisms and the resulting ratios seen on earth, assuming tribimaximal mixing.

Mechanism	*Source*	*Earth*
$\pi + \mu$ decay	1 : 2 : 0	1 : 1 : 1
π decay only	0 : 1 : 0	4 : 7 : 7
neutron decay	1 : 0 : 0	5 : 2 : 2
ν decay, normal hierarchy	Any	4 : 1 : 1
ν decay, inverted hierarchy	Any	0 : 1 : 1
Spacetime foam	Any	1 : 1 : 1

Note: Neutrino decay and spacetime foam (quantum gravity decoherence) are discussed in sections 12.4 and 12.5, respectively. Adapted from [602].

$$p\gamma : \quad \frac{\bar{\nu}_e}{\nu_{\text{tot}}} = \frac{2}{27}, \quad \frac{\bar{\nu}_e}{(\nu_\mu + \bar{\nu}_\mu)} = \frac{2}{9}. \tag{11.10}$$

Deviations from the tribimaximal mixing approximation due to nonzero V_{e3} and nonzero CP phase δ can also in principle be probed [600].

11.7 Neutrinos from Dark Matter Annihilation

Dark matter accounts for about 20% of the energy density of the universe. Particle physics models relate a conserved discrete symmetry to the existence of a stable dark matter particle. A WIMP of mass of order 100 GeV that was produced in the early universe, thermalized, and frozen-out due to the Hubble expansion provides a natural explanation for the observed density of dark matter today. The best motivated DM particle is the Lightest Stable Particle (LSP) of SUSY with R-parity conservation [603, 604]. The LSP is nominally the lightest neutralino (denoted by χ_1^0), a neutral spin-1/2 particle that is a linear combination of gauginos (spin-1/2 SUSY companions of the Spin-1 Standard Model B and W_3 gauge bosons) and higgsinos (spin-1/2 companions of two spin-0 Higgs bosons) [605]. The consequences of neutralino dark matter detection have been the subject of numerous studies; some recent surveys are given in [606–612]. There are three complementary experimental approaches: direct detection via nuclear recoils from WIMP scattering; indirect detection via astrophysics experiments wherein WIMP annihilations give neutrino, gamma ray, positron, antiproton, and antideuteron signals; and collider experiments where the supersymmetric particles undergo cascade decays to final states with two LSPs that give missing energy in the events.

Our focus here is on the DM annihilation signals in neutrino telescopes that result from the annihilations of a broad class of dark matter candidates that have been gravitationally captured by the sun. However, when we provide numerical results we focus on the neutralino. These experiments have the capability to find or limit signals from neutralino DM annihilations in the sun. It is therefore of particular interest to examine the characteristic features imprinted on the neutrino energy spectra of this origin.

Dark Matter Capture and Annihilation

The time dependence of the number of DM particles in the sun is determined by the balance of capture and annihilation. A massive particle can be gravitationally captured by the sun from the galactic halo at a rate $C_\odot$. On the other hand, annihilation at the center of the sun will decrease the number of particles at a rate Γ_A. Thus, the time evolution of the number of DM particles in the sun N is given by the solution to the differential equation [607],

$$\frac{dN}{dt} = C_\odot - C_A N^2, \tag{11.11}$$

where $C_A = 2\Gamma_A/N^2$.

Assuming that the number of collisions DM particles undergo inside the sun during the sun's lifetime is large enough for them to thermalize [613], the number density of DM particles at a distance r from the solar core can be expressed in terms of the DM mass m_χ, the temperature of the sun T and the gravitational potential $2\pi\rho r^2/(3M_{Pl}^2)$:

$$n(r) = n_0 e^{-\frac{2\pi\rho r^2}{3M_{Pl}^2}\frac{m_\chi}{T}}, \tag{11.12}$$

where n_0 is the DM number density at the center of the sun, ρ is the average density of the sun and $M_{Pl} = 1.22 \times 10^{19}$ GeV is the Planck mass. C_A is then a constant that depends only on ρ, T and the annihilation cross section $\langle\sigma_A v\rangle$ averaged over the velocity distribution in the limit $v \to 0$:

$$C_A = \frac{\int n(r)^2 \langle\sigma_A v\rangle d^3r}{(\int n(r) d^3r)^2} = \langle\sigma_A v\rangle \left(\frac{3M_{Pl}^2 T}{m_\chi \rho}\right)^{-\frac{3}{2}}. \tag{11.13}$$

Equation (11.11) admits the solution,

$$N(t) = \sqrt{\frac{C_\odot}{C_A}} \tanh\left(\frac{t}{\tau}\right), \tag{11.14}$$

where $\tau \equiv 1/\sqrt{C_A C_\odot}$ is the characteristic time necessary to reach equilibrium. For $t \sim 4.5$ Gyr and $\langle\sigma_A v\rangle \sim 1$ pb and $\sigma_{SD} \gtrsim 10^{-6}$ pb (the range of interest for IceCube/DeepCore), it is known [607, 613] that $t/\tau \gg 1$ for the range $150\,\text{GeV} < m_\chi < 600\,\text{GeV}$. Since $\Gamma_A = C_A N^2/2 = C_\odot \tanh^2(t/\tau)/2$, one can easily see that when $t \gg \tau$, $\Gamma_A \to \Gamma_{eq} \equiv C_\odot/2$.

This limit simplifies the calculations, as it relates the event rate to the DM-nucleon scattering cross section, thus bypassing all the astrophysical uncertainties related to the solar model. However, the condition of equilibrium is not guaranteed. For example, in [614] it was recently shown that there are regions of mSUGRA parameter space for which the annihilation rate is far below the capture rate.

Moreover, since DM can annihilate to hidden sector particles as well as MSSM particles, the annihilation rate relevant for neutrino detection is scaled by the branching fraction to MSSM decay products, $\Gamma_A^{MSSM} = \Gamma_A B_F^{MSSM}$. These uncertainties can be accounted for by scaling the muon event rates at the detector by the parameter $\xi \equiv \Gamma_A^{MSSM}/\Gamma_{eq} = B_F^{MSSM}\tanh^2(t/\tau)$. We set $\xi = 1$ for our quantitative discussion.

The solar capture rate of neutralinos in the galactic halo is approximately given by [615]

$$C_\odot^{SI} = 4.8 \times 10^{28} s^{-1} \frac{\rho_{0.3}}{\bar{v}_{270} m_\chi} \sum_i F_i f_i \phi_i \frac{\sigma_i^{SI}}{m_{N_i}} S\left(\frac{m_\chi}{m_{N_i}}\right) \quad (11.15)$$

$$C_\odot^{SD} = 1.3 \times 10^{29} s^{-1} \frac{\rho_{0.3}}{\bar{v}_{270} m_\chi} \sigma_H^{SD} S\left(\frac{m_\chi}{m_{N_i}}\right), \quad (11.16)$$

where i sums over the elements with significant abundance in the sun ranging from hydrogen to iron. $\rho_{0.3}$ is the local DM halo density in units $0.3\,\mathrm{GeV/cm^3}$, $\bar{v}_{270}$ is the average DM dispersion velocity in units of 270 km/s, m_{N_i} denotes the mass of the nucleus of the ith element in GeV, and σ_i is the SD/SI scattering cross section in pb. f_i, S, F_i are the mass fraction, kinematic suppression, and form-factor suppression [616] for nucleus i. ϕ_i describes the distributions of the ith element. We refer interested readers to [607] for a detailed discussion and the values for these parameters. For most mSUGRA models consistent with the measured relic density, σ^{SD} is greater than σ^{SI} by two to three orders of magnitude, but does not necessarily dominate the capture rate.

Neutrino Propagation

Here we describe the procedure involving the propagation of neutrinos from the center of the sun to the earth, and the detection at IceCube and DeepCore.

Once the neutrinos are produced at the center of the sun, they propagate through the solar medium and travel to the earth before they can be detected at the neutrino telescope. The appropriate formalism involves the density matrix for the neutrino spectra in the flavor basis [617]. We call this ρ, and indicate matrices in boldface.

Neutrinos of energy E_ν can be propagated from a point r to $r + dr$ inside the sun by solving the Heisenberg equation,

$$\frac{d\rho(E_\nu)}{dr} = -i[\mathbf{H}(E_\nu), \rho(E_\nu)] + \left.\frac{d\rho(E_\nu)}{dr}\right|_{NC} + \left.\frac{d\rho(E_\nu)}{dr}\right|_{CC} + \left.\frac{d\rho(E_\nu)}{dr}\right|_{in}, \quad (11.17)$$

where $\mathbf{H}$ is the Hamiltonian for neutrino oscillations in matter, the term indicated by in is the injection spectrum at the center of the sun and the other two terms represent the matter effects due to NC- and CC-interactions. The Hamiltonian is

$$\mathbf{H} = \frac{1}{2E_\nu}\mathbf{V}\mathrm{diag}(0, \delta m_{21}^2, \delta m_{31}^2)\mathbf{V}^\dagger + \mathrm{diag}(\sqrt{2}G_F N_e, 0, 0), \quad (11.18)$$

where N_e is the radius-dependent density of electrons inside the sun, G_F is the Fermi constant, and E_ν is the energy of the incoming neutrino. We use the following values: $\delta m^2_{21} = 8.1 \times 10^{-5}\,\mathrm{eV}^2$, $\delta m^2_{31} = 2.2 \times 10^{-3}\,\mathrm{eV}^2$, $\theta_{12} = 33.2^\circ$, $\theta_{23} = 45^\circ$, $\theta_{13} = 0$. It has been shown in [617] that the oscillation results are not significantly changed if θ_{13} is small but nonzero. The density profile of the sun is shown in figure 3.3.

The injection term (which is diagonal in the flavor basis) is

$$\left.\frac{d\rho_{ij}}{dr}\right|_{in}(E_\nu) = \delta(r)\delta_{ij}\frac{d\Phi}{dE_\nu}, \tag{11.19}$$

where

$$\frac{d\Phi}{dE_\nu} = \Gamma_A^{MSSM}\sum_k \mathrm{BF}_k\frac{d\Phi_k}{dE_\nu}, \tag{11.20}$$

is the source neutrino/antineutrino flux at the center of the sun; k denotes each annihilation channel, BF_k and $\frac{d\Phi_k}{dE_\nu}$ are the branching fractions and normalized (to each annihilation event) neutrino energy spectra for the kth channel, respectively.

In considering the NC and CC terms, we introduce the following quantity that depends on the deep inelastic scattering (DIS) total cross sections σ:

$$\mathbf{\Gamma}_{NC(CC)}(E_\nu, E') = N_p(r)\mathrm{diag}[\sigma(\nu_l p \to \nu_l'(l)+\mathrm{any})]+N_n(r)\mathrm{diag}[\sigma(\nu_l n \to \nu_l'(l)+\mathrm{any})], \tag{11.21}$$

where $N_p(r)(= N_e(r))$ and $N_n(r)$ are the proton and neutron densities inside the sun, plotted in figure 3.3, E_ν is the incoming neutrino energy, E' the outgoing neutrino (charged-lepton) energy and l labels flavor. Then, the neutral current term is given by

$$\left.\frac{d\rho(E_\nu)}{dr}\right|_{NC} = -\rho(E_\nu)\int_0^{E_\nu}\frac{d\mathbf{\Gamma}_{NC}}{dE'_\nu}(E_\nu, E'_\nu)dE'_\nu + \int_{E_\nu}^{\infty}\frac{d\mathbf{\Gamma}_{NC}}{dE_\nu}(E'_\nu, E_\nu)\rho(E'_\nu)dE'_\nu. \tag{11.22}$$

The CC term is defined in a similar way but it is more complicated for two reasons. Firstly, the CC-DIS cross sections are not the same for all flavors of neutrinos. As a matter of fact the ν_τ-cross sections are suppressed near threshold by the kinematical effects of m_τ. Secondly, one needs to take into account the effects of tau regeneration that couple the propagation of different flavors (different elements of the density matrix) and elements of the neutrino and antineutrino density matrices.

Tau regeneration is an important effect that leads to a reinjection of the neutrinos produced by the decay of the taus that are CC-created by neutrinos of higher energies. The taus produced by energetic neutrinos undergoing CC interactions can decay promptly through various channels, for example $\tau^- \to \nu_\tau + \mathrm{any}$, $\tau^- \to e^-\bar{\nu}_e\nu_\tau$ and $\tau^- \to \mu^-\bar{\nu}_\mu\nu_\tau$, and similarly for the antiparticles. These processes provide additional sources of energetic neutrinos that reenter the flux with lower energies. The probabilities of reinjection are encoded in four functions $f_{\nu_\tau\to\nu_\tau}(u)$, $f_{\bar{\nu}_\tau\to\bar{\nu}_\tau}(u)$, $f_{\nu_\tau\to\bar{\nu}_{e,\mu}}(u)$, $f_{\bar{\nu}_\tau\to\nu_{e,\mu}}(u)$, which depend on the branching ratios of the channels and on $u \equiv E_\nu^{out}/E_\nu^{in}$, where E_ν^{in} is the energy of the tau-neutrino undergoing CC-scattering, and E_ν^{out} is the energy of the lower-energy neutrinos produced by tau decay [617].

The charged-current contribution to the Heisenberg equation is therefore,

$$\left.\frac{d\rho(E_\nu)}{dr}\right|_{CC} = -\frac{\{\mathbf{\Gamma}_{CC}, \rho\}}{2} + \int_{E_\nu}^{\infty} \frac{dE_\nu^{in}}{E_\nu^{in}} \left[\mathbf{\Pi}_\tau \rho_{\tau\tau}(E_\nu^{in})\Gamma_{CC}^{\tau}(E_\nu^{in}) f_{\nu_\tau \to \nu_\tau}\left(\frac{E_\nu}{E_\nu^{in}}\right)\right.$$
$$\left. + \mathbf{\Pi}_{e,\mu}\bar{\rho}_{\tau\tau}(E_\nu^{in})\bar{\Gamma}_{CC}^{\tau}(E_\nu^{in}) f_{\bar{\nu}_\tau \to \nu_{e,\mu}}\left(\frac{E_\nu}{E_\nu^{in}}\right)\right], \tag{11.23}$$

$$\left.\frac{d\bar{\rho}(E_\nu)}{dr}\right|_{CC} = -\frac{\{\bar{\mathbf{\Gamma}}_{CC}, \bar{\rho}\}}{2} + \int_{E_\nu}^{\infty} \frac{dE_\nu^{in}}{E_\nu^{in}} \left[\mathbf{\Pi}_\tau \bar{\rho}_{\tau\tau}(E_\nu^{in})\bar{\Gamma}_{CC}^{\tau}(E_\nu^{in}) f_{\bar{\nu}_\tau \to \bar{\nu}_\tau}\left(\frac{E_\nu}{E_\nu^{in}}\right)\right.$$
$$\left. + \mathbf{\Pi}_{e,\mu}\rho_{\tau\tau}(E_\nu^{in})\Gamma_{CC}^{\tau}(E_\nu^{in}) f_{\nu_\tau \to \bar{\nu}_{e,\mu}}\left(\frac{E_\nu}{E_\nu^{in}}\right)\right], \tag{11.24}$$

where $\mathbf{\Pi}_e = \text{diag}(1, 0, 0)$ are projectors, and similar expressions apply to the other flavors. As is clear from the last term on the right-hand side of equations (11.23) and (11.24), tau regeneration effects couple the two sets of equations.

Once the neutrinos reach the surface of the sun, the propagation to the earth is obtained by the following averaging procedure: rotate the density matrix to the mass basis; drop the off-diagonal terms, and rotate it back to the flavor basis. With our choice of neutrino parameters, the averaging should wash out any observable modulation.

Muon Rates

Upward Events at IceCube. When a muon generated by the CC interactions travels through the rock and ice beneath the detector it loses energy due to ionization, bremsstrahlung, pair production, and photonuclear effects [618]. The average energy loss of the muons that travel a distance dz in a medium of density ρ_{med} is given by:

$$\left\langle \frac{dE}{dz} \right\rangle = -(\alpha + \beta(E)E)\rho_{med}(z), \tag{11.25}$$

where $\alpha = 3.0 \times 10^{-3}\,\text{GeV}\,\text{cm}^2/\text{g}$ is related to ionization, while $\beta(E)$ takes into account bremsstrahlung, pair production, and photonuclear effects. We take $\beta = 3.0 \times 10^{-6}\,\text{cm}^2/\text{g}$ and $\rho_{med} = \rho_{ice} = 0.92\,\text{g/cm}^3$. The results from Muon Monte Carlo [619] are reproduced by choosing these values of α and β [620]. Equation (11.25) can then be easily solved to obtain the final energy $E_{\mu f}$, given initial energy $E_{\mu i}$:

$$E_{\mu f} = -\frac{\alpha}{\beta} + e^{-\beta\rho_{ice}z}\left(E_{\mu i} + \frac{\alpha}{\beta}\right). \tag{11.26}$$

The average range covered by the muon between energies $E_{\mu i}$, $E_{\mu f}$ is then,

$$R_\mu(E_{\mu i}, E_{\mu f}) = \frac{1}{\beta\rho_{ice}} \ln\left(\frac{\alpha + \beta E_{\mu i}}{\alpha + \beta E_{\mu f}}\right). \tag{11.27}$$

Thus, the muon flux at the detector is obtained by a convolution of the following: the probability of the incoming neutrino to CC-scatter with a nucleus in ice; the average range over which energy losses force the muon energy below the detector threshold; and the muon probability of surviving its own decay length. This last effect can be parametrized by the survival probability,

$$P_{sur}(E_{\mu i}, E_{\mu f}) = \left[\frac{E_{\mu f}(\alpha + \beta E_{\mu i})}{E_{\mu i}(\alpha + \beta E_{\mu f})}\right]^{\frac{m_\mu}{c\tau\alpha\rho_{ice}}}, \tag{11.28}$$

which is a solution to the differential equation,

$$\frac{dP_{sur}}{dE_{\mu f}} = \frac{P_{sur}}{E_{\mu f} c\tau\rho_{ice}(\alpha + \beta E_{\mu f})/m_\mu}, \tag{11.29}$$

where τ is the muon lifetime and m_μ its mass. Folding these effects together gives the spectrum of muon events,

$$\begin{aligned}\frac{d\Phi_\mu}{dE_{\mu f}} = &\int_0^{R_\mu(m_X, E_{\mu f})} dz\, e^{\beta\rho_{ice} z} P_{sur}(E_{\mu i}(E_{\mu f}, z), E_{\mu f}) \\ &\times \int_{E_{\mu i}(E_{\mu f}, z)}^{m_X} dE_\nu \left[\frac{d\Phi_\nu}{dE_\nu}\left(\frac{d\sigma_{\nu p}^{CC}}{dE_{\mu i}}(E_\nu)\rho_p + \frac{d\sigma_{\nu n}^{CC}}{dE_{\mu i}}(E_\nu)\rho_n\right)\right. \\ &\left. + \text{(a corresponding contribution for } \bar{\nu})\right],\end{aligned} \tag{11.30}$$

where $E_{\mu i}(E_{\mu f}, z)$ is obtained by inverting equation 11.26, $d\sigma^{CC}/dE_\mu(E_\nu)$ are the differential CC-cross sections to protons (νp) and neutrons (νn), $\rho_p \sim 5/9\ N_A\ \text{cm}^{-3}$ and $\rho_n \sim 4/9\ N_A\ \text{cm}^{-3}$ are the number densities of nucleons in ice expressed in terms of Avogadro's number N_A, and $d\Phi_\nu/dE_\nu$ is the neutrino spectrum at earth.

The event rate for upward events is obtained by convolving equation 11.30 with the muon *effective area* of the detector, $A_{eff}(E_{\mu f})R(\cos\theta)$, which is constituted by a zenith angle-independent part, shown in figure 11.11a, and a factor $R(\cos\theta) = 0.92 - 0.45\cos\theta$ that accounts for the rock bed beneath the ice [621]. We take the average of the effective area over the time of the year that the sun spends below the horizon, namely between the March and September equinoxes. We define the zenith angle θ_z at the South Pole to be the angle centered at the detector with $\theta_z = 0°$ indicating the vertical direction in the sky. θ_z can be parametrized in terms of the time of the year f_y (where $f_y = 0, 1/2$ correspond to the March and September equinoxes, respectively), and the tilt of the earth axis with respect to the perpendicular to the ecliptic plane, $\theta_t = 23°26'$:

$$\theta_z(f_y) = \frac{\pi}{2} + \theta_t \sin(2\pi f_y). \tag{11.31}$$

The event rate reads,

$$N_{events}^{Up} = \xi\Gamma_{eq} \int_0^{m_X} \frac{d\Phi_\mu}{dE_{\mu f}} \langle A_{eff}(E_{\mu f}) R(\cos\theta_z)\rangle dE_{\mu f}, \tag{11.32}$$

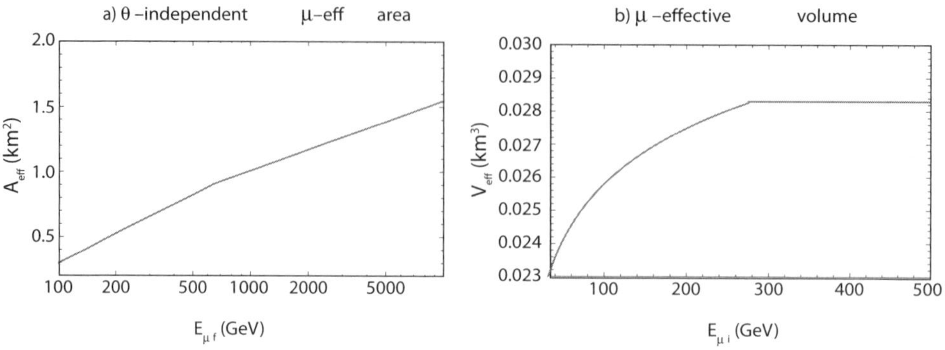

Figure 11.11. a) The IceCube muon effective area with zenith angle dependence factored out. b) Muon effective volume for DeepCore. From [622].

where ξ and Γ_{eq} are the quantities introduced earlier. $\langle A_{eff}(E_{\mu f})R(\cos\theta_z)\rangle$ is the average over the portion of the solid angle (θ_z, ϕ) that corresponds to the time the sun spends below the horizon.

Contained Events at IceCube. For neutrinos that interact within the detector volume, the resulting muons do not lose an appreciable fraction of their energies. Following notation similar to the previous subsection (except that we identify $E_\mu \equiv E_{\mu i}$ since the muons do not propagate) the muon flux for contained events is given by,

$$\frac{d\Phi_\mu}{dE_\mu} = L\int_{E_\mu}^{m_X} dE_\nu \left[\frac{d\Phi_\nu}{dE_\nu}\left(\frac{d\sigma_{\nu p}^{CC}}{dE_\mu}(E_\nu)\rho_p + \frac{d\sigma_{\nu n}^{CC}}{dE_\mu}(E_\nu)\rho_n\right) + (\nu \to \bar{\nu})\right], \quad (11.33)$$

where $L \sim 1$ km is the size of the IceCube detector. The event rate reads,

$$N_{events}^{C} = \xi\Gamma_{eq}\int_{E_{thr}}^{m_X} \frac{d\Phi_\mu}{dE_\mu}(1\ \text{km}^2)dE_\mu, \quad (11.34)$$

where $E_{thr} = 100$ GeV is the energy threshold of the IceCube detector. As previously mentioned, we only consider events observed between the March and September equinoxes.

Events at DeepCore. The great advantage of DeepCore with respect to IceCube is that the outer instrumented volume of IceCube will serve as a veto to atmospheric muon events up to one part in 10^6 [623], so that data can be collected throughout the year, i.e., even when the sun is above the horizon. The rate for contained events at DeepCore can be calculated by convolving equation 11.33 with the muon *effective volume* $V_{eff}(E_{\mu i})$. In the most optimistic estimates the effective volume is constant for muon energies above ~ 300 GeV, and drops significantly at lower energies [624]. For DeepCore we consider the interval $E_{\mu i} > E_{min} = 35$ GeV. We find that in this interval the effective volume in km^3 can be parametrized by

$$V_{eff}(E_{\mu i}) = (0.0056 \log E_{\mu i} + 0.0146)\ \Theta(275 - E_{\mu i}) + 0.0283\ \Theta(E_{\mu i} - 275), \quad (11.35)$$

where Θ is the Heaviside step function and $E_{\mu i}$ is in GeV. The effective volume is plotted in figure 11.11b.

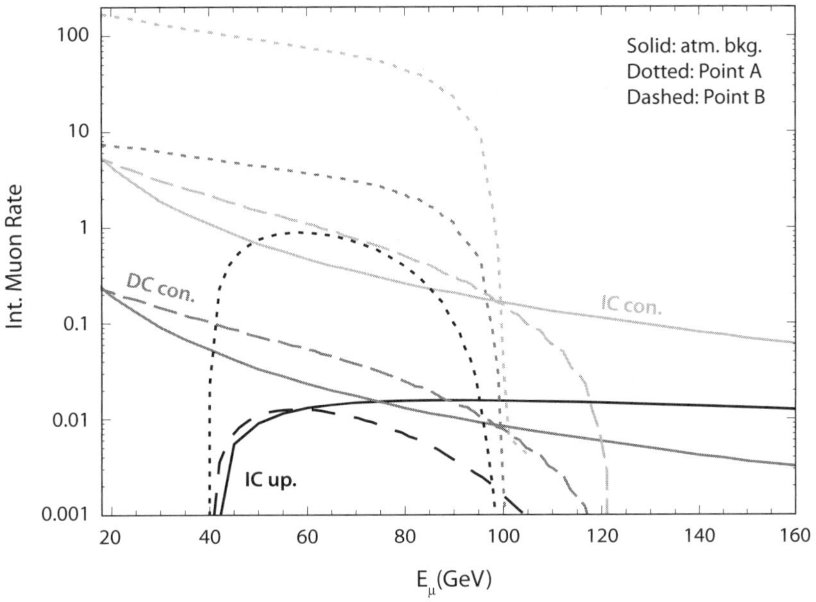

Figure 11.12. Time-integrated muon energy spectra for the atmospheric background and DM signals for IceCube contained (light gray), DeepCore contained (gray), and IceCube upward (black) events. Both signal and background spectra assume an angular cone size of 1°. Adapted from [620].

After convolution one gets

$$N^{DC}_{events} = \xi \Gamma_{eq} \int_{E_{min}}^{m_X} \frac{1}{L} \frac{d\Phi_\mu}{dE_\mu} V_{eff}(E_\mu) dE_\mu. \tag{11.36}$$

Atmospheric Background. The angle-dependent flux of atmospheric neutrinos $d\Phi_\nu^{atm}/(dE_\nu d\cos\theta)$ is given in [625]. A convenient parametrization of this flux is [626]

$$\frac{dN_\nu}{dE_\nu d\Omega} = N_0 E_\nu^{-2.74} \left(\frac{0.018}{1 + 0.024 E_\nu |\cos\theta|} + \frac{0.0069}{1 + 0.00139 E_\nu |\cos\theta|} \right), \tag{11.37}$$

where $N_0 = 1.95 \times 10^{17} (1.35 \times 10^{17})$ GeV^{-1}km^{-2}yr^{-1}sr^{-1} for ν_μ ($\bar{\nu}_\mu$) and θ is the zenith angle. These neutrinos interact with the medium surrounding the detector and produce muons that constitute the background. As in the case of neutrinos from annihilation, the atmospheric background can be divided into upward and contained events. Assuming an acceptance cone-size of 1°, the yearly atmospheric background rate is 6.1 for IceCube upgoing events, 15.6 for IceCube contained events, and 2.5 for DeepCore events; see [622] for a detailed description of the calculation.

Dark Matter Signals

The number of energetic neutrinos above detector threshold determine the prospects for detecting new physics at IceCube/DeepCore. For a solar WIMP signal there are three major factors: (i) annihilation rate; (ii) muon energy threshold vs. WIMP mass; (iii) annihilation channels that produce energetic neutrinos.

The *focus point region* of mSUGRA parameter space that allows a large σ_{SD} coupling is the most popular discovery scenario for mSUGRA. A lower energy threshold at DeepCore greatly enhances the signal with respect to the background and can be crucial for WIMPs with mass $\sim 10^2$ GeV. For illustration we consider a point which yields $m_\chi = 119$ GeV and annihilation BFs to the W^+W^- and ZZ channels of 90% and 8.4%, respectively. We call this "Point A".

The $b\bar{b}$ channel generally produces less and softer neutrinos compared to the WW and ZZ channels. The $\tau^+\tau^-$ channel provides energetic neutrinos but often has a low branching fraction. The neutrino signal from a massive WIMP becomes hard to detect as the $b\bar{b}$ channel dominates. The *A-funnel* region of mSUGRA parameter space is representative of this scenario. We consider a "Point B" with $m_\chi = 123$ GeV and annihilation to $b\bar{b}$ and $\tau^+\tau^-$ with BFs 88% and 12%, respectively.

The differential energy spectra for Points A and B are shown in figure 11.12. With sufficient statistics as for Point A it is also possible to construct the shape of the energy spectrum.

12

Beyond Three Neutrinos

Our focus thus far has largely been on the three-neutrino phenomenology of massive neutrinos and its experimental validation. In addition to this, there are some neutrino experiments, though not conclusive, that may indicate other neutrino phenomena. Moreover, there are numerous theoretical possibilities that go beyond the standard three-neutrino framework. These exotic neutrino phenomena are the subject of this chapter.

12.1 LSND Experiment

The Liquid Scintillator Neutrino Detector (LSND) experiment has provided an indication for the most compelling evidence for new physics beyond the standard three-neutrino picture. New physics models have been proposed to explain the LSND data and other experiments have been carried out to confirm or refute the result. Nonetheless, a definitive statement on the validity of the LSND result cannot yet be made, and currently there is no theoretical framework that can completely explain the LSND data. In this section and the next we summarize the current status of the LSND puzzle and the new questions that have arisen in the attempts to solve it.

The LSND experiment found evidence for $\bar{\nu}_\mu \rightarrow \bar{\nu}_e$ oscillations at 3.3σ significance (oscillation probability $(2.64 \pm 0.67 \pm 0.45) \times 10^{-3}$ [627]) in data on μ^+ decays at rest (DAR) taken from 1993–1998. The antineutrino energy had a maximum value of about 53 MeV, and the distance to the detector was 30 m, giving $L/E \sim 1$ m/MeV. Evidence for $\nu_\mu \rightarrow \nu_e$ oscillations was found at lesser significance from π^+ decay in flight, with oscillation probabilities $(2.6 \pm 1.0 \pm 0.5) \times 10^{-3}$ in the 1993–1995 data [628] and $(1.0 \pm 1.6 \pm 0.4) \times 10^{-3}$ in the 1996–1998 data (see the last paper of [627]).[1] The ν_μ flux in the decay-in-flight experiment peaked at around 60 MeV.

[1] There was a significant difference in the analysis of the decay in flight data in the two time periods due to changes in the neutrino production target, so the two $\nu_\mu \rightarrow \nu_e$ samples were not combined.

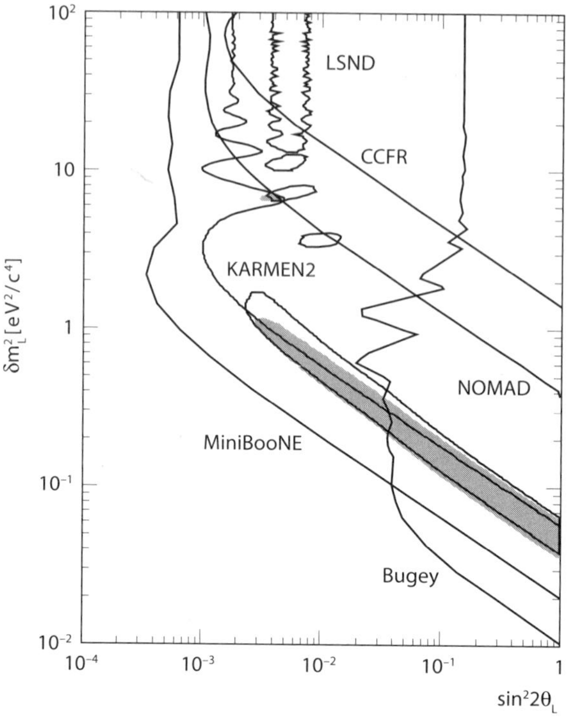

Figure 12.1. The 90% C.L. allowed region (shaded) from a combined fit to $\bar{\nu}_\mu \to \bar{\nu}_e$ data from LSND [627] and KARMEN [629]. The 90% C.L. allowed region from LSND alone is unshaded. The 90% C.L. exclusion regions from KARMEN, Bugey ($\bar{\nu}_e \to \bar{\nu}_e$) [630], CCFR ($\nu_\mu \to \nu_e$) [631], and NOMAD ($\nu_\mu \to \nu_e$) [632] and the expected 90% C.L. sensitivity of the MiniBooNE experiment [156] are also shown. From [633].

In a two-neutrino picture, if the LSND $\bar{\nu}_e$ excess is interpreted as $\bar{\nu}_\mu \to \bar{\nu}_e$ oscillations with oscillation amplitude $\sin^2 2\theta_L$ and mass-squared difference δm_L^2, the allowed parameters fall roughly into two regions. The first region is a band in the $(\sin^2 2\theta_L, \delta m_L^2)$ plane lying along the line described approximately by $\sin^2 2\theta_L(\delta m_L^2)^2 = 0.0025$ eV^4, between $\delta m_L^2 = 0.05$ and 1 eV^2 (see figure 12.1); for higher values of δm_L^2 the suppression of the oscillation probability comes from the small value of $\sin^2 2\theta_L$, while at lower values of δm_L^2 there is large mixing and the suppression comes from the smallness of the oscillation argument $\delta m_L^2 L/(2E)$. The second region is a band where the oscillation argument changes rapidly and the oscillations are effectively averaged over, with $\sin^2 2\theta_L \approx 0.005$ and $\delta m_L^2 > 10\,\text{eV}^2$. There are also some isolated islands in between these two regions.

The KARMEN experiment [629] at the spallation neutron source ISIS used $\bar{\nu}_\mu$ from μ^+ decay at rest to search for $\bar{\nu}_\mu \to \bar{\nu}_e$ oscillations. Like LSND, the antineutrino energies ranged from zero to 53 MeV; the distance from the source to detector was about 18 m. The KARMEN results were consistent with background, which ruled out a large fraction of the LSND allowed region, but still allow a limited region of oscillation parameters [633]. The Bugey reactor experiment [630], which tests the oscillation channel $\bar{\nu}_e \to \bar{\nu}_e$, excludes the part of the LSND region

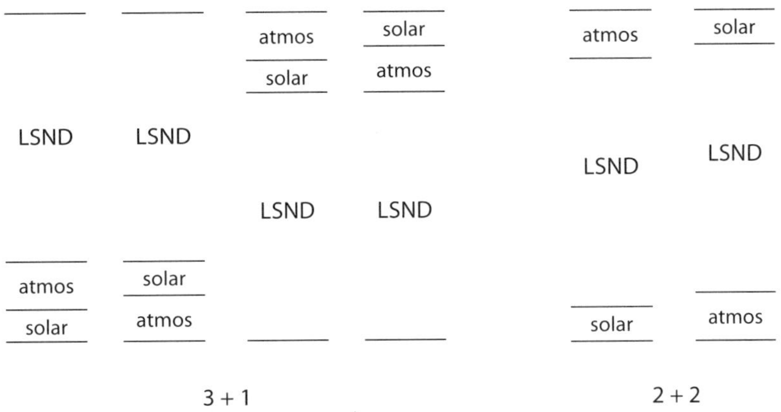

Figure 12.2. The six possible mass spectra in four-neutrino models. From [182].

with $\sin^2 2\theta_L \gtrsim 0.04$. In a two-neutrino parameter space, the indicated oscillation parameters from a combination of LSND and KARMEN data that are consistent with the constraint from Bugey are $\delta m_L^2 \sim 0.2 - 1$ eV^2, $\sin^2 2\theta_L \sim 0.003 - 0.04$, and $\delta m_L^2 \sim 7\, eV^2$, $\sin^2 2\theta_L \sim 0.004$ at the 90% C.L. (see figure 12.1).

The LSND parameters are very different from the oscillation parameters that explain the solar and atmospheric neutrino data; in particular, $|\delta m_L^2| \gg |\delta m_{31}^2|, |\delta m_{21}^2|$. Since a theory with three neutrinos has at most two independent mass-squared difference scales, a third δm^2 scale suggests that there may be a fourth light neutrino participating in neutrino oscillations [634]. The LSND experiment thus cast some doubt on the completeness of the three-neutrino picture. Because measurements of the invisible Z width indicate that there are only three light active neutrinos [119], a fourth light neutrino must be sterile, i.e., it does not participate in the weak interactions [635].

There are two classes of mass spectra possible in four-neutrino models with a single sterile neutrino [636, 637]. In 3 + 1 models, one mass eigenstate is separated from a nearly degenerate triplet of mass eigenstates by δm_L^2; the triplet has a mass ordering like that of a three-neutrino model. The well-separated mass eigenstate can be either lighter or heavier than the other three, and the triplet can have a normal or inverted hierarchy, so there are four possible variations of 3 + 1 models. In 2 + 2 models there is one pair of closely-spaced mass eigenstates separated from another closely-spaced pair by δm_L^2; one pair has a mass-squared difference of δm_{21}^2 and the other δm_{31}^2, and the solar δm^2 can be in either the upper or lower pair. Figure 12.2 shows the six possible mass spectra with four neutrinos.

2 + 2 *Models*

The 2 + 2 models are not a simple extension of a three-neutrino model, since the removal of the sterile neutrino does not leave the standard three-neutrino mass spectrum. If the lower pair of neutrino mass eigenstates are primarily responsible for solar neutrino oscillations, then ν_e is primarily connected to ν_1 and ν_2; similarly, the upper pair of mass eigenstates are primarily responsible for the atmospheric neutrino oscillations, and ν_μ is primarily connected to ν_3 and ν_4.[2] Some examples of explicit $2+2$ models are given in [637, 638].

Since there is a sterile neutrino, the simplest situation would be to have either $\nu_e \to \nu_s$ for solar and KamLAND neutrinos or $\nu_\mu \to \nu_s$ for atmospheric and long-baseline neutrinos. In the most general case, solar neutrinos oscillate to a linear combination of ν_τ and ν_s, and atmospheric neutrinos oscillate to the orthogonal combination [637]:

$$\nu_e \to -\sin\alpha\,\nu_\tau + \cos\alpha\,\nu_s, \tag{12.1}$$

$$\nu_\mu \to \cos\alpha\,\nu_\tau + \sin\alpha\,\nu_s\,. \tag{12.2}$$

Early analyses of the solar neutrino data showed that solar solutions with pure $\nu_e \to \nu_s$ ($\alpha = 0$) did not provide as good a fit to the solar neutrino data (see the second paper of [43]). The main difficulty with pure sterile neutrino solutions is that they give similar predictions for the Chlorine and Super-K experiments, whereas active neutrino solutions give a larger value for Super-K due to neutral current $\nu_\mu e$ and/or $\nu_\tau e$ interactions in the detector, which is in better agreement with the experimental data. Oscillations to sterile neutrinos have different matter effects since the value of δm^2_{21} that gives resonant oscillations is $\delta m^2_{21} = 2\sqrt{2}G_F E_\nu(N_e - \frac{1}{2}N_n)/\cos 2\theta_{12}$ instead of the value $2\sqrt{2}G_F E_\nu N_e/\cos 2\theta_{12}$ for oscillations to active neutrinos (see equation 3.40). Accounting for these matter effects, it was found that there was an upper bound on the fraction of oscillating solar ν_e and KamLAND $\bar{\nu}_e$ that could oscillate to sterile neutrinos [639]:

$$\cos^2\alpha \le 0.25 \qquad 99\%\ \text{C.L.} \tag{12.3}$$

Therefore pure $\nu_e \to \nu_s$ oscillations of solar neutrinos and pure $\bar{\nu}_e \to \bar{\nu}_s$ oscillations of KamLAND neutrinos are excluded.

The opposite extreme is to have pure sterile solutions to the atmospheric neutrino data ($\alpha = \frac{\pi}{2}$). Here there are strong matter effects due to coherent forward scattering in the earth that is present for ν_μ but not ν_s (see equation 3.40), with $N_{eff} = -\frac{1}{2}N_n$; for pure $\nu_\mu \to \nu_\tau$ oscillations, matter effects are small. The Super-K atmospheric data strongly disfavor pure $\nu_\mu \to \nu_s$ oscillations [640, 641], and a fit to Super-K atmospheric neutrino, K2K, and null short-baseline data (KARMEN, CDHS [642], Bugey, CHOOZ, and Palo Verde) found that the fraction of muon neutrinos oscillating to sterile neutrinos in the atmospheric and K2K experiments had an upper bound [639]:

$$\sin^2\alpha \le 0.25 \qquad 99\%\ \text{C.L.} \tag{12.4}$$

The MINOS long-baseline experiment can also measure $\nu_\mu \to \nu_s$ from the number of neutral-current interactions; a deficit from expectation indicates active to sterile oscillations. They have placed the upper bound $\sin^2\alpha < 0.22$ at 90% C.L [643].

[2] The case where the lower pair of mass eigenstates is primarily responsible for atmospheric neutrino oscillations leads to similar conclusions and we therefore do not consider it here.

Although pure $\nu_e \to \nu_s$ solar and pure $\nu_\mu \to \nu_s$ atmospheric neutrino oscillations are ruled out, partial sterile solutions are allowed in each case. However, we must still satisfy $\cos^2\alpha + \sin^2\alpha = 1$, which is clearly disfavored by equations 12.3 and 12.4. An analysis of mixed sterile solutions found that they are excluded at about the 5σ level [639].

3 + 1 *Models*

In contrast to 2 + 2 models, 3 + 1 models are a straightforward extension of a three-neutrino model: the three active neutrinos have mass-squared differences and mixings similar to those in a three-neutrino model, and the sterile neutrino state has only small mixing with active neutrinos. However, 3 + 1 models have trouble accounting for the LSND results and simultaneously obeying the constraints of earlier accelerator and reactor experiments [636, 637]. This can be demonstrated as follows.

Assume a neutrino mass spectrum such that the nearly degenerate triplet of mass eigenstates is lighter than the remaining state and exhibits a normal hierarchy (the first spectrum shown in figure 12.2); similar conclusions can be drawn for the other three 3 + 1 spectra. Then $\delta m^2_{43} \simeq \delta m^2_{42} \simeq \delta m^2_{41} = \delta m^2_L \gg \delta m^2_{32} \simeq \delta m^2_{31} \simeq 2.4 \times 10^{-3}\ \mathrm{eV}^2 \gg \delta m^2_{21} \simeq 7.6 \times 10^{-5}\ \mathrm{eV}^2$, and the oscillation probabilities for the leading oscillation, due to δm^2_L, are

$$P(\bar{\nu}_\mu \to \bar{\nu}_e) \simeq 4|V_{\mu 4}|^2|V_{e4}|^2 \sin^2\Delta_L, \tag{12.5}$$

$$P(\nu_\mu \to \nu_\mu) \simeq 1 - 4|V_{\mu 4}|^2(1 - |V_{\mu 4}|^2)\sin^2\Delta_L, \tag{12.6}$$

$$P(\bar{\nu}_e \to \bar{\nu}_e) \simeq 1 - 4|V_{e4}|^2(1 - |V_{e4}|^2)\sin^2\Delta_L, \tag{12.7}$$

where $\Delta_L \equiv \delta m^2_L L/(4E_\nu)$, analogous to equation 3.5. At L/E_ν values appropriate for atmospheric neutrinos,

$$P(\nu_\mu \to \nu_\mu) \simeq 1 - 4|V_{\mu 3}|^2(1 - |V_{\mu 3}|^2 - |V_{\mu 4}|^2)\sin^2\Delta_{31}, \tag{12.8}$$

and for solar neutrinos

$$P(\nu_e \to \nu_e) \simeq 1 - 4|V_{e1}|^2|V_{e2}|^2 \sin^2\Delta_{21}\,. \tag{12.9}$$

There are very stringent limits on ν_μ disappearance from the CCFR [631], NOMAD [632], CDHS [642], and MiniBooNE/SciBooNE [644] accelerator experiments that constrain $|V_{\mu 4}|$ to be very small or very close to unity via equation 12.6. However, if $|V_{\mu 4}|$ is close to unity then the amplitude of atmospheric neutrino oscillations cannot be as large as required by observation (see equation 12.8); therefore, $|V_{\mu 4}|^2 \ll 1$. Similarly, the Bugey reactor experiment [630] puts severe limits on $\bar{\nu}_e$ disappearance that constrain $|V_{e4}|$ to be very small or very close to unity, and the observation of large angle mixing in solar neutrino oscillations therefore implies $|V_{e4}|$ cannot be close to unity (since V_{e1} cannot be small; see equation 12.9), so $|V_{e4}|^2 \ll 1$.

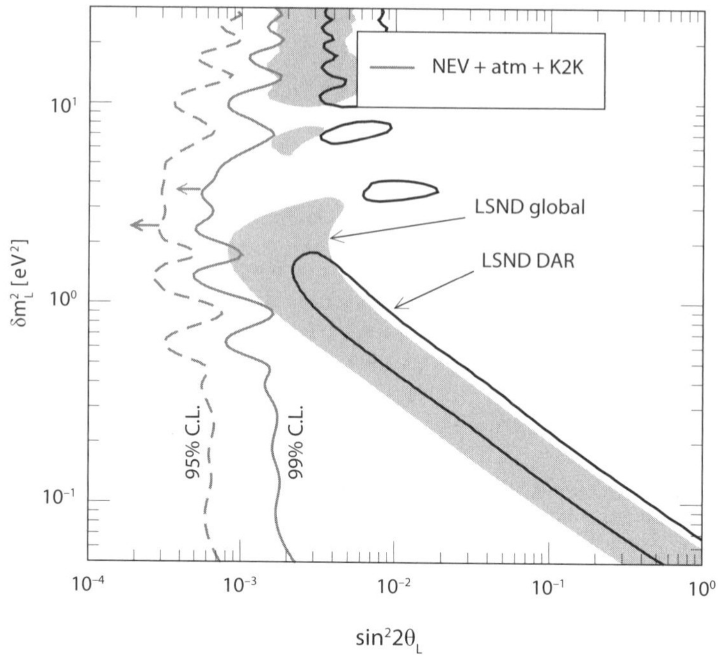

Figure 12.3. Upper bounds at 95% and 99% C.L. on the LSND amplitude in 3 + 1 models ($4|V_{\mu 4}|^2|V_{e4}|^2$) from the accelerator and reactor experiments that show no evidence for oscillations combined with atmospheric and K2K data (curve labeled NEV + atm + K2K). Also shown are the allowed regions from the LSND experiment. Adapted from [639].

Since the oscillation amplitude for the LSND experiment is $4|V_{\mu 4}|^2|V_{e4}|^2$, it has an upper limit of approximately one-fourth of the product of the CDHS and Bugey oscillation amplitude bounds (see equations 12.5–12.7). In practice, the limits on $|V_{\mu 4}|$ and $|V_{e4}|$ depend on δm_L^2, as does the allowed oscillation amplitude from LSND, so a comparison must be made for each value of δm_L^2. Many studies were made to determine whether the 3 + 1 model was consistent with all of the data, with somewhat conflicting conclusions as new data was added to the analyses [636,637,645–649]. Figure 12.3 shows the incompatibility of the combined accelerator and reactor upper limit on the oscillation amplitude in 3 + 1 models and the region allowed by LSND as of 2004. The 3 + 1 models were excluded at about the 3σ level. However, they still provided the best explanation of the data at the time, as 2 + 2 models and the standard three-neutrino scenario were both excluded *relative to the* 3 + 1 *model* at the 3σ level.

12.2 MiniBooNE Experiment

The Mini Booster Neutrino Experiment (MiniBooNE) [156] was designed to test the anomalous LSND result of $\bar{\nu}_e$ appearance in the $\bar{\nu}_\mu$ beam. The average neutrino energy in MiniBooNE is about 800 MeV, and the distance to the detector is approximately 500 m, giving $L/E \sim 1$ m/MeV. The experiment first ran in neutrino

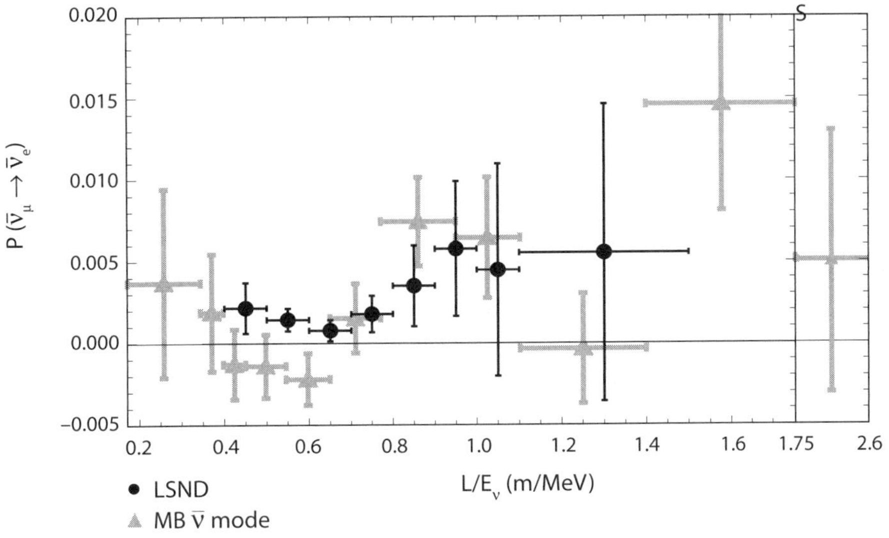

Figure 12.4. $P(\bar{\nu}_\mu \to \bar{\nu}_e)$ versus L/E for LSND and MiniBooNE antineutrino data. From [654].

mode in order to achieve higher statistics, and did not confirm the LSND anomaly for ν_μ in the energy range 475 MeV $< E_\nu <$ 3000 MeV;[3] 408 events were found, with an expectation of 385.9 ± 35.7 without oscillations [650]. Furthermore, a two-neutrino analysis using the LSND, KARMEN, and MiniBooNE data sets found a maximal compatibility of only about 4% [651]. MiniBooNE did, however, observe an anomalous ν_e appearance at low energies, $200 < E_\nu < 475$ MeV, that was not consistent with the oscillation interpretation of the LSND data [652]. This anomaly will be discussed in more detail below. Then MiniBooNE ran with a $\bar{\nu}_\mu$ beam [653, 654] (average energy 600 MeV) and found excess $\bar{\nu}_\mu \to \bar{\nu}_e$ events that gave allowed regions that mostly overlayed those obtained by LSND for antineutrinos, though with lower statistics than for their earlier data with neutrinos; figure 12.4 shows that the inferred oscillation probabilities versus L/E for the LSND and MiniBooNE antineutrino data are very similar.

3 + 2 *Models*

It has been suggested that models with three active and two sterile neutrinos (3 + 2 models) may be able to evade the constraints on 3 + 1 models [655]. Such models have three active neutrinos with the usual mass splittings required to explain solar and atmospheric data, two sterile neutrinos with LSND-type mass splittings from the active neutrinos, and small sterile-active mixing. [4]

Furthermore, CP violation requires at least two oscillation arguments to be non-neglible [78], and at the LSND L/E values δm^2_{31} and δm^2_{21} are not large enough for

[3] If CP and CPT are conserved, $\bar{\nu}_\mu \to \bar{\nu}_e$ and $\nu_\mu \to \nu_e$ oscillations in MiniBooNE would be the same for the same L/E. Then MiniBooNE neutrino beam results on oscillations can be directly compared to the antineutrino beam results of LSND.

[4] Adding an additional sterile neutrino to a 2+2 model still requires sizable oscillations to sterile neutrinos in solar and/or atmospheric experiments, and is therefore excluded.

this to happen; therefore 3 + 1 models necessarily have $P(\nu_\mu \to \nu_e) \equiv P(\bar{\nu}_\mu \to \bar{\nu}_e)$. However, in 3 + 2 models there are at least two large δm^2 scales and significant CP violation is possible; it was speculated that in 3+2 models the difference between the positive antineutrino result in LSND and the null short-baseline experiments might be resolved [656]. The oscillation probabilities in the leading oscillation are

$$P(\nu_\alpha \to \nu_\alpha) = 1 - 4\Big[(1 - |V_{\alpha4}|^2 - |V_{\alpha5}|^2)(|V_{\alpha4}|^2 \sin^2\Delta_{41} + |V_{\alpha5}|^2 \sin^2\Delta_{51}) + |V_{\alpha4}|^2|V_{\alpha5}|^2 \sin^2\Delta_{54}\Big] \quad (12.10)$$

and

$$P(\nu_\mu \to \nu_e) = 4\,|V_{\mu4}|^2|V_{e4}|^2 \sin^2\Delta_{41} + 4|V_{\mu5}|^2|V_{e5}|^2 \sin^2\Delta_{51} + 8|V_{\mu4}||V_{\mu5}||V_{e4}||V_{e5}| \sin\Delta_{41}\sin\Delta_{51}\cos(\Delta_{54} - \phi) \quad (12.11)$$

where $\Delta_{ij} = \delta m^2_{ij} L/(4E)$ and $\phi = arg(V^*_{\mu5} V_{e5} V_{\mu4} V^*_{e4})$ is a CP-violating phase.[5]

After the initial MiniBooNE constraint on ν_μ oscillations was released [650], it was found that the difference in the ν_μ and $\bar{\nu}_\mu$ channels could in fact be explained by CP violation in a 3+2 model; however, the preferred parameters were at odds with short-baseline constraints at about the 3σ level [657].[6] After MiniBooNE released their final ν_μ data and initial indications of $\bar{\nu}_\mu$ oscillations, the tension between different data sets remained; although the overall data were compatible at the 7% level, the neutrino/antineutrino and appearance/disappearance compatibilities were both well below 1% [659]. Therefore, while 3 + 2 models still provide a much better fit to the data than 3 + 1 models (which in turn are better than the null oscillation hypothesis), they do not give a completely consistent description of all the data. A 3 + 1 model with non-standard interactions of neutrinos (see section 12.7) could explain the discrepancy between neutrino and antineutrino results [660], including those of the null experiments, but not the MiniBooNE low-energy anomaly (described in this chapter). Similarly, a 3 + 1 model with energy-dependent neutrino masses and mixings [661] provides a better fit to all data *except* the low-energy anomaly.

It is natural to think that there could be three sterile neutrinos. In one such realization of a 3+3 model the sterile neutrinos are simply the right-handed partners of the left-handed, active neutrinos, with mass of order 1 eV – the light neutrino masses arise from a seesaw mechanism [662]. However, a 3 + 3 model does not fit the data appreciably better than a 3 + 2 model [657]. Finally, if there are one or more sterile neutrinos with a mass of order 1 eV or more that mix strongly enough with ν_e, it could be possible to detect their presence in a KATRIN-like beta decay endpoint experiment [663].

[5] There are two other Dirac CP phases in a 3 + 2 model, but they do not contribute appreciable CP violation when oscillation arguments involving δm^2_{31} and δm^2_{21} are small, such as in the LSND and MiniBooNE experiments.

[6] A limit on ν_μ and $\bar{\nu}_\mu$ disappearance from the MiniBooNE experiment [658] is consistent with the fits of [656], but rule out the best fit point of [657].

Three-neutrino Models with CPT Violation

It was suggested that if CPT were not conserved, then oscillations of three active neutrinos could describe the solar, atmospheric, *and* LSND data simultaneously [664]. In this proposal, the mass matrices (and hence mass-squared differences and mixings) for neutrinos and antineutrinos are different, which violates CPT. It is unclear whether one can construct an acceptable theoretical framework with this property (see [377–379, 665]), but one can still treat this possibility phenomenologically.

In the original versions of CPT-violating models, the neutrino sector had the usual three-neutrino mass spectrum that can account for the oscillation of solar and atmospheric neutrinos, while in the antineutrino sector the mass-squared differences account for the oscillation of antineutrinos in the atmospheric and LSND experiments (the weak indication for $\nu_\mu \to \nu_e$ oscillations in LSND must be ignored). KamLAND data, consistent with oscillations of $\bar{\nu}_e$ at the δm^2_{21} scale, forced a modification of the antineutrino spectrum, so that it describes the oscillation of antineutrinos in LSND and KamLAND (but not in the atmosphere) [666]. Since the atmospheric data does not distinguish between neutrinos and antineutrinos, this latter scenario was not in obvious contradiction to the data; however a detailed analysis of atmospheric data indicated that CPT-violating scenarios are not in good agreement with the atmospheric data [667]. Furthermore, global analyses of all data including KamLAND excludes these CPT-violating scenarios at the 3σ level [667, 668]. Nonetheless, recent very preliminary indications that δm^2_{31} for antineutrinos in MINOS is larger than for neutrinos [273] may revive this scenario [669].

A more speculative possibility is to have a sterile neutrino *and* CPT violation [670]. In 3 + 1 models the strongest constraints on the $\bar{\nu}_\mu \to \bar{\nu}_e$ and $\nu_\mu \to \nu_e$ oscillation amplitudes come primarily from a combination of constraints on $\bar{\nu}_e \to \bar{\nu}_e$ and $\nu_\mu \to \nu_\mu$. Therefore if the neutrino mixing angles are different from those of antineutrinos, the usual constraints from the null short-baseline experiments can be relaxed. A 3 + 1 CPT violating model appears to be consistent with the data; a 3 + 1 structure in the neutrino sector and 2 + 2 in the antineutrino sector also is possible [670].

Low-energy Anomaly

As mentioned, MiniBooNE measured an excess of $\nu_\mu \to \nu_e$ events, 128.8 ± 43.4, for neutrino energies in the range $200 < E < 475\,\text{MeV}$ [652]; see figure 12.5. They did not find a signal in the energy range $475 < E < 1250\,\text{MeV}$, which would have been expected if the there were LSND-type oscillations in the neutrino channel. The best two-neutrino fit for $E > 200\,\text{MeV}$ was $\delta m^2_L = 3.14\,\text{eV}^2$ and $\sin^2 2\theta_L = 0.0017$, outside the LSND allowed region, and for $E > 475$ MeV the best fit was the null oscillation hypothesis. Background processes such as neutral current π^0 production, Δ decays, or neutrino scattering off of carbon atoms could explain the excess if their normalization were larger than expected by factors or 2.0, 2.7, or 2.4, respectively. In the first two cases, photons are misidentified as electrons (or positrons), which makes the background difficult to separate from the signal.

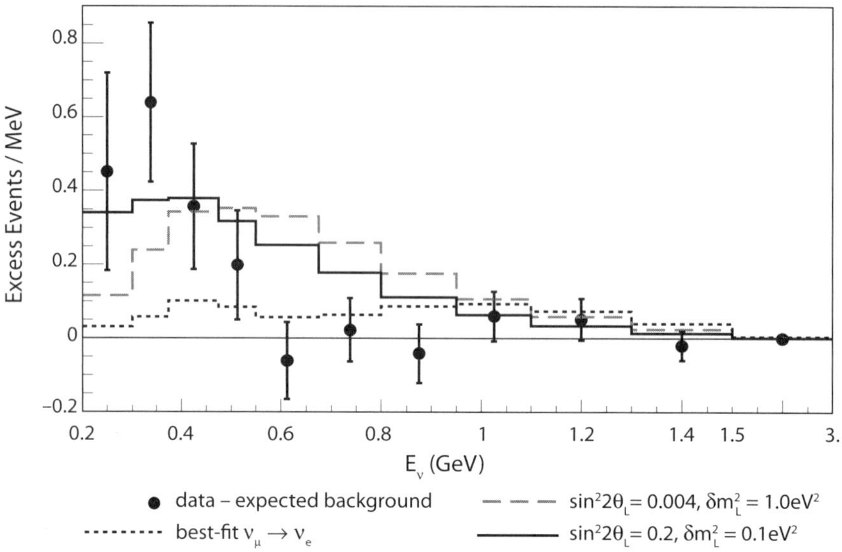

Figure 12.5. Excess of $\nu_\mu \rightarrow \nu_e$ events shown versus reconstructed neutrino energy. The error bars include both statistical and systematic errors. Also shown are the best fit oscillation parameters and two sets of non-optimal oscillation parameters that exhibit a low-energy excess. Adapted from [652].

There have been several attempts to explain the MiniBooNE low-energy anomaly with new physics. In models with extra dimensions, if the active neutrinos lie in a $3 + 1$ dimensional brane and there is a sterile neutrino that lies in the extra-dimensional bulk, then the sterile neutrino can take shortcuts in the bulk, which modifies the usual oscillation formula. Then for oscillations involving the sterile neutrino, the oscillation argument is modified to be proportional to LE, instead of the standard L/E; this means that the effective value of δm^2 is proportional to E^2, and thus there can be a resonant enhancement of oscillations that is energy dependent [671]. Since the neutrino energy in MiniBooNE is significantly higher than in LSND, this in principle could explain any difference between the two experiments. Similarly, the LSND neutrino energies are higher than those of KARMEN, which could explain their different results for $\bar{\nu}_\mu \rightarrow \bar{\nu}_e$.

However, given that MiniBooNE has found $\bar{\nu}_\mu \rightarrow \bar{\nu}_e$ events consistent with LSND, even though the neutrino energies are quite different in the two experiments, energy dependence alone may not be sufficient to explain all the data. More general dispersion relations, where the effective δm^2 has terms proportional to powers of E, allow for more freedom [672]. Such terms could arise in theories with CPT violation [108, 673, 674][7] or extra dimensions [491, 492]. There is also the possibility of neutrino-antineutrino oscillations [635, 675] in the $\nu_\mu \rightarrow \bar{\nu}_e$ channel, since the electrons in the MiniBooNE detector are not distinguished from positrons. Combining non-standard dispersion relations and the neutrino-antineutrino oscillation hypotheses then could potentially describe both the

[7] The CPT violation in these models occurs in the neutrino interactions, not in the mass matrix; see section 12.6.

low-energy excess in the neutrino channel at MiniBooNE and the oscillation signals in both LSND and MiniBooNE in the antineutrino channel [676].

The GALLEX and SAGE solar neutrino detectors were exposed to neutrinos from intense, artificial ^{52}Cr and ^{37}Ar radioactive sources, as a test of those detectors [677]. Combining the results of all of these experiments, the ratio of measured to expected events was [678] 0.86 ± 0.05. This deficit could be the due to an overestimation of transition rates to excited states, but it also has been speculated that this "Gallium anomaly" and the MiniBooNE low-energy excess could both be due to oscillations between ν_e and ν_μ in a 3+1 model with one sterile neutrino [679]. The two-neutrino parameters associated with this oscillation are $\delta m^2 = 2.24\ eV^2$ and $\sin^2 2\theta = 0.46$, although $\delta m^2 \gtrsim 0.2\ eV^2$ and $\sin^2 2\theta \gtrsim 0.1$ are allowed at the 3σ level. However, the apparent oscillation of $\bar{\nu}_\mu \to \bar{\nu}_e$ in LSND and MiniBooNE would either have to be a statistical fluctuation or have a different explanation.

Another possibility for explaining the low-energy excess is via the decay of a heavy sterile neutrino with a mass around 500 MeV [680]. The decay rate is proportional to m_ν^3, however, and because the antineutrino beams in MiniBooNE and, especially LSND, have lower energies than the MiniBooNE neutrino beam, this process would not be able to explain the antineutrino signal. Such a heavy sterile neutrino could also appear in the decay $D_s^+ \to \mu^+\nu$, which could account [681] for the discrepancy between the predicted $D_s^+ \to \mu^+\nu_\mu$ decay rate and theoretical expectation. There is an alternative neutrino decay model in which a heavy sterile ν_h is produced by the reaction $\nu_\mu + {}^{12}\mathrm{C} \to \nu_h + n + {}^{11}\mathrm{C}$ and the photon in the $\nu_h \to \nu\gamma$ decay and a 2.2 MeV photon from neutron capture are misidentified as electrons; this model can account for all of the LSND, KARMEN, and MiniBooNE data, although there is some tension in the fits to LSND and MiniBooNE [682]. It has also been suggested [683] that a Wess-Zumino-Witten term coupling the photon, Z boson, and ω meson may be responsible for the MiniBooNE low-energy excess.

Summary of LSND and MiniBooNE Anomalies

The MiniBooNE confirmation of the LSND antineutrino signal validates a conventional explanation with the oscillation argument proportional to L/E, but may present a challenge for models with a different dispersion relation due to the energy difference in the two experiments. On the other hand, conventional oscillations cannot consistently explain both the signal in antineutrinos and the MiniBooNE low-energy anomaly. It seems likely that a single phenomenon will not be able to explain all of the data, and that at least one non-standard process will be required, such as CPT violation and/or neutrino-antineutrino oscillations. It is possible that one or more of the experimental effects may go away with higher statistics. MiniBooNE will take more antineutrino data, and the future MicroBooNE detector [155], a liquid argon TPC, will be able to separate photons from electrons/positrons and therefore the signal can be better discriminated from the background. Another proposal is to move MiniBooNE closer to the source (to a distance of 200 m), which will give a much higher flux and allow for a cancellation of systematic uncertainties [684]. It has also been suggested [685] that the LSND experiment be repeated using the Super-K detector doped with Gadolinium, which would enhance the detection of the inverse beta decay cross section. The effects of sterile neutrinos can also be probed at neutrino factories [686]. Thus there is hope that a clearer picture will emerge in the future.

12.3 Mass-varying Neutrinos

Abundant cosmological data indicate that the expansion of our universe is in an accelerating phase caused by a negative pressure component called dark energy. Dark energy is troubling because the acceleration of the universe is a very recent phenomenon in its expansion history. This "cosmic coincidence" problem can be expressed as follows: Why are the dark matter and dark energy densities comparable today even though their ratio scales as $\sim 1/a^3$ (where a is the expansion scale factor)?

The coincidence that the scale of dark energy $(2 \times 10^{-3}\ \text{eV})^4$ is similar to the scale of neutrino mass-squared differences $(0.01\ \text{eV})^2$ has been proposed as a solution to the coincidence problem [687]. The authors of [687] considered the possibility of coupling neutrinos to dark energy by supposing that the dark energy density is a function of neutrino mass and imposing the condition that the total energy density of neutrinos and dark energy remain stationary under variations in neutrino mass. Then neutrino masses vary in such a way that the neutrino energy density and the dark energy density are related over a wide range of the scale factor a.

A simple way [353] to make the dark energy density neutrino-mass-dependent is to introduce a Yukawa coupling between a sterile neutrino s and a light scalar field ϕ (similar to quintessence) called the acceleron. At scales below the sterile neutrino mass, a Lagrangian of the form

$$-\mathcal{L} = m_D \nu s + \lambda \phi s s + V_0(\phi), \tag{12.12}$$

where ν is a Standard Model left-handed neutrino, leads to an effective potential for the acceleron (if neutrinos are nonrelativistic) given by

$$V = \frac{m_D^2}{\lambda \phi} n_\nu + V_0(\phi)\,. \tag{12.13}$$

Thus, the effective potential of the acceleron at late times receives a contribution equal to $m_\nu n_\nu$, where $m_\nu = m_D^2/(\lambda\phi)$ and n_ν are the active neutrino mass and number density, respectively. More elaborate supersymmetric models of neutrino dark energy have been constructed in [688].

Model-independent tests of neutrino dark energy are cosmological [687, 689]. A strict relationship between the equation of state of the combined dark energy-neutrino fluid $w = p_{nde}/\rho_{nde}$ (where nde denotes neutrino dark energy) and neutrino mass is predicted [687]

$$w = -1 + \frac{m_\nu n_\nu}{V}. \tag{12.14}$$

Further, since neutrino masses are predicted to scale with redshift approximately as a^3 in the nonrelativistic regime, cosmological and terrestrial probes of neutrino mass could give conflicting results.

It has been argued in [690] that it is natural to expect couplings of the acceleron to quarks and charged leptons to be generated radiatively. Moreover, Yukawa couplings of the acceleron to visible matter could be low-energy manifestations of nonrenormalizable operators arising from quantum gravity. If the acceleron couples both to neutrinos and matter, it may be possible to investigate this scenario through neutrino oscillations [690, 691]. However, the coupling to matter is model-dependent. The effective neutrino mass in matter is altered by the interactions via the scalar, which in turn modifies neutrino oscillations.

At low redshifts, the contribution to the neutrino mass caused by the interactions of the acceleron with electrons and neutrinos is of the form [692]

$$M = \frac{\lambda_\nu}{m_\phi^2}(\lambda_e n_e + \lambda_\nu(n_\nu^{C\nu B} + \frac{m_\nu}{E_\nu} n_\nu^{rel})), \tag{12.15}$$

where λ_ν (λ_e) is the Yukawa coupling of the acceleron to ν (the electron). In principle, ϕ has a mass, m_ϕ, that depends on n_e and the n_ν. This dependence is weak since the underlying assumption in obtaining equation 12.15 is that ϕ evolves adiabatically and remains at the minimum of its potential. The number density of the cosmic neutrino background in one generation of neutrinos and antineutrinos is $n_\nu^{C\nu B} \sim 112\ \text{cm}^{-3} \sim 10^{-12}\ \text{eV}^3$, the number density of relativistic neutrinos in the background frame is n_ν^{rel}, and the electron number density is n_e. In this model m_ν are neutrino masses in a background dominated environment.

In terrestrial environments and even for applications to solar neutrinos, the dominant contribution to the mass shift arises from the $\lambda_e n_e$ term. Then, one can adopt a matter dependence of the form [692]

$$M(n_e) = M^0 \left(\frac{n_e}{n_e^0}\right)^k, \tag{12.16}$$

where M^0 is the value at some reference density n_e^0 and k parametrizes a power law dependence of the neutrino mass on density. In principle, M is expected to depend linearly on n_e, but one may allow k to deviate from unity as a phenomenological parameter. The choice of reference density is arbitrary. If the environment that neutrinos traverse has a constant density (e.g., for passage through the earth's crust), then that density could be taken to be the reference density. If neutrino propagation is adiabatic (as in the sun), the reference density could be taken to be the density at which the neutrinos are produced. Implicit in the form of equation 12.16 is the assumption that the neutrino number density has a negligible effect on neutrino masses. Thus, it applies only in the current epoch when the cosmic neutrino background number density ($\mathcal{O}(10^{-12})\ \text{eV}^3$) is tiny. At earlier epochs, the neutrino number density is orders of magnitude larger and must be taken into account. For example, in the era of Big Bang Nucleosynthesis (BBN), the neutrino number density is $\mathcal{O}(10^{30})\ \text{eV}^3$. For the compatibility of mass-varying neutrinos (MaVaNs) with BBN see [693].

A simplifying assumption is that the heaviest neutrino has a mass of $\mathcal{O}(0.05)$ eV in the present epoch. As a result of their non-negligible velocities, the neutrino overdensity in the Milky Way from gravitational clustering can be neglected [694]. Then m_ν represents the masses of terrestrial neutrinos in laboratory experiments like those that could in principle be measured in tritium beta decay. We note that cosmological bounds on the sum of neutrino masses of $\mathcal{O}(1)$ eV are inapplicable to MaVaNs. Consequently, the usual relationship between neutrino dark matter and $0\nu\beta\beta$-decay [297] is also rendered inapplicable. Moreover, it was pointed out that if the acceleron couples to highly nonrelativistic neutrino eigenstates, then neutrino dark energy is unstable [695]. The assumption that the background neutrino masses are small circumvents the latter problem.[8]

For such light neutrinos, only model-dependent (neutrino oscillation) tests of the MaVaN scenario are viable because the model-independent (cosmological) tests become inoperable. There are two reasons for this: (1) The dark energy behaves almost exactly as a cosmological constant today. (2) If these light neutrinos do not cluster sufficiently, the local neutrino mass is the same as the background value, which is below the sensitivity of tritium beta decay experiments. Then, high-redshift cosmological data (which should show no evidence for neutrino mass) and data from tritium beta decay experiments are guaranteed to be consistent.

It has been shown in [692] that oscillations of mass-varying neutrinos (that result in exotic matter effects of the same size as standard matter effects) can lead to an improved agreement (relative to conventional oscillations) with solar neutrino data while remaining compatible with KamLAND, CHOOZ, K2K, and atmospheric data. MaVaN oscillations are perfectly compatible with solar data because the survival probability can change from a higher-than-vacuum value (at low energies) to $\sin^2\theta$ (at high energies) over a very narrow range of energies. It is noteworthy that non-standard interactions that modify the kinetic part of the neutrino evolution equations lead to a prediction for the survival probability of *pp* neutrinos that is different from that for standard vacuum oscillations or from non-standard oscillations that modify the potential contribution to the neutrino evolution equations via changes to the vector part of neutrino-matter interactions. A global analysis of solar and KamLAND data within the context of this scenario concludes that the fit in the LMA-II region is improved considerably [697].

As shown in [698], other tests in reactor and long-baseline experiments emerge when the former scheme is embedded in a comprehensive model that can explain all extant neutrino oscillation data including the LSND anomaly and a null MiniBooNE result. Specifically, the large values of $\sin^2 2\theta_{13}$ required in this model ($0.10 \lesssim \sin^2 2\theta_{13} \lesssim 0.30$),[9] lead to the expectation that θ_{13} should be detectable in ongoing and proposed reactor experiments with expected sensitivity $\sin^2 2\theta_{13} \geq 0.01$ where most of the neutrino path is in earth matter, such as Angra or Daya Bay. However, Double Chooz [80], which should be sensitive to $\sin^2 2\theta_{13} \geq 0.03$, would see a null result since most of the neutrino path is in air. The MINOS experiment, which is sensitive to $\sin^2 2\theta_{13} \geq 0.05$ at the 90% C.L., should also see a positive signal in the $\nu_\mu \to \nu_e$ appearance channel.

The idea of using reactor experiments with different fractions of air and Earth matter along the neutrino path to study MaVaN oscillations has been further

[8] The instability is also avoided for sufficiently weak coupling of the neutrinos to the acceleron during the relevant cosmological era [696].

[9] The CHOOZ reactor constraint on $\bar{\nu}_e \to \bar{\nu}_e$ oscillations at the atmospheric scale ($L/E_\nu \simeq 250$ m/MeV) does not apply since the neutrino path in the CHOOZ experiment was primarily in air; it is instead replaced by the weaker Palo Verde constraint at a somewhat smaller L/E_ν value.

explored in [699]. It has been demonstrated that for $\sin^2 2\theta_{13} \gtrsim 0.04$, two reactor experiments with baselines of at least 1.5 km, one of whose neutrino path is in air and the other in matter, can constrain an oscillation effect that is different in air and matter at the level of a few percent.

Neutrino superbeam experiments may probe mass-varying neutrinos in a controlled environment if the effects are large enough. The sensitivity of long-baseline experiments to non-standard matter effects in MaVaN oscillations has been examined in [700].

12.4 Neutrino Decay

Active Neutrino Decay

In the enhanced standard model with neutrino masses, a heavier neutrino can decay into a lighter neutrino plus photon with decay rate [701]

$$\Gamma(\nu_i \rightarrow \nu_j \gamma) \simeq \frac{9\alpha G_F^2 m_i^5}{1024\pi^4} \left| \sum_\alpha V_{\alpha j}^* V_{\alpha i} \left(\frac{m_\alpha}{m_W}\right)^2 \right|^2, \qquad (12.17)$$

where V is the MNS mixing matrix, $\alpha = e, \mu, \tau$ and m_α represents a charged-lepton mass. For light neutrinos, the lifetime is extremely long compared to the age of the universe.

A heavier neutrino can also decay into three lighter neutrinos if there are flavor-changing neutral-current couplings between the neutrino mass eigenstates. In a theory with only three light neutrinos, the GIM mechanism will lead to a cancellation of the flavor-changing neutral currents (FCNC), but if the light neutrino masses are due to a seesaw mechanism, the 3×3 mixing matrix of the light neutrinos is not exactly unitary and the GIM (Glashow-Iliopoulos-Maiani) cancellation is not exact; FCNC exist of order $M_D(M_R^{-1})^2 M_D^T$ (see equation 9.10) [702]. Again, for light neutrinos these are extremely long-lived.

Fast neutrino decays could occur at tree-level due to a massless, spinless scalar particle J

$$\nu_i \rightarrow \bar{\nu}_j + J, \qquad (12.18)$$

where i, j are mass-eigenstates that may be mixtures of active and sterile flavors. The couplings of J to ν_μ and ν_e are experimentally constrained by π and K meson decays [703], but the bounds still allow fast decays. The J-particle could be cosmologically relevant. Possible candidates for J include a Majoron [704] or a familon (flavor-changing axion) [705].

The general oscillation probability for ν_μ if mass eigenstate ν_2 decays into $\bar{\nu}_3 J$ is [64]

$$P(\nu_\mu \rightarrow \nu_\mu) = \sin^4\theta + \cos^4\theta e^{-\alpha L/E} + 2\sin^2\theta\cos^2\theta e^{-\alpha L/2E} \cos\left(\frac{\delta m_{23}^2 L}{2E}\right), \qquad (12.19)$$

where $\nu_\mu = \cos\theta\ \nu_2 + \sin\theta\ \nu_3$ and $\alpha = m_2/\tau_2$. A more general treatment can be found in [706].

If δm_{23}^2 is sufficiently large, of order $1\,\text{eV}^2$, then the oscillation term averages to zero and

$$P(\nu_\mu \to \nu_\mu) = \sin^4\theta + \cos^4\theta\, e^{-\alpha L/E}. \tag{12.20}$$

However, the L/E dependence of the oscillations disfavors this possibility [707–709]. If instead $\nu_2 \to \bar{\nu}_4 J$, and $\nu_4 \simeq \nu_s$, then δm_{23}^2 can be very small and [65]

$$P(\nu_\mu \to \nu_\mu) = (\sin^2\theta + \cos^2\theta\, e^{-\alpha L/2E})^2. \tag{12.21}$$

However, the L/E dependence of Super-K atmospheric neutrinos rules out this case at more than the 3σ level [251]. Also, in such models ν_e must mix predominantly with ν_1, and small mixing with the other neutrinos then prefers the SMA solar solution that is strongly disfavored by the data. The best fits occur for $\delta m_{23}^2 \approx 3\times 10^{-3}\,\text{eV}^2$ when the primary feature is an oscillation and the decay contribution is subdominant [709], and when combined with data from long-baseline experiments there is no evidence for active neutrino decays [710].

Solar neutrino data are not well-described by neutrino decay [711] and can place limits on neutrino decay parameters [712]. Also, neutrino decay cannot simultaneously explain the solar and KamLAND results [56]. However, these data do not preclude the possibility that neutrinos from distant astrophysical sources may decay [713].

Detection of supernova relic neutrinos can place further limits on neutrino lifetimes [714], and decays of supernova neutrinos can affect interpretations of supernova dynamics [715].

Sterile Neutrino Decay

Sterile neutrinos have been invoked to explain the observed velocity "kicks" of pulsars [111]. Neutrino oscillations involving active-sterile transitions could be important in supernova dynamics [716]. It has been suggested that sterile neutrinos with masses of several keV could make up most of the dark matter [717]. Attempts have also been made to explain the LSND anomaly via a decaying sterile neutrino [718].

A sterile neutrino of mass $m_s < 2m_e$ could decay into active neutrinos via the exchange of a virtual Z-boson [719]. However, the lifetime would be longer than the age of the universe. A sterile neutrino can decay radiatively to an active neutrino and a photon, $\nu_s \to \nu_\alpha\gamma$, via a one-loop diagram [719, 720]. The radiative lifetime is

$$\frac{1}{\tau} = \left(6.8\times 10^{-33} s^{-1}\right)\left[\frac{\sin^2 2\theta}{10^{-10}}\right]\left[\frac{m_s}{\text{keV}}\right]^5, \tag{12.22}$$

where θ is the sterile-active neutrino mixing angle. If sterile neutrinos of about 5 keV mass should constitute the dark matter, the signature of the decays would be a gamma ray line spectrum [721]. Limits on such a spectral feature have been placed with x-ray data from a variety of sources [722, 723] (see also figure 12.6). Future x-ray observations of dwarf galaxies can also test this possibility [730].

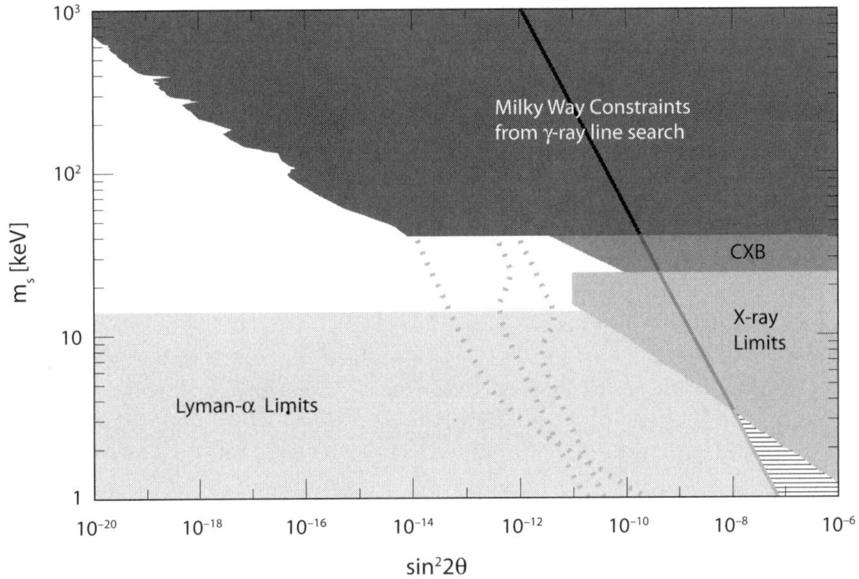

Figure 12.6. Excluded regions in sterile neutrino mass and mixing from Milky Way gamma-ray line emission [724], cosmic x-ray background (CXB) [725], other x-ray sources [726], and Lyman-α forest data [727]. The Lyman-α constraint may vary according to the sterile neutrino production model [728]. The solid line represents the model of [717]; the dotted lines are model predictions of [729]. From [724].

12.5 Neutrino Decoherence

One possible effect of quantum gravity is that pure states could evolve into mixed states over time [731] by interacting with the spacetime foam at the Planck scale [732]. Such decoherence is distinct from and in addition to the kinematical decoherence caused by the finite size of the source and imperfect detector resolution described in section 3.6. The earliest studies of possible quantum gravity decoherence were made in the kaon system [733, 734] and for neutrons [733, 735]. Many studies of possible neutrino decoherence have also been made for solar neutrinos [736, 737], atmospheric neutrinos [67, 737, 738] and long-baseline neutrino experiments [739]. More general studies have also been made [740].

The neutrino density matrix evolves according to [741]

$$\frac{d\rho}{dt} = -i[H, \rho] - \gamma[D, [D, \rho]], \tag{12.23}$$

where the second, dissipative term is the source of the decoherence, D is the dimensionless operator that describes the decoherence effect, and γ is the strength of the effect, with dimensions of energy, representing the inverse of the decoherence length. The resulting survival probability for two-neutrino oscillations is [66]

$$P(\nu_\alpha \to \nu_\alpha) = 1 - \frac{1}{2}\sin^2 2\theta\left[1 - e^{-\gamma L}\cos\left(\frac{\delta m^2 L}{2E}\right)\right]. \tag{12.24}$$

Without a full theory of quantum gravity the form of γ is not known, but it is common to parametrize it as a power law in energy, $\gamma = \gamma_0(E/E_0)^n$, where γ_0 is a constant with dimensions of energy and E_0 is an arbitrary reference energy [66, 742]. There have been several suggestions for the exponent n: (i) $n = 0$ (energy independent), (ii) $n = -1$ (Lorentz invariant [66]), and (iii) $n = 2$ (motivated by dimensional analysis and assuming the effect is suppressed by the Planck scale, $\gamma{\sim}E^2/M_{Pl}$ [743]). Decoherence could also occur due to stochastic density fluctuations as neutrinos propagate through matter [744], rather than by quantum gravity effects, in which case $n = -2$ [742]. Furthermore, it has been shown that $\gamma \sim (\delta m^2)^2 L/E^2$ if one assumes Gaussian uncertainties in the source position and detector energy resolution [745].

The energy dependence of the $\bar{\nu}_e \rightarrow \bar{\nu}_e$ oscillation in the KamLAND experiment excludes the case of pure decoherence, i.e., massless neutrinos, for $n = -1$ [56,746]. A later study [742] used solar and KamLAND data to show that decoherence must be at most a subdominant effect and placed upper bounds on γ_0 given approximately by

$$\gamma_0 < 10^{1.94n-24.18}\ \text{GeV}\ (95\%\ \text{C.L.}), \tag{12.25}$$

for $E_0 = 1\,\text{GeV}$. Pure decoherence is excluded at more than the 3σ level for atmospheric neutrinos for $n = -1$ from the L/E dependence of the oscillations [251] (see also figure 5.3). Attempts have also been made to explain the LSND anomaly using decoherence [747].

Neutrino decoherence can also be manifested in high-energy astrophysical neutrinos and, because of the much larger distances involved, place much stricter upper bounds on γ (or, alternatively, lower bounds on the decoherence length) [748]. Decoherence of supernova neutrinos due to matter density fluctuations can in principle affect supernova dynamics [749]. Also, for large matter densities the standard kinematical decoherence is not necessarily complete and synchronized oscillations can occur [750].

12.6 Lorentz Invariance Violation

General relativity and standard quantum field theory assume Lorentz invariance. The unification of gravity with the physics of the standard model is expected to occur at the Planck scale and some proposed candidates for this unification can have a small amount of Lorentz invariance violation (LIV), e.g., string theory [751] and quantum gravity [752]. The size of the LIV is expected to be suppressed by the inverse of the Planck mass. The detailed dynamics at the Planck scale are not known, but the LIV effects can be expressed as non-Standard Model terms in an effective field theory. The most general framework for describing these violations is the Standard Model Extension (SME) [673], which includes all possible LIV interaction terms that can be written as observer scalars and that involve Standard Model fields.

The largest LIV effect on neutrinos is in their propagation. In the limit of highly relativistic neutrinos the effective Hamiltonian for neutrino propagation in the minimal SME (with only renormalizable terms) becomes [674]

$$(H_{eff})_{ij} = E\delta_{ij} + \frac{(m^2)_{ij}}{2E} + \frac{1}{E}\left[a^{\mu}p_{\mu} - c^{\mu\nu}p_{\mu}p_{\nu}\right]_{ij}, \tag{12.26}$$

where i, j are flavor indices, $p_{\mu} = (E, -E\hat{p})$ is the neutrino 4-momentum, $\hat{p}$ the neutrino direction, and E the neutrino energy. The first term can be ignored since oscillations are insensitive to terms proportional to the identity, and the second term is the usual one due to neutrino mass. The terms involving the a^{μ} coefficients are energy independent and violate CPT, while the $c^{\mu\nu}$ terms are proportional to energy and conserve CPT. Note, however, that the CPT violation in the SME cannot explain the CPT violation discussed in section 12.2 that has different mass matrices for neutrinos and antineutrinos. Since H_{eff} is Hermitian and terms proportional to the identity do not contribute to oscillations, there are eight real independent parameters for each a^{μ} and $c^{\mu\nu}$. Therefore there are 16 parameters that are direction independent and 96 that do not obey rotational invariance. For antineutrinos, $a^{\mu}_{ij} \to -a^{\mu}_{ij}$.

In principle LIV of neutrinos could be related to that of the charged leptons, for which there are some strong constraints [753], but it is also possible to have LIV for neutrinos but not charged leptons [754].

It has been determined that atmospheric neutrinos have an oscillation argument that goes predominantly like $1/E$ over several decades in energy [755, 756], which would seem to imply that the LIV terms are subleading at best. However, it is possible to construct an effective Hamiltonian with *massless* neutrinos and LIV that mimics this behavior: in the bicycle model [757] the only nonzero LIV terms are $c^{00}_{ee} \equiv c$ and $a^{\mu}_{e\mu} = a^{\mu}_{e\tau} \equiv a/\sqrt{2}$ for $\mu = 0, 1, 2$, or 3, and at high neutrino energies there is a seesaw mechanism that leads to $\nu_{\mu} - \nu_{\tau}$ oscillations with an oscillation argument proportional to $1/E$ and an effective mass-squared difference proportional to a^2/c. If the nonzero a^{μ} is space-like, the δm^2_{eff} will have directional dependence.

A detailed examination of the neutrino phenomenology [758] indicates that the bicycle model cannot simultaneously explain the solar, atmospheric, long-baseline, and KamLAND data. Models using LIV combined with massive neutrinos have been proposed [759] to explain all neutrino oscillation phenomena including the LSND effect. If LIV is induced by dark energy, only the energy-independent LIV terms in equation 12.26 are present [760]. Also, Lorentz invariance violation with more complicated neutrino dispersion relations have been studied in the context of explaining the LSND anomaly, although significant finetuning is required [761]. Models with neutrino mass and higher dimensional LIV operators (involving oscillation arguments with higher powers of energy) may be able to explain the MiniBooNE low-energy anomaly [762].

Since the LIV terms in equation 12.26 are proportional to higher powers of energy than the neutrino mass terms, measurements of high-energy astrophysical neutrinos can also place limits on LIV effects in neutrinos [763]. Interference between the CPT-violating and neutrino mass terms can lead to a resonant enhancement of oscillations in matter simultaneously for neutrinos and antineutrinos [764], unlike the usual case where one is suppressed if the other is enhanced. Searches for direction dependence have found null results [765, 766], although a recent study suggests that the MiniBooNE low-energy excess in the neutrino channel could be explained by direction-independent LIV [766].

12.7 Non-standard Neutrino Interactions

Although neutrino mass and mixing will require some new physics beyond the minimal Standard Model, most neutrino oscillation analyses are done assuming Standard Model neutrino interactions. However, in many cases new physics will also introduce additional terms to the neutrino interaction Lagrangian.[10] The usual ansatz is to parametrize these non-standard interactions (NSI) of neutrinos [35, 768, 769] in terms of dimension-6 operators in an effective Lagrangian

$$\mathcal{L}_{eff} = -2\sqrt{2} \sum_{f=e,u,d} \left[\bar{\nu}_\alpha \gamma_\mu P_L \nu_\beta\right] \left[\bar{f} \gamma_\mu \left(\epsilon_{\alpha\beta}^{fL} P_L + \epsilon_{\alpha\beta}^{fR} P_R\right) f\right], \qquad (12.27)$$

where $P_R, P_L = \frac{1}{2}(1 \pm \gamma_5)$. The size of the $\epsilon_{\alpha\beta}^{fP}$ ($P = L$ or R) are expected to be of order $(m_W/\Lambda)^2$, where Λ is the scale of new physics. Supersymmetry could give rise to NSI of neutrinos; in the MSSM with right-handed neutrinos NSI are suppressed [770], but they may be observable in SUSY with R-parity violation [771].

There are numerous experimental constraints on the NSI of neutrinos [772]. Elastic $\nu_e e$ scattering measured in LSND [773] and $\bar{\nu}_e e$ scattering measured in the MUNU experiment [774] imply bounds on ϵ_{ee}^{eP} and $\epsilon_{e\tau}^{eP}$, and elastic $\nu_\mu e$ scattering measured in the CHARM II detector [775] leads to limits on $\epsilon_{\mu\mu}^{eP}$ and $\epsilon_{\mu\tau}^{eP}$. Similarly $\nu_e q$ and $\nu_\mu q$ scattering in CHARM [776] and NuTeV [777], respectively, give limits on $\epsilon_{e\beta}^{uP}$ and $\epsilon_{\mu\beta}^{dP}$. The contribution of $\epsilon_{\tau\tau}^{eP}$ to $e^+e^- \rightarrow \nu\bar{\nu}\gamma$ at LEP constrains that parameter [778]. These limits are summarized in tables 12.1 and 12.2, assuming only one parameter is nonzero; in some cases the limits relax if more than one parameter is allowed to vary [778], but inclusion of all the data makes the bounds fairly robust [779]. Loop diagrams with $\nu_\tau \nu_\tau q q$ vertices contribute to the invisible width of the Z boson, but constraints on $\epsilon_{\tau\tau}^{qP}$ are rather weak, of order unity or a little larger [772] (but see the following discussion on their effects on atmospheric neutrinos). Future neutrino scattering experiments on electrons [781], coherent neutrino scattering on nuclei [782], measurements of rare K and D decays [783], or Borexino measurements of the shape of the low-energy solar neutrino spectrum [784] can improve these limits.

In principle there are also strong constraints on NSI of neutrinos from flavor-changing decays such as $\mu \rightarrow eee$ if the effective $\nu\nu\ell\ell$ coupling is related to the $\ell\ell\ell\ell$ coupling, as might be expected from $SU(2)_L$ gauge invariance. There is one dimension-6 and one dimension-8 operator that can lead to NSI of neutrinos without contributing to processes such as $\mu \rightarrow eee$ at tree-level [785]. The corresponding NSI will contribute to $\mu \rightarrow eee$ at the one-loop level, which leads to the limits $|\epsilon_{e\mu}^{eP}| < 5 \times 10^{-4}$ [772]. Similarly, there is a bound $|\epsilon_{e\mu}^{qP}| < 7.7 \times 10^{-4}$ from the rate of $\mu - e$ conversion on nuclei [772, 786].

[10] Standard Model radiative corrections can also give interactions that differ slightly from those at tree-level, although these effects are small [767].

TABLE 12.1
Compilation of 90% C.L. allowed ranges for NSI parameters $\epsilon_{\alpha\beta}^{eP}$ defined in equation 12.27. If a single number is given it refers to the bound on the absolute value. Bounds are from [779] unless otherwise noted. For bounds on $\epsilon_{e\mu}^{eP}$, see the discussion in the text.

$\alpha\beta$	$\epsilon_{\alpha\beta}^{eL}$	$\epsilon_{\alpha\beta}^{eR}$
ee	$(-0.03, 0.08)$	$(0.004, 0.08)$ [779,780]
$\mu\mu$	0.03 [772]	0.03 [772]
$\tau\tau$	$(-0.46, 0.24)$	$(-0.25, 0.43)$
$e\tau$	0.33	0.18 [780]
$\mu\tau$	0.1 [772]	0.1 [772]

TABLE 12.2
Compilation of 90% C.L. allowed ranges for NSI parameters $\epsilon_{\alpha\beta}^{qP}$ defined in equation 12.27. Where a single number is given it refers to the bound on the absolute value. For bounds on $\epsilon_{\tau\tau}^{qP}$ and $\epsilon_{e\mu}^{qP}$, see the discussion in the text. From [772].

$\alpha\beta$	$\epsilon_{\alpha\beta}^{uL}$	$\epsilon_{\alpha\beta}^{uR}$	$\epsilon_{\alpha\beta}^{dL}$	$\epsilon_{\alpha\beta}^{dR}$
ee	$(-1.0, 0.3)$	$(-0.4, 0.7)$	$(-0.3, 0.3)$	$(-0.6, 0.5)$
$\mu\mu$	0.003	$(-0.008, 0.003)$	0.003	$(-0.008, 0.015)$
$e\tau$	0.5	0.5	0.5	0.5
$\mu\tau$	0.05	0.05	0.05	0.05

However, it has been shown [787] that if $SU(2)_L$ gauge invariance is assumed in the effective theory leading to the couplings in equation 12.27, then there is a cancellation of the NSI loop contributions in the massless limit and the limits are relaxed by a factor m_W^2/m_ℓ^2 for the dimension-6 operator, where ℓ is a charged lepton. For the dimension-8 operator, one can make model-dependent naturalness assumptions and recover the stringent bounds on the $\epsilon_{e\mu}^{fP}$, but the model independent bound is similar to that of ϵ_{ee}^{eP}, i.e., about 0.13 [779]. Nevertheless, saturating this bound would require significant fine tuning [787].

One interesting possibility is that NSI can cause neutrino oscillations in matter via coherent forward scattering [35, 768], which in principle could provide an explanation for the solar neutrino problem without neutrino mass [788]. The contribution of NSI to neutrino propagation in matter depends only on the vector part of the interaction, which leads to an additional term in equation 3.30 of

$$i\left(\frac{d\nu_\alpha}{dL}\right)_{NSNI} = \sqrt{2}G_F \sum_\beta \left(\sum_{f=e,u,d}\sum_{P=L,R} N_f \epsilon_{\alpha\beta}^{fP}\right)\nu_\beta, \qquad (12.28)$$

where N_f is the number density of fermion f.

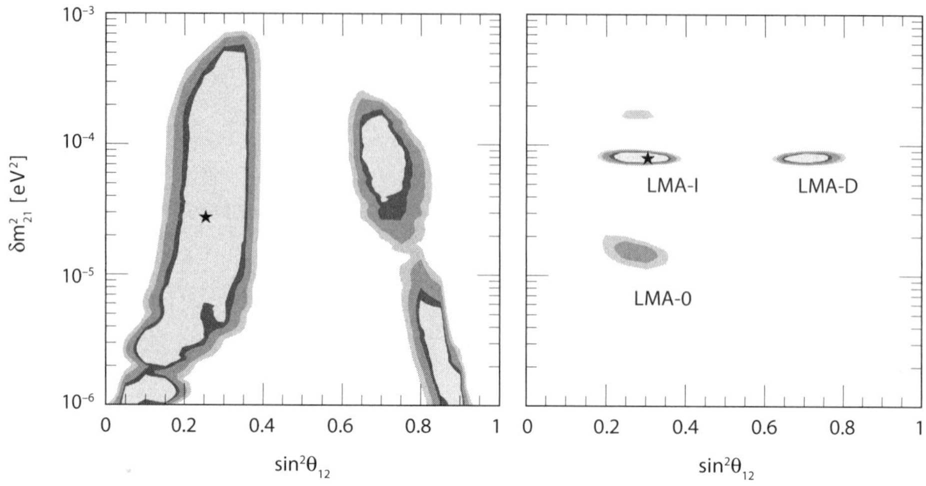

Figure 12.7. The 90%, 95%, 99%, and 99.73% C.L. allowed regions for $\sin^2\theta_{12}$ and δm^2_{21} for neutrino oscillations due to neutrino mass plus NSI (the NSI parameters have been marginalized). The left panel shows the fit for solar neutrino data only; the right panel is for a fit to solar and KamLAND data. The fits in the LMA-I and LMA-D regions are of similar quality; the LMA-0 region is allowed only at 97% C.L. Adapted from [790].

Many studies have been made of the effect of NSI on the propagation of solar neutrinos [788, 789]; with the addition of KamLAND data, a pure NSI solution of the solar neutrino problem is excluded [216]. However, a hybrid solution with neutrino mass *and* NSI is possible with $\sin^2\theta_{12} > \frac{1}{2}$ [790]; see figure 12.7. Taking into account NSI in both propagation and detection, solar and KamLAND data place a constraint $-0.036 < \epsilon^{eL}_{\alpha\beta} < 0.063$ [791], which is comparable to other bounds on this parameter. By including CHARM data on neutrino-quark scattering, tighter bounds can be obtained [792].

There have also been studies of the effect of neutrino NSI on atmospheric neutrinos [793]. Pure NSI of neutrinos as an explanation of the atmospheric neutrino oscillations are excluded at more than 99% C.L. [794]. If there are NSI only in the $\nu_\mu - \nu_\tau$ sector, the preferred values of the oscillation parameters are unaffected by any NSI contribution [795], unlike the case of solar neutrinos. Also, the atmospheric data place contraints on the *vector* NSI couplings $|\epsilon^{qV}_{\mu\tau}| \lesssim 0.01$ and $|\epsilon^{qV}_{\mu\mu} - \epsilon^{qV}_{\tau\tau}| \lesssim 0.03$ [795,796], where $\epsilon^{fV}_{\alpha\beta} = \epsilon^{fL}_{\alpha\beta} + \epsilon^{fR}_{\alpha\beta}$. However, if NSI are also present for ν_e, the best-fit value of θ_{23} can change substantially and much larger NSI effects ($\epsilon^{fV}_{\alpha\beta} \sim 1$) are allowed by the data [797].

NSI of neutrinos could also be probed in superbeam experiments [798–800] and neutrino factories [801–812]. Effects of NSI at the source can be distinguished from oscillations by a detector near the source [798, 804, 810, 812], while ambiguities between NSI and θ_{13} can benefit from multiple detectors away from the source [799, 803, 806, 811]. NSI effects can also affect the measurement of CP violation [801, 805, 808, 809, 811], both by modifying matter effects and by intrinsic CP violation in the NSI, and can cause CP violation even in the limits $\delta = 0, \pi$ and/or $\theta_{13} = 0$ [808]. The optimization of the neutrino factory configuration does not change if NSI of neutrinos are considered [807]. NSI of neutrinos can be confused

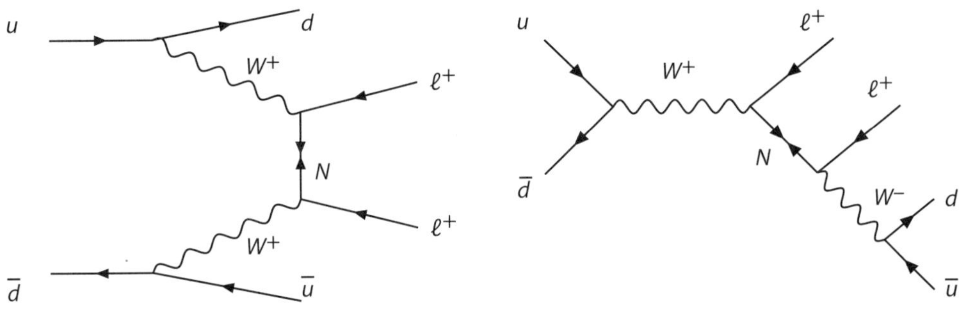

Figure 12.8. Feynman diagrams for $\Delta L = 2$ processes involving a Majorana neutrino N in hadron colliders. The W bosons can be either a Standard Model W with left-handed couplings or a W_R in a left-right model. From [818].

with a θ_{13} contribution to oscillations of solar and KamLAND neutrinos [813]. The effect of NSI of neutrinos in supernovae has also been considered [814].

A mapping of vacuum neutrino oscillation formulas to the corresponding expressions in matter with NSI is given in [815]. Also, the expansion in small parameters of oscillation probabilities in equations 3.31, 3.32, 3.38, and 3.39 have been generalized to the case of oscillations plus NSI [816]. The resulting expressions reveal a new parameter degeneracy between the oscillation and NSI parameters that involves both solar and atmospheric neutrino oscillations [816]; some degeneracies involving NSI are not easily resolved, even if the spectrum shape is well-measured [769].

In summary, NSI of neutrinos could affect the production, propagation, and detection of neutrinos, and in practice can be difficult to disentangle from other effects, such as neutrino oscillations due to neutrino masses and mixing. An important aspect of any future neutrino program will be to identify or constrain these interactions.

12.8 Heavy Majorana Neutrinos at Colliders

The unambiguous proof of the Majorana nature of a heavy neutrino would be the observation of a lepton-number (L) violating process. If Majorana neutrinos N exist with GeV to TeV masses, they may be discovered at colliders via L nonconservation. The fact that N decays equally to leptons and antileptons leads to $\Delta L = 2$ processes at colliders. This was first pointed out [817] in the context of the $SU(2)_L \times SU(2)_R \times U(1)$ model, where the production of right-handed neutrinos N occurs via the right-handed gauge boson W_R. Feynman diagrams that give like-sign leptons in this model are shown in figure 12.8.

A simple extension of the SM is to append right-handed singlet Majorana neutrinos. The prospects for their discovery at the LHC was first studied in [818]. The decay branching fractions of N are $BF(N \to \ell^- W^+) = BF(N \to \ell^+ W^-) \approx BF(N \to \nu Z) \approx BF(N \to \nu H) \approx \frac{1}{4}$. Here the W, Z, and H may be real or virtual, depending on the mass of N. The Feynman diagrams that give $\Delta L = 2$ processes in singlet models are similar to those in the left-right model (see figure 12.8). The same-sign dilepton event rates are proportional to the square of the mixing matrix element $V_{\ell 4}$, which for $\ell = \mu$ is constrained by experiments to be $|V_{\mu 4}|^2 \lesssim 10^{-4}$.

Nonetheless, it was found that masses of m_N from 10 to 180 GeV could be probed [818, 819].

Right-handed neutrinos could be produced in pairs in new physics models with an extra Z boson via its $Z' \to NN$ decays. These models and others with seesaw generation of neutrino masses may be testable by the associated $\Delta L = 2$ signals [820]. Tri-lepton signals such as $\ell^+\ell^+\ell^-\bar{\nu}$ may also be observable [821].

12.9 Neutrino Magnetic Moment

The fundamental neutrino magnetic moments are properly associated with neutrino mass eigenstates. In the minimal extension of the Standard Model that includes neutrino mass, the magnetic moment for a Dirac neutrino is [822]

$$\mu_\nu = \frac{3eG_F m_\nu}{8\sqrt{2}\pi^2} = \frac{3G_F m_e m_\nu}{4\sqrt{2}\pi^2}\mu_B\,, \tag{12.29}$$

where $\mu_B = e/2m_e$ is the Bohr magneton. A Majorana neutrino can have only a transition magnetic moment (unless CPT is not conserved [823]); a Dirac neutrino can have both diagonal and off-diagonal magnetic moments, although in the Standard Model the off-diagonal moments are suppressed relative to the diagonal moments [720, 824]. For $m_\nu \gtrsim 0.05$ eV, the magnetic moment from equation equation (12.29) is $\mu_\nu \gtrsim 1.5 \times 10^{-20}\ \mu_B$. Model-independent calculations show that if new physics at the TeV scale contributes to a neutrino magnetic moment and it is found that $\mu_\nu \gtrsim 10^{-15}\ \mu_B$, then neutrinos must be Majorana, not Dirac, particles [825]. For a detailed discussion of the electromagnetic properties of neutrinos, see [826].

A neutrino magnetic moment affects $\nu_e e$ and $\bar{\nu}_e e$ scattering. The differential cross-section is given by [827]

$$\frac{d\sigma}{dT} = \frac{G_F^2 m_e}{2\pi}\left[(g_V + g_A)^2 + (g_V - g_A)^2\left(1 - \frac{T}{E_\nu}\right) + (g_A^2 - g_V^2)\frac{m_e T}{E_\nu^2}\right] + \frac{\pi\alpha^2\mu_\nu^2}{m_e^2}\left(\frac{1}{T} - \frac{1}{E_\nu}\right), \tag{12.30}$$

where T is the electron (positron) recoil kinetic energy, $g_V = 2x_W + 1$, $g_A = +\frac{1}{2}(-\frac{1}{2})$ for ν_e ($\bar{\nu}_e$) and μ_ν is expressed in units of μ_B. If there is neutrino mixing, the effective magnetic moment for a particular neutrino flavor is a complicated function of both magnetic and electric diagonal and transition moments [828, 829].

The predicted distortion of the low-energy spectrum in solar and reactor experiments allows upper limits in the range $\mu_{\nu_e} \lesssim (2 - 11) \times 10^{-11}\ \mu_B$ [51, 774, 830–832] to be set. After accounting for neutrino oscillations, the Borexino data can also be interpreted [833] as setting limits on the muon and tau neutrino magnetic moments, $\mu_{\nu_\mu} < 1.5 \times 10^{-10}\ \mu_B$ and $\mu_{\nu_\tau} < 1.9 \times 10^{-10}\ \mu_B$. Also, a photon traveling in a medium acquires an effective mass, enabling the so-called plasmon decay, $\gamma^* \to \nu\bar{\nu}$. In stars, a large enough neutrino magnetic moment can enhance the plasmon decay rate and the energy loss in neutrinos leads to faster cooling, delaying helium ignition. The lack of evidence for this effect gives the constraint $\mu_\nu \leq 3 \times 10^{-12}\ \mu_B$ [834].

If the neutrino has a magnetic moment, it will precess in a magnetic field, causing a $\nu_L \to \nu_R$ transition. In combination with matter effects, this can lead to resonant spin-flavor precession (RSFP) [835] in the Sun. If neutrinos are Dirac particles the transitions are to sterile neutrinos, but for Majorana neutrinos the transitions are to antineutrinos [836]. The Super-K [837] and SNO [838] experiments find that the flux of solar antineutrinos is at most about 1% of the flux of neutrinos (in their ranges of sensitivity), while KamLAND has placed a stronger limit of 0.03% on $\nu_e \to \bar{\nu}_e$ conversion [839] (both limits are quoted at 90% C.L.), which in principle put strong constraints on the size of the neutrino magnetic moment for Majorana neutrinos [840].

The MSW resonance and the spin-flavor precession resonance generally occur at different positions in the sun [826, 841]. To calculate the effects of RSFP [842] on solar neutrinos, one must have detailed information about solar magnetic fields, which are not well understood [843]. Since KamLAND antineutrinos will not have sizeable RSFP effects, spin-flavor precession cannot be the primary solution to the solar neutrino problem [217], although it may contribute to solar neutrino transitions [844] and cause time variations of the solar neutrino flux [845]. In some extensions of the Standard Model the tau neutrino can have an enhanced magnetic moment [846].

12.10 Fourth Generation Neutrino

Although extra sterile neutrinos have been suggested as a solution to the LSND anomaly, it is also possible to have extra heavy neutrinos that have the standard electroweak interactions. The existence of a sequential fourth generation of quarks and leptons (with SM $V - A$ gauge couplings, sometimes called SM4) has been entertained for a long time [847]. There has been a revival of interest in an additional generation of fermions [848,849], including extensions to fourth generation particles with vector gauge coupling [850], as part of models with mirror symmetry (with $V + A$ couplings) [846, 851], or with mixed gauge couplings [852].

Investigations [853] have shown that sequential fourth generation particles are not ruled out by precision electroweak data if they are heavy, with masses at or above about 100 GeV. The experimental constraints on a fourth-generation neutrino come from LEP2 and the Tevatron [854]. It is assumed that the heavy neutrino decays via $N \to W^{\pm}\ell^{\mp}$, with $\ell = e, \mu$, or τ. The current lower bound on the mass of a neutrino decaying to eW or μW is about 80 GeV; the limit for τW decay is about 62 GeV. One possible reason the fourth-generation neutrino is heavy (and hence has not already been discovered) is that it obtains its mass via the seesaw mechanism, but the right-handed neutrino mass for the fourth generation is much lower than for the first three generations, which results in a weaker suppression for the left-handed neutrino mass.

A fourth generation could also provide possible explanations for the Higgs naturalness problem and the fermion mass hierarchy, and provide a source for CP violation effects that cannot be explained by the Standard Model [855]. Since partial-wave unitarity places an upper bound on fourth-generation fermion masses of about 1 TeV [856], the LHC likely can either discover or rule out their existence [857].

✳13✳

Summary and Outlook

In recent years we have witnessed a revolution in the physics of neutrinos. The observation of neutrino oscillations demonstrated both the quantum mechanical nature of neutrinos and that neutrinos are massive particles. These discoveries came from astrophysics sources – the neutrinos created by interactions of cosmic rays with the earth's atmosphere and the neutrinos released by the fusion reactions in the core of the sun. The same oscillation phenomena were then found in terrestrial experiments, using detectors at long baselines from accelerators and detectors at about 180 km distances from nuclear reactors.

How convenient that nature provided neutrino mass-squared differences that could be probed at L/E scales in both extra-terrestrial and terrestrial experiments, with their vastly different lengths and energies! In the atmosphere, the lever arm of oscillations is the diameter of the earth, $L \sim 13 \times 10^3$ km, and typical neutrino energies are $E \sim 10$ GeV, leading to values of L/E up to 10^3 km/GeV. In a long-baseline terrestrial experiment in which a detector is located at a distance $L \sim 10^3$ km from an accelerator that produces a neutrino beam of energy $E \sim 5$ GeV, the L/E is again of order 10^3 km/GeV. Both are thus germane to neutrino oscillations associated with a neutrino mass-squared difference of order 10^{-3} eV^2. For solar neutrino oscillations, it is not as straight-forward to infer the effective L/E because of the importance of matter effects on the exiting neutrinos. However, from phenomenological analyses of the solar neutrino data, we know that the mass-squared difference needed to describe the survival probability versus energy of solar neutrinos is of order 10^{-4} eV^2. A detector placed at a distance $L \sim 200$ km from an ensemble of reactors emitting antineutrinos of energy a few MeV gives similar L/E sensitivity as the solar neutrinos.

The observations of atmospheric, accelerator, solar, and reactor neutrino observations fit neatly into a three-neutrino framework. The disappearance of the atmospheric muon-neutrinos is ascribed to their conversion to tau neutrinos, with the latter still to be definitively established, although one event of this kind has been seen in an accelerator experiment at long baselines. The nonobservation of a similar disappearance of atmospheric ν_e is due to a suppressed mixing of ν_e with the other two flavors at this atmospheric mass-squared difference scale. The solar neutrino disappearance and the reactor antineutrino disappearance both involve the interchanges of electron-type neutrinos with the other two neutrino flavors at the solar mass-squared difference scale, with CPT symmetry relating the probabilities for the neutrino and antineutrino probabilities.

The mixings among the 3 neutrino flavors can be represented by a 3×3 unitary matrix in the basis that the charged-lepton matrix is chosen to be diagonal. This mixing matrix is specified by three mixing angles and one complex phase. In the conventional representation of the matrix, the 23 angle, θ_{23}, is consistent with 45 degrees (corresponding to maximal muon-neutrino, tau-neutrino mixing), the 12 angle, θ_{12}, is 34 degrees (it is the best determined of the neutrino mixings) but consistent with 35 degrees, and the value of the 13 angle, θ_{13}, is about 9 degrees. For angles of 45 degrees, 35 degrees, and 0 degrees the mixing matrix takes the simple tribimaximal form. With $\theta_{13} \neq 0$ the matrix depends on the complex phase and there is hope for detecting CP violation in the oscillation probabilities. Also, the atmospheric and solar neutrino oscillation phenomena are not completely decoupled. Why the mixing matrix has this form remains a mystery that is a focus of theoretical constructions.

Now that neutrino oscillation experiments have shown with certainty that neutrinos have mass and substantial knowledge has been acquired about neutrino mass-squared differences and mixings, the path forward to a deeper level of discovery and fundamental understanding is evident. The first step is to precisely determine the 13 angle, θ_{13}, of the mixing matrix. Matter effects in the propagation of neutrinos through the earth and CP violation are both dependent on the value of this angle. The predictions of various theoretical models for θ_{13} range from the maximal presently allowed by experiment to very small values. Thus, a θ_{13} measurement has considerable potential to discriminate theoretical models.

One way forward is reactor experiments designed with an appropriate L/E to probe $\bar{\nu}_e$ disappearance oscillations connected to the atmospheric mass-squared difference. Three such reactor experiments are underway with this goal. Another way forward is accelerator experiments to detect $\nu_\mu \to \nu_e$ appearance, with detectors at baselines of 250 km (Japan) or 750 km (U.S.) and narrow band beams that have a nearly unique neutrino energy. These medium-baseline experiments could be competitive with the time-frame of the reactor experimental programs.

Now that a nonzero value of θ_{13} is established, attention will turn to experiments with long enough baselines (1000 km or more) to be sensitive to matter effects on neutrinos propagating through the earth's mantle. These experiments will determine the sign of the atmospheric mass-squared difference, presently unknown, that corresponds to a normal or inverted neutrino mass hierarchy. In the normal hierarchy, the masses of the neutrinos associated with solar neutrino oscillations are smaller than the mass of the neutrino that drives the atmospheric neutrino oscillations, and vice versa for the inverted mass hierarchy. The 1300 km baseline from Fermilab to the DUSEL underground laboratory is under consideration, as is a similarly long baseline from Japan to Korea. In long-baseline experiments, there may be an eight-fold degeneracy of the determined values of the oscillation parameters θ_{13} and δ, the sign of the atmospheric mass-squared difference, and the quadrant of θ_{23}. To resolve these degeneracies, a wide-band neutrino beam is advantageous. At DUSEL the neutrinos would be detected in three or more 100 kt water Cherenkov modules or in Liquid Argon detector modules, if the latter can be successfully scaled up from the present 0.6 kt to the 100 kt size. A reactor measurement of θ_{13} has no degeneracy ambiguity and thus will simplify the solution to the physical parameters.

Neutrino oscillations do not provide information on the absolute values of neutrino masses. Moreover, the sub-eV neutrino mass range is likely below the reach of beta decay experiments in the near future. Neutrinoless double beta decay experiments can probe down to about 0.05 eV, if the neutrino mass hierarchy is inverted. Fortunately, cosmology provides sensitivity to very small neutrino masses, which affect the cosmic microwave background and the formation of large-scale structure. These may allow the sum of neutrino masses to be determined, even if it is only 0.05 eV. Then, combined with the mass-squared difference values from oscillation experiments, the individual neutrino masses will be known.

A crucial question to be answered is whether neutrinos are Majorana particles (for which the neutrino is its own antineutrino) or Dirac particles, which may be resolved by $0\nu\beta\beta$-decay experiments. This process is forbidden if neutrinos are Dirac but is allowed if neutrinos are Majorana. The $0\nu\beta\beta$-decay rate is proportional to the *ee* element of the neutrino mass matrix, and so also probes the absolute neutrino mass. However, in the $0\nu\beta\beta$-decay matrix element, the neutrino masses are weighted by the square of the V_{ei} mixing matrix elements, and consequently small θ_{13} suppresses the rate when the neutrino mass hierarchy is normal. A succession of increasingly sensitive experiments are underway and in the planning stages to do the very challenging $0\nu\beta\beta$-decay measurements for more than one nucleus.

Although it may take a decade or more to successfully complete the neutrino experimental program as outlined here the road map to accomplish these goals is straightforward in design. That full information is vital to the pursuit of the fundamental theory of neutrinos.

In the process of following this conventional road map, other new physics effects may well be found and lead to new roads to be followed. Such interesting possibilities include sub-dominant oscillations to sterile neutrinos, non-standard neutrino interactions, environmental dependence of the neutrino mass, and the existence of new neutrinos with TeV scale masses. The connections of neutrinos with GUT scale physics models may also become evident. It is possible that neutrinos are connected with the dark energy in the universe. The prospect for fundamental discoveries along the neutrino pathway may thus be especially bright.

The advent of neutrino telescopes opens a new window for the experimental study of neutrinos. At TeV to PeV energies, neutrinos may help to reveal the sources of cosmic rays. Since the magnetic moments of neutrinos are constrained to be very small by solar neutrino data, neutrinos are unaffected by magnetic fields as they propagate from distant sources. Then observed excesses of events above atmospheric neutrino backgrounds in certain directions could identify their sources if the directions point back to AGN and/or past supernovae. Another important issue that cosmic neutrinos could resolve is whether the observed cut-off of cosmic ray events of order 10^{21} eV is due to the GZK mechanism, wherein the absorption of the cosmic rays would necessarily give rise to ultra high-energy neutrinos, or because of an energy cut-off associated with the production engine. The nature of the neutrino sources can be identified through the oscillation-averaged neutrino flavor ratios that are observed at earth.

Neutrinos from the sun may also probe the nature of dark matter. Dark matter particles may be gravitationally attracted by the sun. If the DM particle has spin-dependent interactions with nuclei, it will be captured by the sun and migrate to the solar core. Then, if the DM is a Majorana particle, it can self-annihilate. Over the age of the sun, the captures and decays of the dark matter will have equilibrated and led to an annihilation rate today that is 1/2 of the capture rate, since two dark matter particles give one annihilation event. The pairs of unstable SM particles ($\tau\bar{\tau}$, $b\bar{b}$, W^+W^-, $t\bar{t}$, etc.) produced in the annihilation events then give neutrinos through their weak decays. These neutrinos can be detected in the DeepCore subdetector of IceCube as events that point back to the sun, within the angular resolution of the detector. Such events are background free due to the vetoing of backgrounds by the surrounding IceCube detector. The upper end point of the neutrino spectrum would determine the dark matter mass. The shape of the spectrum gives information about the participating SM particles. The spin-dependent cross section inferred from the rate of neutrinos detected would give information about the strength of the annihilation process.

These are salient examples of the power that the new large scale neutrino detectors bring to bear on unsolved problems in neutrino physics. After a decade of this exploration of the astrophysics neutrino frontier, new chapters on the associated physics discoveries can be anticipated.

The information that can be gleaned about neutrinos from particle physics, nuclear physics, astrophysics, and cosmology is highly complementary and all of these areas are contributing to the advancement of knowledge of neutrino physics. The past history of neutrino physics has been full of surprises and has contributed importantly to our fundamental understanding of elementary particles. It is quite possible that the results of future exploration of neutrino physics will also exceed our optimistic expectations.

- References -

[1] W. Pauli, letter to *Tübingen Conference*, December 4, 1930, translated in L. M. Brown, *Phys. Today* **23**, Sept. 1978.

[2] J. Chadwick, *Nature* **129**, 312 (1932).

[3] E. Fermi, *Z. Phys.* **88**, 161 (1934), translated in F. L. Wilson, *Am. J. Phys.* **36**, 1150 (1960).

[4] C. L. Cowan, F. Reines, F. B. Harrison, H. W. Kruse, and A. D. McGuire, *Science* **124**, 103 (1956).

[5] T. D. Lee and C. N. Yang, *Phys. Rev.* **104**, 254 (1956).

[6] C. S. Wu, E. Ambler, R. W. Hayward, D. D. Hoppes, and R. P. Hudson, *Phys. Rev.* **105**, 1413 (1957).

[7] E.C.G. Sudarshan and R. E. Marshak, *Proc. of the Conference on Mesons and Newly-Discovered Particles*, ed. by N. Zanichelli (Bologna, 1957); *Phys. Rev.* **109**, 1860 (1958).

[8] R. P. Feynman and M. Gell-Mann, *Phys. Rev.* **109**, 193 (1958).

[9] M. Goldhaber, L. Grodzins, and A. W. Sunyar, *Phys. Rev.* **109**, 1015 (1958).

[10] G. Danby, J. M. Gaillard, K. A. Goulianos, L. M. Lederman, N. B. Mistry, M. Schwartz, and J. Steinberger, *Phys. Rev. Lett.* **9**, 36 (1962).

[11] J. H. Christenson, J. W. Cronin, V. L. Fitch, and R. Turlay, *Phys. Rev. Lett.* **13**, 138 (1964).

[12] A. D. Sakharov, *Pisma Zh. Eksp. Teor. Fiz.* **5**, 32 (1967) [*JETP Lett.* **5**, 24 (1967)] [*Sov. Phys. Usp.* **34**, 392 (1991)] [*Usp. Fiz. Nauk* **161**, 61 (1991)].

[13] B. Aubert et al. [BaBar Collaboration], *Phys. Rev. Lett.* **93**, 131801 (2004) [arXiv:hep-ex/0407057].

[14] Y. Chao et al. [Belle Collaboration], *Phys. Rev. Lett.* **93**, 191802 (2004) [arXiv:hep-ex/0408100].

[15] V. M. Lobashev et al., *Nucl. Phys. Proc. Suppl.* **91**, 280 (2001); *Phys. Lett. B* **460**, 227 (1999).

[16] C. Kraus et al., *Eur. Phys. J. C* **40**, 447 (2005) [arXiv:hep-ex/0412056]; C. Weinheimer et al., *Phys. Lett. B* **460**, 219 (1999).

[17] B. Pontecorvo, *Sov. Phys. JETP* **6**, 429 (1957) [*Zh. Eksp. Teor. Fiz.* **33**, 549 (1957)].

[18] Z. Maki, M. Nakagawa, and S. Sakata, *Prog. Theor. Phys.* **28**, 870 (1962).

[19] M. Kobayashi and T. Maskawa, *Prog. Theor. Phys.* **49**, 652 (1973).

[20] M. L. Perl et al., *Phys. Rev. Lett.* **35**, 1489 (1975).

[21] K. Kodama et al. [DONUT Collaboration], *Phys. Lett. B* **504**, 218 (2001) [arXiv:hep-ex/0012035].

[22] R. Davis, *Phys. Rev. Lett.* **12**, 303 (1964).

[23] R. J. Davis, D. S. Harmer, and K. C. Hoffman, *Phys. Rev. Lett.* **20**, 1205 (1968).

[24] J. N. Bahcall, *Phys. Rev. Lett.* **12**, 300 (1964); J. N. Bahcall, N. A. Bahcall, and G. Shaviv, *Phys. Rev. Lett.* **20**, 1209 (1968).

[25] R. Davis, L. C. Rogers, and V. Rodeka, *Bull. Amer. Phys. Soc.* **16**, 631 (1971).

[26] B. Pontecorvo, *Sov. Phys. JETP* **26**, 984 (1968) [*Zh. Eksp. Teor. Fiz.* **53**, 1717 (1967)]; V. N. Gribov and B. Pontecorvo, *Phys. Lett.* **28B**, 493 (1969).

[27] J. N. Abdurashitov et al. [SAGE Collaboration], *J. Exp. Theor. Phys.* **95**, 181 (2002) [*Zh. Eksp. Teor. Fiz.* **122**, 211 (2002)] [arXiv:astro-ph/0204245].

[28] W. Hampel et al. [GALLEX Collaboration], *Phys. Lett. B* **447**, 127 (1999).

[29] M. Altmann et al. [GNO Collaboration], *Phys. Lett. B* **490**, 16 (2000) [arXiv:hep-ex/0006034].
[30] Y. Fukuda et al. [Super-Kamiokande Collaboration], *Phys. Rev. Lett.* **81**, 1158 (1998) [Erratum-ibid. **81**, 4279 (1998)] [arXiv:hep-ex/9805021]; *Phys. Rev. Lett.* **82**, 2430 (1999) [arXiv:hep-ex/9812011].
[31] S. Fukuda et al. [Super-Kamiokande Collaboration], *Phys. Rev. Lett.* **86**, 5651 (2001) [arXiv:hep-ex/0103032]; *Phys. Lett. B* **539**, 179 (2002) [arXiv:hep-ex/0205075].
[32] M. B. Smy et al. [Super-Kamiokande Collaboration], *Phys. Rev. D* **69**, 011104 (2004) [arXiv:hep-ex/0309011].
[33] B. Aharmim et al. [SNO Collaboration], *Phys. Rev. C* **75**, 045502 (2007) [arXiv:nucl-ex/0610020]; Q. R. Ahmad et al. [SNO Collaboration], *Phys. Rev. Lett.* **89**, 011301 (2002) [arXiv:nucl-ex/0204008]; *Phys. Rev. Lett.* **89**, 011302 (2002) [arXiv:nucl-ex/0204009]; *Phys. Rev. Lett.* **87**, 071301 (2001) [arXiv:nucl-ex/0106015].
[34] B. Aharmim et al. [SNO Collaboration], *Phys. Rev. Lett.* **101**, 111301 (2008) [arXiv:0806.0989 [nucl-ex]].
[35] L. Wolfenstein, *Phys. Rev. D* **17**, 2369 (1978).
[36] V. D. Barger, K. Whisnant, S. Pakvasa, and R. J. Phillips, *Phys. Rev. D* **22**, 2718 (1980).
[37] S. P. Mikheyev and A. Yu. Smirnov, *Yad. Fiz.* **42**, 1441 (1985); *Nuovo Cim.* **9** C, 17 (1986).
[38] H. A. Bethe, *Phys. Rev. Lett.* **56**, 1305 (1986).
[39] V. D. Barger, R.J.N. Phillips, and K. Whisnant, *Phys. Rev. D* **34**, 980 (1986).
[40] S. J. Parke, *Phys. Rev. Lett.* **57**, 1275 (1986); S. J. Parke and T. P. Walker, *Phys. Rev. Lett.* **57**, 2322 (1986) [Erratum-ibid. **57**, 3124 (1986)].
[41] S. P. Rosen and J. M. Gelb, *Phys. Rev. D* **34**, 969 (1986).
[42] W. C. Haxton, *Phys. Rev. Lett.* **57**, 1271 (1986); V. D. Barger, R. J. Phillips, and K. Whisnant, *Phys. Rev. D* **34**, 980 (1986).
[43] S. A. Bludman, N. Hata, D. C. Kennedy, and P. G. Langacker, *Phys. Rev. D* **47**, 2220 (1993) [arXiv:hep-ph/9207213]; N. Hata and P. Langacker, *Phys. Rev. D* **50**, 632 (1994) [arXiv:hep-ph/9311214]; G. L. Fogli, E. Lisi, and D. Montanino, *Phys. Rev. D* **49**, 3626 (1994); *Phys. Rev. D* **54**, 2048 (1996) [arXiv:hep-ph/9605273]; J. N. Bahcall, P. I. Krastev, and A. Yu. Smirnov, *Phys. Rev. D* **58**, 096016 (1998) [arXiv:hep-ph/9807216].
[44] P. I. Krastev and S. T. Petcov, *Phys. Lett. B* **299**, 99 (1993); A. J. Baltz and J. Weneser, *Phys. Rev. D* **50**, 5971 (1994) [Addendum-ibid. D **51**, 3960 (1995)]; J. N. Bahcall and P. I. Krastev, *Phys. Rev. C* **56**, 2839 (1997) [arXiv:hep-ph/9706239].
[45] A. Friedland, *Phys. Rev. Lett.* **85**, 936 (2000) [arXiv:hep-ph/0002063]; A. de Gouvea, A. Friedland, and H. Murayama, *Phys. Lett. B* **490**, 125 (2000) [arXiv:hep-ph/0002064]; G. L. Fogli, E. Lisi, D. Montanino, and A. Palazzo, *Phys. Rev. D* **62**, 113004 (2000) [arXiv:hep-ph/0005261].
[46] V. D. Barger, K. Whisnant, and R. J. Phillips, *Phys. Rev. D* **24**, 538 (1981); S. L. Glashow and L. M. Krauss, *Phys. Lett. B* **190**, 199 (1987).
[47] Y. Fukuda et al. [Super-Kamiokande Collaboration], *Phys. Rev. Lett.* **82**, 1810 (1999) [arXiv:hep-ex/9812009]; J. Hosaka et al. [Super-Kamkiokande Collaboration], *Phys. Rev. D* **73**, 112001 (2006) [arXiv:hep-ex/0508053].
[48] B. Aharmim et al. [SNO Collaboration], *Phys. Rev. C* **72**, 055502 (2005) [arXiv:nucl-ex/0502021]; S. N. Ahmed et al. [SNO Collaboration], *Phys. Rev. Lett.* **92**, 181301 (2004) [arXiv:nucl-ex/0309004].
[49] G. Alimonti [Borexino Collaboration], *Nucl. Instrum. Meth. A* **600**, 568 (2009) [arXiv:0806.2400 [physics.ins-det]]; *Astropart. Phys.* **16**, 205 (2002) [arXiv:hep-ex/0012030].
[50] C. Arpesella et al. [The Borexino Collaboration], *Phys. Lett. B* **658**, 101 (2008) [arXiv:0708.2251 [astro-ph]].

[51] C. Arpesella et al. [The Borexino Collaboration], *Phys. Rev. Lett.* **101**, 091302 (2008) [arXiv:0805.3843 [astro-ph]].

[52] G. Bellini et al. [The Borexino Collaboration], arXiv:1104.1816 [hep-ex].

[53] G. Bellini et al. [The Borexino Collaboration], *Phys. Rev. D* **82**, 033006 (2010) [arXiv:0808.2868 [astro-ph]].

[54] A. Piepke [KamLAND Collaboration], *Nucl. Phys. Proc. Suppl.* **91**, 99 (2001).

[55] K. Eguchi et al. [KamLAND Collaboration], *Phys. Rev. Lett.* **90**, 021802 (2003) [arXiv:hep-ex/0212021].

[56] T. Araki et al. [KamLAND Collaboration], *Phys. Rev. Lett.* **94**, 081801 (2005) [arXiv:hep-ex/0406035].

[57] S. Abe et al. [KamLAND Collaboration], *Phys. Rev. Lett.* **100**, 221803 (2008) [arXiv:0801.4589 [hep-ex]].

[58] K. Hirata et al. [KAMIOKANDE-II Collaboration], *Phys. Rev. Lett.* **58**, 1490 (1987).

[59] C. B. Bratton et al. [IMB Collaboration], *Phys. Rev. D* **37**, 3361 (1988).

[60] K. S. Hirata et al. [KAMIOKANDE-II Collaboration], *Phys. Lett. B* **205**, 416 (1988); *Phys. Lett. B* **280**, 146 (1992).

[61] D. Casper et al. *Phys. Rev. Lett.* **66**, 2561 (1991); R. Becker-Szendy et al. *Phys. Rev. Lett.* **69**, 1010 (1992).

[62] J. G. Learned, S. Pakvasa, and T. J. Weiler, *Phys. Lett. B* **207**, 79 (1988); V. D. Barger and K. Whisnant, *Phys. Lett. B* **209**, 365 (1988); K. Hidaka, M. Honda, and S. Midorikawa, *Phys. Rev. Lett.* **61**, 1537 (1988).

[63] Y. Fukuda et al. [Super-Kamiokande Collaboration], *Phys. Lett. B* **433**, 9 (1998) [arXiv:hep-ex/9803006]; *Phys. Lett. B* **436**, 33 (1998) [arXiv:hep-ex/9805006]; *Phys. Rev. Lett.* **81**, 1562 (1998) [arXiv:hep-ex/9807003]; *Phys. Rev. Lett.* **82**, 2644 (1999) [arXiv:hep-ex/9812014]; *Phys. Lett. B* **467**, 185 (1999) [arXiv:hep-ex/9908049].

[64] V. D. Barger, J. G. Learned, S. Pakvasa, and T. J. Weiler, *Phys. Rev. Lett.* **82**, 2640 (1999) [arXiv:astro-ph/9810121].

[65] V. D. Barger, J. G. Learned, P. Lipari, M. Lusignoli, S. Pakvasa, and T. J. Weiler, *Phys. Lett. B* **462**, 109 (1999) [arXiv:hep-ph/9907421].

[66] E. Lisi, A. Marrone, and D. Montanino, *Phys. Rev. Lett.* **85**, 1166 (2000) [arXiv:hep-ph/0002053].

[67] G. L. Fogli, E. Lisi, A. Marrone, and D. Montanino, *Phys. Rev. D* **67**, 093006 (2003) [arXiv:hep-ph/0303064].

[68] M. C. Sanchez et al. [Soudan 2 Collaboration], *Phys. Rev. D* **68**, 113004 (2003) [arXiv:hep-ex/0307069].

[69] M. Ambrosio et al. [MACRO Collaboration], *Eur. Phys. J. C* **36**, 323 (2004).

[70] M. H. Ahn et al. [K2K Collaboration], *Phys. Rev. D* **74**, 072003 (2006) [arXiv:hep-ex/0606032].

[71] MINOS Collaboration, Fermilab Report No. NuMI-L-375 (1998).

[72] M. Apollonio et al. [CHOOZ Collaboration], *Phys. Lett. B* **420**, 397 (1998) [arXiv:hep-ex/9711002]; *Phys. Lett. B* **466**, 415 (1999) [arXiv:hep-ex/9907037]; *Eur. Phys. J. C* **27**, 331 (2003) [arXiv:hep-ex/0301017].

[73] F. Boehm et al., *Phys. Rev. Lett.* **84**, 3764 (2000) [arXiv:hep-ex/9912050]; *Phys. Rev. D* **62**, 072002 (2000) [arXiv:hep-ex/0003022]; *Phys. Rev. D* **64**, 112001 (2001) [arXiv:hep-ex/0107009].

[74] R. Acquafredda et al. [OPERA Collaboration], *New J. Phys.* **8**, 303 (2006) [arXiv:hep-ex/0611023].

[75] N. Agafonova et al. [OPERA Collaboration], *Phys. Lett. B* **691**, 138 (2010) [arXiv:1006.1623 [hep-ex]].

[76] B. W. Lee, S. Pakvasa, R. E. Shrock, and H. Sugawara, *Phys. Rev. Lett.* **38**, 937 (1977) [Erratum-ibid. **38**, 1230 (1977)]; N. Cabibbo, *Phys. Lett. B* **72**, 333 (1978).

[77] V. D. Barger, K. Whisnant, and R. J. Phillips, Phys. Rev. D **22**, 1636 (1980).

[78] V. D. Barger, K. Whisnant, and R. J. Phillips, *Phys. Rev. Lett.* **45**, 2084 (1980); S. Pakvasa, in *High Energy Physics - 1980, AIP Conf. Proc.* No. 68, ed. by L. Durand and L. G. Pondrom (AIP, New York, 1981).
[79] K. Nakamura et al. [Particle Data Group], *J. Phys. G* **37**, 075021 (2010).
[80] F. Ardellier et al. [Double Chooz Collaboration], arXiv:hep-ex/0606025; arXiv:hep-ex/0405032; S. Berridge et al., arXiv:hep-ex/0410081; Y. Abe et al., arXiv: 1112.6353v1 [hep-ex].
[81] X. Guo et al. [Daya Bay Collaboration], arXiv:hep-ex/0701029.
[82] J. K. Ahn et al. [RENO Collaboration], arXiv:1003.1391 [hep-ex].
[83] Y. Itow et al. [The T2K Collaboration], arXiv:hep-ex/0106019; T. Le [T2K Collaboration], arXiv:0910.4211 [hep-ex].
[84] D. S. Ayres et al. [NOνA Collaboration], arXiv:hep-ex/0503053.
[85] V. Barger, D. Marfatia, and K. Whisnant, *Phys. Rev. D* **65**, 073023 (2002) [arXiv:hep-ph/0112119].
[86] E. Majorana, *Nuovo Cim.* **14**, 171 (1937).
[87] L. Wolfenstein, *Phys. Lett. B* **107**, 77 (1981).
[88] S. M. Bilenky, J. Hosek, and S. T. Petcov, *Phys. Lett. B* **94**, 495 (1980); M. Doi, T. Kotani, H. Nishiura, K. Okuda, and E. Takasugi, *Phys. Lett. B* **102**, 323 (1981).
[89] D. Beavis et al., E889 Collaboration, BNL preprint BNL-52459, April 1995.
[90] A. Para and M. Szleper, arXiv:hep-ex/0110032.
[91] B. Richter, arXiv:hep-ph/0008222.
[92] V. D. Barger, S. Geer, R. Raja, and K. Whisnant, *Phys. Rev. D* **63**, 113011 (2001) [arXiv:hep-ph/0012017].
[93] D. Beavis et al., arXiv:hep-ex/0205040; M. Diwan et al., arXiv:hep-ex/0211001; *Phys. Rev. D* **68**, 012002 (2003) [arXiv:hep-ph/0303081].
[94] V. Barger, M. Dierckxsens, M. Diwan, P. Huber, C. Lewis, D. Marfatia, and B. Viren, *Phys. Rev. D* **74**, 073004 (2006) [arXiv:hep-ph/0607177].
[95] M. C. Sanchez [LBNE DUSEL Collaboration], *AIP Conf. Proc.* **1222**, 479 (2010).
[96] C. K. Jung, *AIP Conf. Proc.* **533**, 29 (2000) [arXiv:hep-ex/0005046].
[97] D. B. Cline, F. Sergiampietri, J. G. Learned, and K. McDonald, *Nucl. Instrum. Meth. A* **503**, 136 (2003) [arXiv:astro-ph/0105442].
[98] P. Zucchelli, *Phys. Lett. B* **532**, 166 (2002).
[99] S. Geer, *Phys. Rev. D* **57**, 6989 (1998) [Erratum-ibid. D **59**, 039903 (1999)] [arXiv:hep-ph/9712290].
[100] B. Autin et al., CERN-SPSC-98-30; A. De Rujula, M. B. Gavela, and P. Hernandez, *Nucl. Phys. B* **547**, 21 (1999) [arXiv:hep-ph/9811390]; K. Dick, M. Freund, M. Lindner, and A. Romanino, *Nucl. Phys. B* **562**, 29 (1999) [arXiv:hep-ph/9903308]; M. Campanelli, A. Bueno, and A. Rubbia, arXiv:hep-ph/9905240; V. D. Barger, S. Geer, and K. Whisnant, *Phys. Rev. D* **61**, 053004 (2000) [arXiv:hep-ph/9906487].
[101] W. Hu, D. J. Eisenstein, and M. Tegmark, *Phys. Rev. Lett.* **80**, 5255 (1998) [arXiv:astro-ph/9712057].
[102] E. Komatsu et al. [WMAP Collaboration], *Astrophys. J. Suppl.* **192**, 18 (2011) [arXiv:1001.4538 [astro-ph.CO]].
[103] P. F. Harrison, D. H. Perkins, and W. G. Scott, *Phys. Lett. B* **530**, 167 (2002) [arXiv:hep-ph/0202074].
[104] E. Ma and G. Rajasekaran, *Phys. Rev. D* **64**, 113012 (2001) [arXiv:hep-ph/0106291].
[105] See, e.g., C. H. Albright and S. Geer, *Phys. Rev. D* **65**, 073004 (2002) [arXiv:hep-ph/0108070], and references therein.
[106] P. Minkowski, *Phys. Lett. B* **67** (1977) 421; T. Yanagida, in *Proceedings of the Workshop on the Unified Theory and the Baryon Number in the Universe*, eds. O. Sawada et al., (KEK Report 79-18, Tsukuba, 1979), p. 95; M. Gell-Mann, P. Ramond, and R. Slansky, in *Supergravity*, eds. P. van Nieuwenhuizen et al.,

(North-Holland, 1979), p. 315; S. L. Glashow, in *Quarks and Leptons*, Cargèse, eds. M. Lévy et al., (Plenum, 1980), p. 707; R. N. Mohapatra and G. Senjanović, *Phys. Rev. Lett.* **44** (1980) 912.

[107] R. Fardon, A. E. Nelson, and N. Weiner, *JCAP* **0410**, 005 (2004) [arXiv:astro-ph/0309800]; D. B. Kaplan, A. E. Nelson, and N. Weiner, *Phys. Rev. Lett.* **93**, 091801 (2004) [arXiv:hep-ph/0401099].

[108] D. Colladay and V. A. Kostelecky, *Phys. Rev. D* **55**, 6760 (1997) [arXiv:hep-ph/9703464].

[109] S. R. Coleman and S. L. Glashow, *Phys. Rev. D* **59**, 116008 (1999) [arXiv:hep-ph/9812418].

[110] V. D. Barger, B. Kayser, J. Learned, T. J. Weiler, and K. Whisnant, *Phys. Lett. B* **489**, 345 (2000) [arXiv:hep-ph/0008019].

[111] A. Kusenko and G. Segre, *Phys. Lett. B* **396**, 197 (1997) [arXiv:hep-ph/9701311]; A. Kusenko, *Phys. Rept.* **481**, 1 (2009) [arXiv:0906.2968 [hep-ph]].

[112] A. Achterberg et al. [IceCube Collaboration], *Astropart. Phys.* **26**, 155 (2006) [arXiv:astro-ph/0604450]; [IceCube Collaboration], arXiv:0711.0353 [astro-ph]; J. Ahrens et al. [IceCube Collaboration], *Astropart. Phys.* **20**, 507 (2004) [arXiv:astro-ph/0305196]; F. Halzen and S. Klein, *Rev. Sci. Instrum.* **81**, 081101 (2010) [arXiv:1007. 1247 [astro-ph: HE]].

[113] E. Aslanides et al. [ANTARES Collaboration], arXiv:astro-ph/9907432.

[114] F. Halzen, *Science* **315**, 66 (2007).

[115] T. J. Weiler, W. A. Simmons, S. Pakvasa, and J. G. Learned, arXiv:hep-ph/9411432.

[116] G. Arnison et al. [UA1 Collaboration], *Phys. Lett. B* **166**, 484 (1986); J. A. Appel et al. [UA2 Collaboration], *Z. Phys. C* **30**, 1 (1986).

[117] W. L. van Neerven, J.A.M. Vermaseren K.J.F. Gaemers, V. D. Barger, A. D. Martin, and R.J.N. Phillips, *Z. Phys. C* **21**, 99 (1983); J. Smith, W. L. van Neerven, and J.A.M. Vermaseren, *Phys. Rev. Lett.* **50**, 1738 (1983).

[118] V. Barger and R.J.N. Phillips, *Collider Physics*, Updated Edition, Westview Press (1996).

[119] [ALEPH and DELPHI and L3 and OPAL and SLD and LEP Electroweak Working Group and SLD Electroweak Group and SLD Heavy Flavour Group Collaborations], Phys. Rept. **427**, 257–454 (2006) [arXiv:hep-ex/0509008].

[120] G. Steigman, D. N. Schramm, and J. E. Gunn, *Phys. Lett.* **66B**, 202 (1977).

[121] For a review of BBN see G. Steigman, arXiv:astro-ph/0307244.

[122] Y. I. Izotov and T. X. Thuan, *Astrophys. J.* **710**, L67 (2010) [arXiv:1001.4440 [astro-ph.CO]].

[123] G. Steigman, arXiv:1008.4765 [astro-ph.CO]; J. Hamann, S. Hannestad, G. G. Raffelt, I. Tamborra, and Y.Y.Y. Wong, *Phys. Rev. Lett.* **105**, 181301 (2010) [arXiv:1006.5276 [hep-ph]].

[124] E. Aver, K. A. Olive, and E. D. Skillman, *JCAP* **1005**, 003 (2010) [arXiv:1001.5218 [astro-ph.CO]].

[125] E. Aver, K. A. Olive, and E. D. Skillman, arXiv:1012.2385 [astro-ph.CO].

[126] G. Steigman, *Ann. Rev. Nucl. Part. Sci.* **57**, 463 (2007) [arXiv:0712.1100 [astro-ph]].

[127] See e.g., A. Ringwald, *Nucl. Phys. A* **827**, 501C (2009) [arXiv:0901.1529 [astro-ph]].

[128] B. F. Shvartsman, V. B. Braginsky, S. S. Gershtein, Y. B. Zeldovich, and M. Y. Khlopov, *JETP Lett.* **36**, 277 (1982) [*Pisma Zh. Eksp. Teor. Fiz.* **36**, 224 (1982)]; P. F. Smith and J. D. Lewin, *Phys. Lett. B* **127**, 185 (1983).

[129] J. M. Irvine and R. Humphreys, *J. Phys. G* **9**, 847 (1983); A. G. Cocco, G. Mangano, and M. Messina, *JCAP* **0706**, 015 (2007) [*J. Phys. Conf. Ser.* **110**, 082014 (2008)] [arXiv:hep-ph/0703075]; R. Lazauskas, P. Vogel, and C. Volpe, *J. Phys. G* **35**, 025001 (2008) [arXiv:0710.5312 [astro-ph]]; M. Blennow, *Phys. Rev. D* **77**, 113014 (2008) [arXiv:0803.3762 [astro-ph]].

[130] T. J. Weiler, *Phys. Rev. Lett.* **49**, 234 (1982); *Astrophys. J.* **285**, 495 (1984); *Astropart. Phys.* **11**, 303 (1999) [arXiv:hep-ph/9710431]; E. Roulet, *Phys. Rev. D* **47**, 5247 (1993); S. Yoshida, H. Y. Dai, C. C. Jui, and P. Sommers, *Astrophys. J.* **479**, 547 (1997) [arXiv:astro-ph/9608186]; D. Fargion, B. Mele, and A. Salis, *Astrophys. J.* **517**, 725 (1999) [arXiv:astro-ph/9710029].

[131] B. Eberle, A. Ringwald, L. Song, and T. J. Weiler, *Phys. Rev. D* **70**, 023007 (2004) [arXiv:hep-ph/0401203]; G. Barenboim, O. M. Requejo, and C. Quigg, *Phys. Rev. D* **71**, 083002 (2005) [arXiv:hep-ph/0412122]; J. C. D'Olivo, L. Nellen, S. Sahu, and V. Van Elewyck, *Astropart. Phys.* **25**, 47 (2006) [arXiv:astro-ph/0507333]; A. Ringwald and L. Schrempp, *JCAP* **0610**, 012 (2006) [arXiv:astro-ph/0606316].

[132] Yu. Efremenko and W. R. Hix, *J. Phys. Conf. Ser.* **173**, 012006 (2009) [arXiv:0807.2801 [nucl-ex]].

[133] S. I. Dutta, M. H. Reno, and I. Sarcevic, *Phys. Rev. D* **62**, 123001 (2000) [arXiv:hep-ph/0005310].

[134] P. Lipari, *Astropart. Phys.* **1**, 195 (1993).

[135] V. Barger, W. Y. Keung, G. Shaughnessy, and A. Tregre, *Phys. Rev. D* **76**, 095008 (2007) [arXiv:0708.1325 [hep-ph]].

[136] K. S. Krane, *Introductory Nuclear Physics*, John Wiley and Sons (2006).

[137] E. J. Konopinski, *Rev. Mod. Phys.* **15**, 209 (1943).

[138] W. J. Marciano and Z. Parsa, *J. Phys. G* **29**, 2629 (2003) [arXiv:hep-ph/0403168].

[139] V. D. Barger and D. V. Nanopoulos, *Nucl. Phys. B* **124**, 426 (1977).

[140] For a recent treatment, including the effects of a nonzero charged lepton mass, see E. A. Paschos and J. Y. Yu, *Phys. Rev. D* **65**, 033002 (2002) [arXiv:hep-ph/0107261].

[141] A. A. Aguilar-Arevalo et al. [MiniBooNE Collaboration], *Phys. Rev. D* **81**, 092005 (2010) [arXiv:1002.2680 [hep-ex]].

[142] J. Nieves, I. R. Simo, and M.J.V. Vacas, arXiv:1106.5374 [hep-ph].

[143] V. Lyubushkin et al. [NOMAD Collaboration], *Eur. Phys. J. C* **63**, 355 (2009) [arXiv:0812.4543 [hep-ex]].

[144] E. A. Paschos, J. Y. Yu, and M. Sakuda, *Phys. Rev. D* **69**, 014013 (2004) [arXiv:hep-ph/0308130].

[145] V. D. Barger, S. Geer, and K. Whisnant, *Phys. Rev. D* **61**, 053004 (2000) [arXiv:hep-ph/9906487].

[146] A. Strumia and F. Vissani, arXiv:hep-ph/0606054.

[147] For a comprehensive discussion of the tau neutrino cross section, see Y. S. Jeong and M. H. Reno, arXiv:1007.1966 [hep-ph].

[148] J. Edsjo, arXiv:hep-ph/9704384.

[149] P. Adamson et al. [MINOS Collaboration], *Phys. Rev. D* **81**, 072002 (2010) [arXiv:0910.2201 [hep-ex]].

[150] K. Kodama et al. [DONUT Collaboration], *Phys. Rev. D* **78**, 052002 (2008) [arXiv:0711.0728 [hep-ex]].

[151] A. A. Aguilar-Arevalo et al. [SciBooNE Collaboration], arXiv:hep-ex/0601022.

[152] D. Drakoulakos et al. [MINERνA Collaboration], arXiv:hep-ex/0405002.

[153] M. Soderberg [ArgoNeuT Collaboration], arXiv:0910.3433 [physics.ins-det].

[154] L. Bugel et al. [FINeSSE Collaboration], arXiv:hep-ex/0402007.

[155] H. Chen et al. [MicroBooNE Collaboration], "Proposal for a New Experiment Using the Booster and NUMI Neutrino Beamlines: MicroBooNE," http://www-microboone.fnal.gov/Documents.html.

[156] A. A. Aguilar-Arevalo et al. [MiniBooNE Collaboration], *Nucl. Instrum. Meth. A* **599**, 28 (2009) [arXiv:0806.4201 [hep-ex]]; *Phys. Rev. D* **79**, 072002 (2009) [arXiv:0806.1449 [hep-ex]].

[157] A. A. Aguilar-Arevalo et al. [MiniBooNE Collaboration], *Phys. Rev. Lett.* **103**, 081801 (2009) [arXiv:0904.3159 [hep-ex]]; *Phys. Rev. D* **81**, 013005 (2010)

[arXiv:0911.2063 [hep-ex]]; *Phys. Rev. D* **82**, 092005 (2010) [arXiv:1007.4730 [hep-ex]]; *Phys. Rev. D* **83**, 052007 (2011) [arXiv:1011.3572 [hep-ex]]; *Phys. Rev. D* **83**, 052009 (2011) [arXiv:1010.3264 [hep-ex]].

[158] S. H. Ahn et al. [K2K Collaboration], *Phys. Lett. B* **511**, 178 (2001) [arXiv:hep-ex/0103001]; M. H. Ahn et al. [K2K Collaboration], *Phys. Rev. Lett.* **90**, 041801 (2003) [arXiv:hep-ex/0212007]; *Phys. Rev. Lett.* **93**, 051801 (2004) [arXiv:hep-ex/0402017]; E. Aliu et al. [K2K Collaboration], *Phys. Rev. Lett.* **94**, 081802 (2005) [arXiv:hep-ex/0411038].

[159] P. Lipari, M. Lusignoli, and F. Sartogo, *Phys. Rev. Lett.* **74**, 4384 (1995) [arXiv:hep-ph/9411341].

[160] J. N. Bahcall, E. Lisi, D. E. Alburger, L. De Braeckeleer, S. J. Freedman, and J. Napolitano, *Phys. Rev. C* **54**, 411 (1996) [arXiv:nucl-th/9601044].

[161] J. N. Bahcall and R. K. Ulrich, *Rev. Mod. Phys.* **60**, 297 (1988).

[162] J. N. Bahcall, *Phys. Rev. C* **56**, 3391 (1997) [arXiv:hep-ph/9710491].

[163] Y. Ashie et al. [Super-Kamiokande Collaboration], *Phys. Rev. D* **71**, 112005 (2005) [arXiv:hep-ex/0501064].

[164] V. Barger et al., arXiv:0705.4396 [hep-ph].

[165] K. Nakamura, *Int. J. Mod. Phys. A* **18**, 4053 (2003).

[166] A. de Bellefon et al., arXiv:hep-ex/0607026.

[167] C. Alberini et al. [LVD Collaboration], *Nuovo Cim.* C **9**, 237 (1986); C. Bari et al. [LVD Collaboration], *Nucl. Instrum. Meth. A* **264**, 5 (1988); G. Bari et al. [LVD Collaboration], *Nucl. Instrum. Meth. A* **277**, 11 (1989).

[168] J. G. Learned, S. T. Dye, and S. Pakvasa, arXiv:0810.4975 [hep-ex].

[169] B. D. Fields and K. A. Hochmuth, *Earth, Moon, Planets* **99**, 155 (2006) [arXiv:hep-ph/0406001]; T. Marrodan Undagoitia, F. von Feilitzsch, M. Goder-Neff, K. A. Hochmuth, L. Oberauer, W. Potzel, and M. Wurm, *Prog. Part. Nucl. Phys.* **57**, 283 (2006) [*J. Phys. Conf. Ser.* **39**, 278 (2006)] [arXiv:hep-ph/0605229].

[170] A. Rubbia for the ICARUS Collaboration, talk at *Skandinavian NeutrinO Workshop (SNOW)*, Uppsala, Sweden (February 2001), *Phys. Scripta T* **93**, 70 (2001).

[171] B. Baibussinov et al., *Astropart. Phys.* **29**, 174 (2008) [arXiv:0704.1422 [hep-ph]].

[172] D. B. Cline, F. Raffaelli, and F. Sergiampietri, *JINST* **1**, T09001 (2006) [arXiv:astro-ph/0604548].

[173] S. T. Dye et al., "A White Paper for Large Liquid Scintillation Detectors at DUSEL," http://www.phys.hawaii.edu/~jgl/post/A White Paper for Large Liquid Scintillation Detectors at DUSEL.pdf.

[174] D. Indumathi [INO Collaboration], *Pramana* **63**, 1283 (2004); M. S. Athar et al. [INO Collaboration], http://www.imsc.res.in/~ino/OpenReports/INOReport.pdf.

[175] P. Vahle, private communication.

[176] S. Geer, *Ann. Rev. Nucl. Part. Sci.* **59**, 347 (2009).

[177] J. Bernabeu, J. Burguet-Castell, C. Espinoza, and M. Lindroos, *JHEP* **0512**, 014 (2005) [arXiv:hep-ph/0505054].

[178] C. Giunti, *AIP Conf. Proc.* **1026**, 3–19 (2008) [arXiv:0801.0653 [hep-ph]].

[179] S. M. Bilenky, M. D. Mateev, *Phys. Part. Nucl.* **38**, 117–128 (2007) [arXiv:hep-ph/0604044].

[180] Y. Grossmann and H. J. Lipkin, *Phys. Rev. D* **55**, 2760–2767 (1997) [arXiv:hep-ph/9607201]; L. Stodolsky, *Phys. Rev. D* **58**, 036006 (1998) [arXiv:hep-ph/9802387]; H. J. Lipkin, [arXiv:hep-ph/0212093]; H. J. Lipkin, [arXiv:hep-ph/0312292].

[181] B. Kayser, J. Kopp, R.G.H. Robertson, and P. Vogel, *Phys. Rev. D* **82**, 093003 (2010) [arXiv:1006.2372 [hep-ph]].

[182] "Progress in the Physics of Massive Neutrinos", V. Barger, D. Marfatia, and K. Whisnant, *Int. J. Mod. Phys.*, Vol. **12**, No. 5, pp. 569–647, ©2003 World Scientific.

[183] L. L. Chau and W. Y. Keung, *Phys. Rev. Lett.* **53**, 1802 (1984); C. Jarlskog, *Z. Phys. C* **29**, 491 (1985); *Phys Rev. D* **35**, 1685 (1987).

[184] P. Langacker, J. P. Leveille, and J. Sheiman, *Phys. Rev. D* **27**, 1228 (1983); For an early review on neutrino oscillations in matter, see T. K. Kuo and J. Pantaleone, *Rev. Mod. Phys.* **61**, 937 (1989).

[185] J. N. Bahcall, M. H. Pinsonneault, and S. Basu, *Astrophys. J.* **555**, 990 (2001) [arXiv:astro-ph/0010346].

[186] S. Toshev, *Phys. Lett. B* **196**, 170 (1987); S. T. Petcov, *Phys. Lett. B* **200**, 373 (1988).

[187] L. D. Landau, *Physik Z. Sowjetunion* **2**, 46 (1932); C. Zener, *Proc. Roy. Soc. A* **137**, 696 (1932); E.C.G. Stueckelberg, *Helv. Phys. Acta* **5**, 369 (1932).

[188] W. C. Haxton, *Phys. Rev. D* **35**, 2352 (1987).

[189] P. I. Krastev and S. T. Petcov, *Phys. Lett. B*207, 64 (1988) [Erratum-ibid. B **214**, 661 (1988)]; M. Bruggen, W. C. Haxton and Y. Z. Qian, *Phys. Rev. D* **51**, 4028 (1995); E. Torrente Lujan, *Phys. Rev. D* **53**, 4030 (1996) [arXiv:hep-ph/9505209]; S. T. Petcov, *Phys. Lett. B* **406**, 355 (1997) [arXiv:hep-ph/9910335].

[190] A. B. Balantekin and J. F. Beacom, *Phys. Rev. D* **54**, 6323 (1996) [arXiv:hep-ph/9606353].

[191] T. K. Kuo and J. Pantaleone, *Phys. Rev. Lett.* **57**, 1805 (1986); *Phys. Rev. D* **35**, 3432 (1987).

[192] J. Burguet-Castell, M. B. Gavela, J. J. Gomez-Cadenas, P. Hernandez, and O. Mena, *Nucl. Phys. B* **608**, 301 (2001) [arXiv:hep-ph/0103258].

[193] M. Freund, *Phys. Rev. D* **64**, 053003 (2001) [arXiv:hep-ph/0103300].

[194] M. Koike and J. Sato, *Phys. Rev. D* **61**, 073012 (2000); [Erratum-ibid. D **62**, 079903 (2000)] [arXiv:hep-ph/9909469].

[195] P. F. Harrison and W. G. Scott, *Phys. Lett. B* **535**, 229 (2002) [arXiv:hep-ph/0203021].

[196] H. W. Zaglauer and K. H. Schwarzer, *Z. Phys. C* **40**, 273 (1988); K. Kimura, A. Takamura, and H. Yokomakura, *Phys. Lett. B* **537**, 86 (2002) [arXiv:hep-ph/0203099]; *Phys. Rev. D* **66**, 073005 (2002) [arXiv:hep-ph/0205295]; P. F. Harrison, W. G. Scott, and T. J. Weiler, *Phys. Lett. B* **565**, 159 (2003) [arXiv:hep-ph/0305175]; O. Yasuda, arXiv:0704.1531 [hep-ph].

[197] M. Freund and T. Ohlsson, *Mod. Phys. Lett. A* **15**, 867 (2000) [arXiv:hep-ph/9909501]; T. Ota and J. Sato, *Phys. Rev. D* **63**, 093004 (2001) [arXiv:hep-ph/0011234]; P. M. Fishbane and P. Kaus, *J. Phys. G* **27**, 2405 (2001) [arXiv:hep-ph/0101013]; T. Ohlsson and H. Snellman, *Eur. Phys. J. C* **20**, 507 (2001) [arXiv:hep-ph/0103252]; T. Ohlsson, *Phys. Lett. B* **522**, 280 (2001) [arXiv:hep-ph/0109003]; L. Y. Shan, B. L. Young, and X. M. Zhang, *Phys. Rev. D* **66**, 053012 (2002) [arXiv:hep-ph/0110414]; G. L. Fogli, G. Lettera, and E. Lisi, arXiv:hep-ph/0112241; H. Yokomakura, K. Kimura, and A. Takamura, *Phys. Lett. B* **544**, 286 (2002) [arXiv:hep-ph/0207174]; T. Ota and J. Sato, *Phys. Rev. D* **67**, 053003 (2003) [arXiv:hep-ph/0211095]; L. Y. Shan, Y. F. Wang, C. G. Yang, X. M. Zhang, F. T. Liu, and B. L. Young, *Phys. Rev. D* **68**, 013002 (2003) [arXiv:hep-ph/0303112]; E. Kozlovskaya, J. Peltoniemi, and J. Sarkamo, arXiv:hep-ph/0305042.

[198] F. N. Loreti and A. B. Balantekin, *Phys. Rev. D* **50**, 4762 (1994) [arXiv:nucl-th/9406003]; F. N. Loreti, Y. Z. Qian, G. M. Fuller, and A. B. Balantekin, *Phys. Rev. D* **52**, 6664 (1995) [arXiv:astro-ph/9508106]; A. B. Balantekin, J. M. Fetter, and F. N. Loreti, *Phys. Rev. D* **54**, 3941 (1996) [arXiv:astro-ph/9604061]; B. Jacobsson, T. Ohlsson, H. Snellman, and W. Winter, *Phys. Lett. B* **532**, 259 (2002) [arXiv:hep-ph/0112138].

[199] V. D. Barger, N. Deshpande, P. B. Pal, R. J. Phillips, and K. Whisnant, *Phys. Rev. D* **43**, 1759 (1991).

[200] See, e.g., S. Nussinov, *Phys. Lett. B* **63**, 201 (1976); B. Kayser, *Phys. Rev. D* **24**, 110 (1981); C. Giunti, C. W. Kim, J. A. Lee, and U. W. Lee, *Phys. Rev. D* **48**, 4310 (1993) [arXiv:hep-ph/9305276]; K. Kiers, S. Nussinov, and N. Weiss, *Phys.*

Rev. D **53**, 537 (1996) [arXiv:hep-ph/9506271]; W. Grimus and P. Stockinger, *Phys. Rev. D* **54**, 3414 (1996) [arXiv:hep-ph/9603430]; W. Grimus, P. Stockinger and S. Mohanty, *Phys. Rev. D* **59**, 013011 (1999) [arXiv:hep-ph/9807442]; M. Beuthe, *Phys. Rept.* **375**, 105 (2003) [arXiv:hep-ph/0109119]; C. Giunti, *JHEP* **0211**, 017 (2002) [arXiv:hep-ph/0205014]; F. R. Torres and M. M. Guzzo, *Braz. J. Phys.* **37**, 1273 (2007); E. K. Akhmedov and A. Y. Smirnov, *Phys. Atom. Nucl.* **72**, 1363 (2009) [arXiv:0905.1903 [hep-ph]]; E. K. Akhmedov and J. Kopp, *JHEP* **1004**, 008 (2010) [arXiv:1001.4815 [hep-ph]].

[201] C. Giunti, C. W. Kim, and U. W. Lee, *Phys. Rev. D* **44**, 3635 (1991); *Phys. Lett. B* **274**, 87 (1992); *Phys. Lett. B* **421**, 237 (1998) [arXiv:hep-ph/9709494]; M. Beuthe, *Phys. Rev. D* **66**, 013003 (2002) [arXiv:hep-ph/0202068].

[202] K. Kiers and N. Weiss, *Phys. Rev. D* **57**, 3091 (1998) [arXiv:hep-ph/9710289].

[203] B. Kayser and J. Kopp, arXiv:1005.4081 [hep-ph].

[204] For discussions, see J. T. Goldman, *Mod. Phys. Lett. A* **25**, 479 (2010) [arXiv:hep-ph/9604357]; M. Nauenberg, *Phys. Lett. B* **447**, 23 (1999) [Erratum-ibid. B **452**, 434 (1999)] [arXiv:hep-ph/9812441]; A. G. Cohen, S. L. Glashow, and Z. Ligeti, *Phys. Lett. B* **678**, 191 (2009) [arXiv:0810.4602 [hep-ph]]; R.G.H. Robertson, arXiv:1004.1847 [hep-ph]; D. V. Ahluwalia and S. P. Horvath, arXiv:1006.1710 [hep-ph]; B. Kayser, J. Kopp, R.G.H. Robertson, and P. Vogel, arXiv:1006.2372 [hep-ph].

[205] A. Einstein, B. Podolsky, and N. Rosen, *Phys. Rev. D* **47**, 777 (1935).

[206] E. K. Akhmedov and A. Y. Smirnov, arXiv:1008.2077 [hep-ph].

[207] See, e.g., J. N. Bahcall and R. K. Ulrich, *Rev. Mod. Phys.* **60**, 297 (1988).

[208] J. N. Bahcall, A. M. Serenelli, and S. Basu, *Astrophys. J.* **621**, L85 (2005) [arXiv:astro-ph/0412440]; J. N. Bahcall and A. M. Serenelli, *Astrophys. J.* **626**, 530 (2005) [arXiv:astro-ph/0412096]; J. N. Bahcall, M. H. Pinsonneault, and S. Basu, *Astrophys. J.* **555**, 990 (2001) [arXiv:astro-ph/0010346].

[209] C. Pena-Garay and A. Serenelli, arXiv:0811.2424 [astro-ph].

[210] J. P. Cravens et al. [Super-Kamiokande Collaboration], *Phys. Rev. D* **78**, 032002 (2008) [arXiv:0803.4312 [hep-ex]].

[211] B. T. Cleveland et al., *Astrophys. J.* **496**, 505 (1998).

[212] B. Aharmim et al. [SNO Collaboration], *Phys. Rev. C* **81**, 055504 (2010) [arXiv:0910.2984 [nucl-ex]].

[213] A. J. Baltz and J. Weneser, *Phys. Rev. D* **35**, 528 (1987); *Phys. Rev. D* **37**, 3364 (1988); J. N. Bahcall, P. I. Krastev, and A. Y. Smirnov, *Phys. Rev. D* **62**, 093004 (2000) [arXiv:hep-ph/0002293]; V. D. Barger, D. Marfatia, K. Whisnant, and B. P. Wood, *Phys. Rev. D* **64**, 073009 (2001) [arXiv:hep-ph/0104095]; M. Maris and S. T. Petcov, *Phys. Lett. B* **534**, 17 (2002) [arXiv:hep-ph/0201087].

[214] M. Ikeda, *Precise Measurement of Solar Neutrinos with Super-Kamiokande III*, doctoral thesis, 2009.

[215] A. M. Gago, M. M. Guzzo, P. C. de Holanda, H. Nunokawa, O. L. Peres, V. Pleitez, and R. Z. Funchal, *Phys. Rev. D* **65**, 073012 (2002) [arXiv:hep-ph/0112060]; S. Bergmann, M. M. Guzzo, P. C. de Holanda, P. I. Krastev, and H. Nunokawa, *Phys. Rev. D* **62**, 073001 (2000) [arXiv:hep-ph/0004049].

[216] See, e.g., the appendix of M. Guzzo, P. C. de Holanda, M. Maltoni, H. Nunokawa, M. A. Tortola, and J.W.F. Valle, *Nucl. Phys. B* **629**, 479 (2002) [arXiv:hep-ph/0112310].

[217] J. Barranco, O. G. Miranda, T. I. Rashba, V. B. Semikoz, and J. W. Valle, *Phys. Rev. D* **66**, 093009 (2002) [arXiv:hep-ph/0207326].

[218] A. B. Balantekin and C. Volpe, *Phys. Rev. D* **72**, 033008 (2005) [arXiv:hep-ph/0411148]; D. Yilmaz, arXiv:0810.1037 [hep-ph].

[219] M. Gasperini, *Phys. Rev. D* **38**, 2635 (1988); *Phys. Rev. D* **39**, 3606 (1989); A. Halprin and C. N. Leung, *Phys. Rev. Lett.* **67**, 1833 (1991); J. T. Pantaleone, A. Halprin, and C. N. Leung, *Phys. Rev. D* **47**, 4199 (1993) [arXiv:hep-ph/9211214]; J. N. Bahcall,

P. I. Krastev, and C. N. Leung, *Phys. Rev. D* **52**, 1770 (1995) [arXiv:hep-ph/9410353]; A. Halprin, C. N. Leung, and J. T. Pantaleone, *Phys. Rev. D* **53**, 5365 (1996) [arXiv:hep-ph/9512220]; S. L. Glashow, A. Halprin, P. I. Krastev, C. N. Leung, and J. T. Pantaleone, *Phys. Rev. D* **56**, 2433 (1997) [arXiv:hep-ph/9703454]; A. Halprin and C. N. Leung, *Phys. Lett. B* **416**, 361 (1998) [arXiv:hep-ph/9707407]; R. Foot, C. N. Leung, and O. Yasuda, *Phys. Lett. B* **443**, 185 (1998) [arXiv:hep-ph/9809458].

[220] G. Bellini et al. [The Borexino Collaboration], arXiv:1104.2150 [hep-ex].

[221] V. Barger, D. Marfatia, and K. Whisnant, *Phys. Lett. B* **617**, 78 (2005) [arXiv:hep-ph/0501247].

[222] L. Oberauer [Borexino Collaboration], *J. Phys. Conf. Ser.* **203**, 012081 (2010).

[223] M. C. Gonzalez-Garcia, M. Maltoni, and J. Salvado, *JHEP* **1005**, 072 (2010) [arXiv:0910.4584 [hep-ph]].

[224] J. N. Bahcall and C. Pena-Garay, *JHEP* **0311**, 004 (2003) [arXiv:hep-ph/0305159].

[225] Y. Kishimoto [KamLAND Collaboration], *J. Phys. Conf. Ser.* **120**, 052010 (2008).

[226] C. Kraus [SNO+ Collaboration], *Prog. Part. Nucl. Phys.* **57**, 150 (2006).

[227] V. N. Kornoukhov [LENS Collaboration], *Phys. Atom. Nucl.* **65**, 2161 (2002) [Yad. Fiz. **65**, 2224 (2002)].

[228] R. Hazama, H. Ejiri, J. Engel, P. Krastev, N. Kudomi, M. Nomachi, and R. G. Robertson, *AIP Conf. Proc.* **610**, 959 (2002).

[229] R. Luscher et al., *Nucl. Phys. Proc. Suppl.* **110**, 423 (2002).

[230] Y. D. Kim [XMASS Collaboration], *Phys. Atom. Nucl.* **69**, 1970 (2006); K. Ueshima et al. [XMASS Collaboration], *Nucl. Instrum. Meth. A* **594**, 148 (2008) [arXiv:0803.2888 [physics.ins-det]];

[231] D. N. McKinsey and J. M. Doyle, *Journal of Low Temperature Physics* **118**, 153 (2000); D. N. McKinsey and K. J. Coakley, *Astropart. Phys.* **22**, 355 (2005) [arXiv:astro-ph/0402007]; D. N. McKinsey [Mini-CLEAN Collaboration], *Nucl. Phys. Proc. Suppl.* **173**, 152 (2007).

[232] K. Arisaka et al., *Astropart. Phys.* **31**, 63 (2009) [arXiv:0808.3968 [astro-ph]].

[233] A. Kopylov, I. Orekhov, V. Petukhov, A. Solomatin, and M. Arnoldov, *Phys. Atom. Nucl.* **67**, 1182 (2004) [*Yad. Fiz.* **67**, 1204 (2004)] [arXiv:hep-ph/0310163].

[234] For discussions of future reactor experiments to measure θ_{12}, see A. Bandyopadhyay, S. Choubey, and S. Goswami, *Phys. Rev. D* **67**, 113011 (2003) [arXiv:hep-ph/0302243]; A. Bandyopadhyay, S. Choubey, S. Goswami, and S. T. Petcov, *Phys. Rev. D* **72**, 033013 (2005) [arXiv:hep-ph/0410283]; H. Minakata, H. Nunokawa, W.J.C. Teves, and R. Z. Funchal, *Nucl. Phys. Proc. Suppl.* **145**, 45 (2005) [arXiv:hep-ph/0501250]; S. T. Petcov and T. Schwetz, *Phys. Lett. B* **642**, 487 (2006) [arXiv:hep-ph/0607155].

[235] G. Eder, *Nucl. Phys.* **78**, 657 (1966).

[236] L. M. Krauss, S. L. Glashow, and D. N. Schramm, *Nature* **310**, 191 (1984).

[237] See, e.g., the reference model in F. Mantovani, L. Carmignani, G. Fiorentini, and M. Lissia, *Phys. Rev. D* **69**, 013001 (2004) [arXiv:hep-ph/0309013].

[238] For recent discussions, see V. R. Murthy, W. van Westrenen, and Y. Fei, *Nature* **423**, 163 (2003); K. M. Lee and R. Jeanloz, *Geophys. Res. Lett.* **30**, 2212 (2003).

[239] R. S. Raghavan, S. Schonert, S. Enomoto, J. Shirai, F. Suekane, and A. Suzuki, *Phys. Rev. Lett.* **80**, 635 (1998).

[240] C. G. Rothschild, M. C. Chen, and F. P. Calaprice, *Geophys. Res. Lett.* **25**, 1083 (1998) [arXiv:nucl-ex/9710001].

[241] G. L. Fogli, E. Lisi, A. Palazzo, and A. M. Rotunno, arXiv:hep-ph/0405139.

[242] G. Fiorentini, F. Mantovani, and B. Ricci, *Phys. Lett. B* **557**, 139 (2003) [arXiv:nucl-ex/0212008].

[243] S. Abe et al. [KamLAND Collaboration], *Phys. Rev. Lett.* **100**, 221803 (2008) [arXiv:0801.4589 [hep-ex]].

[244] G. Fiorentini, T. Lasserre, M. Lissia, B. Ricci, and S. Schonert, *Phys. Lett. B* **558**, 15 (2003) [arXiv:hep-ph/0301042].

[245] G. Fiorentini, M. Lissia, F. Mantovani, and B. Ricci, *Phys. Lett. B* **629**, 77 (2005) [arXiv:hep-ph/0508048].

[246] G. Bellini et al., *Phys. Lett. B* **687**, 299 (2010) [arXiv:1003.0284 [hep-ex]].

[247] G. L. Fogli, E. Lisi, A. Palazzo, and A. M. Rotunno, arXiv:1006.1113 [hep-ph].

[248] V. V. Sinev, arXiv:1007.2526 [hep-ph].

[249] T. K. Gaisser and M. Honda, *Ann. Rev. Nucl. Part. Sci.* **52**, 153 (2002) [arXiv:hep-ph/0203272]; G. D. Barr, T. K. Gaisser, P. Lipari, S. Robbins, and T. Stanev, *Phys. Rev. D* **70**, 023006 (2004) [arXiv:astro-ph/0403630]; M. Honda, T. Kajita, K. Kasahara, and S. Midorikawa, *Phys. Rev. D* **70**, 043008 (2004) [arXiv:astro-ph/0404457].

[250] R. Enberg, M. H. Reno, and I. Sarcevic, *Phys. Rev. D* **78**, 043005 (2008) [arXiv:0806.0418 [hep-ph]].

[251] Y. Ashie et al. [Super-Kamiokande Collaboration], *Phys. Rev. Lett.* **93**, 101801 (2004) [arXiv:hep-ex/0404034];

[252] W.W.M. Allison et al. [Soudan-2 collaboration], *Nucl. Instrum. Meth.* A **381**, 385 (1996); *Phys. Lett. B* **391**, 491 (1997) [arXiv:hep-ex/9611007]; *Phys. Lett. B* **449**, 137 (1999) [arXiv:hep-ex/9901024]; W.W.M. Allison et al. [Soudan-2 Collaboration], *Phys. Rev. D* **72**, 052005 (2005) [arXiv:hep-ex/0507068].

[253] M. Ambrosio et al. [MACRO Collaboration], *Phys. Lett. B* **434**, 451 (1998) [arXiv:hep-ex/9807005]; *Phys. Lett. B* **478**, 5 (2000) [arXiv:hep-ex/0001044]; *Phys. Lett. B* **517**, 59 (2001) [arXiv:hep-ex/0106049]; *Nucl. Instrum. Meth.* A **486**, 663 (2002); *Phys. Lett. B* **566**, 35 (2003) [arXiv:hep-ex/0304037].

[254] J. Hosaka et al. [Super-Kamiokande Collaboration], *Phys. Rev. D* **74**, 032002 (2006) [arXiv:hep-ex/0604011].

[255] K. Abe et al. [Super-Kamiokande Collaboration], *Phys. Rev. Lett.* **97**, 171801 (2006) [arXiv:hep-ex/0607059].

[256] J. Conrad, A. de Gouvea, S. Shalgar, and J. Spitz, *Phys. Rev. D* **82**, 093012 (2010) [arXiv:1008.2984 [hep-ph]].

[257] V. K. Ermilova, V. A. Tsarev, and V. A. Chechin, *Short Notices Lebedev Inst.* **5**, 26 (1986); E. K. Akhmedov, *Yad. Fiz.* **47**, 475 (1988) [*Sov. J. Nucl. Phys.* **47**, 301 (1988)]; P. I. Krastev and A. Y. Smirnov, *Phys. Lett. B* **226**, 341 (1989); E. K. Akhmedov, *Nucl. Phys. B* **538**, 25 (1999) [arXiv:hep-ph/9805272]; E. K. Akhmedov, A. Dighe, P. Lipari, and A. Y. Smirnov, *Nucl. Phys. B* **542**, 3 (1999) [arXiv:hep-ph/9808270]; E. K. Akhmedov, arXiv:hep-ph/9903302.

[258] S. T. Petcov, *Phys. Lett. B* **434**, 321 (1998) [arXiv:hep-ph/9805262]; M. V. Chizhov and S. T. Petcov, *Phys. Rev. Lett.* **83**, 1096 (1999) [arXiv:hep-ph/9903399].

[259] M. C. Banuls, G. Barenboim, and J. Bernabeu, *Phys. Lett. B* **513**, 391 (2001) [arXiv:hep-ph/0102184].

[260] E. K. Akhmedov, M. Maltoni, and A. Y. Smirnov, *JHEP* **0806**, 072 (2008) [arXiv:0804.1466 [hep-ph]].

[261] O.L.G. Peres and A. Y. Smirnov, *Phys. Lett. B* **456**, 204–213 (1999) [arXiv:hep-ph/9902312]; M. C. Gonzalez-Garcia, M. Maltoni, and A. Y. Smirnov, *Phys. Rev. D* **70**, 093005 (2004) [arXiv:hep-ph/0408170]; S. Choubey and P. Roy, *Phys. Rev. D* **73**, 013006 (2006) [arXiv:hep-ph/0509197].

[262] J. Bernabeu, S. Palomares Ruiz, and S. T. Petcov, *Nucl. Phys. B* **669**, 255 (2003) [arXiv:hep-ph/0305152].

[263] S. T. Petcov and T. Schwetz, *Nucl. Phys. B* **740**, 1–22 (2006) [arXiv:hep-ph/0511277].

[264] D. Indumathi and M.V.N. Murthy, *Phys. Rev. D* **71**, 013001 (2005) [arXiv:hep-ph/0407336].

[265] R. Gandhi, P. Ghoshal, S. Goswami, P. Mehta, and S. Uma Sankar, *Phys. Rev. D* **73**, 053001 (2006) [arXiv:hep-ph/0411252]; R. Gandhi, P. Ghoshal, S. Goswami,

P. Mehta, and S. Uma Sankar, and S. Shalgar, *Phys. Rev. D* **76**, 073012 (2007) [arXiv:0707.1723 [hep-ph]]; R. Gandhi, P. Ghoshal, S. Goswami, and S. Uma Sankar, *Phys. Rev. D* **78**, 073001 (2008) [arXiv:0807.2759 [hep-ph]].

[266] P. Huber, M. Maltoni, and T. Schwetz, *Phys. Rev. D* **71**, 053006 (2005) [arXiv:hep-ph/0501037].

[267] M. C. Gonzalez-Garcia, M. Maltoni, and J. Salvado [arXiv:1103.4365 [hep-ph]].

[268] A. Friedland, C. Lunardini, and M. Maltoni, *Phys. Rev. D* **70**, 111301 (2004) [arXiv:hep-ph/0408264]; A. Friedland, and C. Lunardini, *Phys. Rev. D* **72**, 053009 (2005) [arXiv:hep-ph/0506143].

[269] M. C. Gonzalez-Garcia, F. Halzen, M. Maltoni, and H.K.M. Tanaka, *Phys. Rev. Lett.* **100**, 061802 (2008) [arXiv:0711.0745 [hep-ph]].

[270] J. Kopp, and M. Lindner, *Phys. Rev. D* **76**, 093003 (2007) [arXiv:0705.2595 [hep-ph]].

[271] S. Yamamoto et al. [K2K Collaboration], *Phys. Rev. Lett.* **96**, 181801 (2006) [arXiv:hep-ex/0603004].

[272] P. Adamson et al. [The MINOS Collaboration], [arXiv:1103.0340 [hep-ex]]; *Phys. Rev. Lett.* **101**, 131802 (2008) [arXiv:0806.2237 [hep-ex]].

[273] P. Adamson et al. [MINOS Collaboration] arXiv:1104.0344 [hep-ex].

[274] P. Adamson et al. [The MINOS Collaboration], *Phys. Rev. D* **82**, 051102 (2010) [arXiv:1006.0996 [hep-ex]]; *Phys. Rev. Lett.* **103**, 261802 (2009) [arXiv:0909.4996 [hep-ex]].

[275] K. Abe et al. [T2K Collaboration], arXiv:1106.2822 [hep-ex].

[276] G. L. Fogli, E. Lisi, A. Marrone, and A. Palazzo, *Prog. Part. Nucl. Phys.* **57**, 742 (2006) [arXiv:hep-ph/0506083].

[277] J. E. Roa, D. C. Latimer, and D. J. Ernst, *Phys. Rev. C* **81**, 015501 (2010) [arXiv:0904.3930 [nucl-th]].

[278] T. Schwetz, M. A. Tortola, and J.W.F. Valle, *New J. Phys.* **10**, 113011 (2008) [arXiv:0808.2016 [hep-ph]]; version 3 of the preprint, dated Feb. 11, 2010, presented an updated global analysis.

[279] M. Maltoni and T. Schwetz, *PoS* **IDM2008**, 072 (2008) [arXiv:0812.3161 [hep-ph]].

[280] M. C. Gonzalez-Garcia, M. Maltoni, and J. Salvado, *JHEP* **1004**, 056 (2010) [arXiv:1001.4524 [hep-ph]].

[281] G. L. Fogli, E. Lisi, A. Marrone, A. Palazzo, and A. M. Rotunno, *Phys. Rev. Lett.* **101**, 141801 (2008) [arXiv:0806.2649 [hep-ph]]; *J. Phys. Conf. Ser.* **203**, 012103 (2010).

[282] G. L. Fogli, E. Lisi, A. Marrone, A. Palazzo, and A. M. Rotunno, arXiv:1106.6028 [hep-ph].

[283] A. B. Balantekin and D. Yilmaz, *J. Phys. G* **35**, 075007 (2008) [arXiv:0804.3345 [hep-ph]].

[284] G. L. Fogli, E. Lisi, A. Marrone, A. Palazzo, and A. M. Rotunno, arXiv:0809.2936 [hep-ph].

[285] G. L. Fogli, E. Lisi, A. Marrone, A. Palazzo, and A. M. Rotunno, *J. Phys. Conf. Ser.* **203**, 012103 (2010).

[286] A. Gando et al., arXiv:1009.4771 [hep-ex].

[287] B. H. McKellar, *Phys. Lett. B* **97**, 93 (1980); E. Holzschuh, *Rept. Prog. Phys.* **55**, 1035 (1992).

[288] A. Osipowicz et al. [KATRIN Collaboration], arXiv:hep-ex/0109033.

[289] W. Hu and D. J. Eisenstein, *Astrophys. J.* **498**, 497 (1998) [arXiv:astro-ph/9710216]; D. J. Eisenstein and W. Hu, *Astrophys. J.* **511**, 5 (1997) [arXiv:astro-ph/9710252].

[290] B. A. Reid et al., *Mon. Not. Roy. Astron. Soc.* **404**, 60 (2010) [arXiv:0907.1659 [astro-ph.CO]].

[291] U. Seljak, A. Slosar, and P. McDonald, *JCAP* **0610**, 014 (2006) [arXiv:astro-ph/0604335].

[292] M. Tegmark, A. Vilenkin, and L. Pogosian, *Phys. Rev. D* **71**, 103523 (2005) [arXiv:astro-ph/0304536].

[293] W. Hu and M. Tegmark, *Astrophys. J. Lett.* **514**, 65 (1999) [arXiv:astro-ph/9811168]; K. N. Abazajian and S. Dodelson, *Phys. Rev. Lett.* **91**, 041301 (2003) [arXiv:astro-ph/0212216]; M. Kaplinghat, L. Knox, and Y. S. Song, *Phys. Rev. Lett.* **91**, 241301 (2003) [arXiv:astro-ph/0303344]; J. Lesgourgues, L. Perotto, S. Pastor, and M. Piat, *Phys. Rev. D* **73**, 045021 (2006) [arXiv:astro-ph/0511735]. S. Hannestad, H. Tu, and Y.Y.Y. Wong, *JCAP* **0606**, 025 (2006) [arXiv:astro-ph/0603019]; T. D. Kitching, A. F. Heavens, L. Verde, P. Serra, and A. Melchiorri, *Phys. Rev. D* **77**, 103008 (2008) [arXiv:0801.4565 [astro-ph]]; R. de Putter, O. Zahn, and E. V. Linder, *Phys. Rev. D* **79**, 065033 (2009) [arXiv:0901.0916 [astro-ph.CO]].

[294] V. D. Barger, T. J. Weiler, and K. Whisnant, *Phys. Lett. B* **442**, 255 (1998) [arXiv:hep-ph/9808367].

[295] Y. Chikashige, R. N. Mohapatra, and R. D. Peccei, *Phys. Rev. Lett.* **45**, 1926 (1980); *Phys. Lett. B* **98**, 265 (1981).

[296] B. Kayser and A. S. Goldhaber, *Phys. Rev. D* **28**, 2341 (1983); B. Kayser, *Phys. Rev. D* **30**, 1023 (1984); V. D. Barger and K. Whisnant, *Phys. Lett. B* **456**, 194 (1999) [arXiv:hep-ph/9904281]; F. Vissani, *JHEP* **9906**, 022 (1999) [arXiv:hep-ph/9906525]; S. M. Bilenky, C. Giunti, W. Grimus, B. Kayser, and S. T. Petcov, *Phys. Lett. B* **465**, 193 (1999) [arXiv:hep-ph/9907234]; M. Czakon, J. Gluza, and M. Zralek, arXiv:hep-ph/0003161; H. V. Klapdor-Kleingrothaus, H. Pas, and A. Y. Smirnov, *Phys. Rev. D* **63**, 073005 (2001) [arXiv:hep-ph/0003219]; F. Feruglio, A. Strumia, and F. Vissani, *Nucl. Phys. B* **637**, 345 (2002) [Addendum-ibid. B **659**, 359 (2003)] [arXiv:hep-ph/0201291]; T. Hambye, *Eur. Phys. J. direct C* **4**, 13 (2002) [arXiv:hep-ph/0201307]; S. Pascoli and S. T. Petcov, *Phys. Lett. B* **544**, 239 (2002) [arXiv:hep-ph/0205022]; H. Minakata and H. Sugiyama, *Phys. Lett. B* **567**, 305 (2003) [arXiv:hep-ph/0212240]; S. Pascoli, S. T. Petcov, and T. Schwetz, *Nucl. Phys. B* **734**, 24 (2006) [arXiv:hep-ph/0505226]; S. Choubey and W. Rodejohann, *Phys. Rev. D* **72**, 033016 (2005) [arXiv:hep-ph/0506102]; S. M. Bilenky, A. Faessler, T. Gutsche, and F. Simkovic, *Phys. Rev. D* **72**, 053015 (2005) [arXiv:hep-ph/0507260]; M. Lindner, A. Merle, and W. Rodejohann, *Phys. Rev. D* **73**, 053005 (2006) [arXiv:hep-ph/0512143]; S. Pascoli and S. T. Petcov, *Phys. Rev. D* **77**, 113003 (2008) [arXiv:0711.4993 [hep-ph]].

[297] V. Barger, S. L. Glashow, D. Marfatia, and K. Whisnant, *Phys. Lett. B* **532**, 15 (2002) [arXiv:hep-ph/0201262].

[298] S. Pascoli, S. T. Petcov, and L. Wolfenstein, *Phys. Lett. B* **524**, 319 (2002) [arXiv:hep-ph/0110287]; S. Pascoli, S. T. Petcov, and W. Rodejohann, *Phys. Lett. B* **549**, 177 (2002) [arXiv:hep-ph/0209059].

[299] V. Barger, S. L. Glashow, P. Langacker, and D. Marfatia, *Phys. Lett. B* **540**, 247 (2002) [arXiv:hep-ph/0205290].

[300] Z. Z. Xing, *Phys. Rev. D* **68**, 053002 (2003) [arXiv:hep-ph/0305195].

[301] H. V. Klapdor-Kleingrothaus et al., *Eur. Phys. J. A* **12**, 147 (2001) [arXiv:hep-ph/0103062].

[302] See, e.g., the recent calculations in M. Kortelainen, and J. Suhonen, *Phys. Rev. C* **75**, 051303 (2007) [arXiv:0705.0469 [nucl-th]]; *Phys. Rev. C* **76**, 024315 (2007) [arXiv:0708.0115 [nucl-th]]; F. Simkovic, A. Faessler, V. Rodin, P. Vogel, J. Engel, *Phys. Rev. C* **77**, 045503 (2008) [arXiv:0710.2055 [nucl-th]]; J. Menendez, A. Poves, E. Caurier, and F. Nowacki, *Nucl. Phys. A* **818**, 139–151 (2009) [arXiv:0801.3760 [nucl-th]]; F. Simkovic, A. Faessler, and P. Vogel, *Phys. Rev. C* **79**, 015502 (2009) [arXiv:0812.0348 [nucl-th]].

[303] F. T. Avignone, III, S. R. Elliott, and J. Engel, *Rev. Mod. Phys.* **80**, 481–516 (2008) [arXiv:0708.1033 [nucl-ex]].

[304] H. V. Klapdor-Kleingrothaus, A. Dietz, H. L. Harney, and I. V. Krivosheina, *Mod. Phys. Lett. A* **16**, 2409 (2001) [arXiv:hep-ph/0201231].

[305] C. E. Aalseth et al., *Mod. Phys. Lett. A* **17**, 1475 (2002) [arXiv:hep-ex/0202018]; H. V. Klapdor-Kleingrothaus, arXiv:hep-ph/0205228.

[306] C. Arnaboldi et al. [CUORE Collaboration], *Nucl. Instrum. Meth. A* **518**, 775-798 (2004) [arXiv:hep-ex/0212053]; *Astropart. Phys.* **20**, 91 (2003) [arXiv:hep-ex/0302021] C. Arnaboldi et al. [CUORICINO Collaboration], *Phys. Rev. C* **78**, 035502 (2008); [arXiv:0802.3439 [hep-ex]].

[307] K. Wamba [EXO Collaboration], *eConf* **C020620**, THAP11 (2002) [arXiv:hep-ph/0210186].

[308] H. V. Klapdor-Kleingrothaus et al. [GENIUS Collaboration], arXiv:hep-ph/9910205.

[309] C. E. Aalseth et al. [Majorana Collaboration], arXiv:hep-ex/0201021; R. Gaitskell et al. [Majorana Collaboration], arXiv:nucl-ex/0311013.

[310] H. Ohsumi [NEMO and SuperNEMO Collaboration], *J. Phys. Conf. Ser.* **120**, 052054 (2008).

[311] I. Mocioiu and R. Shrock, *JHEP* **0111**, 050 (2001) [arXiv:hep-ph/0106139].

[312] V. D. Barger, D. Marfatia, and K. Whisnant, in *Proc. of the APS/DPF/DPB Summer Study on the Future of Particle Physics (Snowmass 2001)* ed. N. Graf, *eConf* **C010630**, E102 (2001) [arXiv:hep-ph/0108090].

[313] T. Kajita, H. Minakata, and H. Nunokawa, *Phys. Lett. B* **528**, 245 (2002) [arXiv:hep-ph/0112345].

[314] P. Huber, M. Lindner, and W. Winter, *Nucl. Phys. B* **645**, 3 (2002) [arXiv:hep-ph/0204352].

[315] P. Lipari, *Phys. Rev. D* **61**, 113004 (2000) [arXiv:hep-ph/9903481]; I. Mocioiu and R. Shrock, *Phys. Rev. D* **62**, 053017 (2000) [arXiv:hep-ph/0002149]; V. D. Barger, S. Geer, R. Raja, and K. Whisnant, *Phys. Lett. B* **485**, 379 (2000) [arXiv:hep-ph/0004208]; M. Koike, T. Ota, and J. Sato, *Phys. Rev. D* **65**, 053015 (2002) [arXiv:hep-ph/0011387]; H. Minakata and H. Nunokawa, *JHEP* **0110**, 001 (2001) [arXiv:hep-ph/0108085].

[316] V. D. Barger, S. Geer, R. Raja, and K. Whisnant, *Phys. Rev. D* **62**, 013004 (2000) [arXiv:hep-ph/9911524].

[317] G. L. Fogli and E. Lisi, *Phys. Rev. D* **54**, 3667 (1996) [arXiv:hep-ph/9604415].

[318] H. Minakata, H. Nunokawa, and S. Parke, *Phys. Rev. D* **66**, 093012 (2002) [arXiv:hep-ph/0208163].

[319] V. Barger, D. Marfatia, and K. Whisnant, *Phys. Rev. D* **66**, 053007 (2002) [arXiv:hep-ph/0206038].

[320] P. Huber and W. Winter, *Phys. Rev. D* **68**, 037301 (2003) [arXiv:hep-ph/0301257].

[321] A. Asratyan, G. Davidenko, A. Dolgolenko, V. Kaftanov, M. Kubantsev, and V. Verebryusov, arXiv:hep-ex/0303023.

[322] A description of the NuMI beam is provided in J. Hylen et al., Fermilab-TM-2018, September 1997, http://lss.fnal.gov/archive/test-tm/2000/fermilab-tm-2018.pdf.

[323] G. Barenboim, A. De Gouvea, M. Szleper, and M. Velasco, arXiv:hep-ph/0204208.

[324] P. Huber, M. Lindner, T. Schwetz, and W. Winter, *JHEP* **0911**, 044 (2009) [arXiv:0907.1896 [hep-ph]].

[325] Y. Kozlov, L. Mikaelyan, and V. Sinev, *Phys. Atom. Nucl.* **66**, 469 (2003) [*Yad. Fiz.* **66**, 497 (2003)] [arXiv:hep-ph/0109277]; V. Martemyanov, L. Mikaelyan, V. Sinev, V. Kopeikin, and Y. Kozlov, *Phys. Atom. Nucl.* **66**, 1934 (2003) [*Yad. Fiz.***66**, 1982 (2003)] [arXiv:hep-ex/0211070].

[326] H. Minakata, H. Sugiyama, O. Yasuda, K. Inoue, and F. Suekane, *Phys. Rev. D* **68**, 033017 (2003) [Erratum-ibid. D **70**, 059901 (2004)] [arXiv:hep-ph/0211111].

[327] P. Huber, M. Lindner, T. Schwetz, and W. Winter, *Nucl. Phys. B* **665**, 487 (2003) [arXiv:hep-ph/0303232].
[328] F. Suekane, K. Inoue, T. Araki, and K. Jongok, arXiv:hep-ex/0306029.
[329] M. H. Shaevitz and J. M. Link, arXiv:hep-ex/0306031.
[330] K. Anderson et al., "White paper report on using nuclear reactors to search for a value of theta(13)," arXiv:hep-ex/0402041.
[331] C. Bemporad, G. Gratta, and P. Vogel, *Rev. Mod. Phys.* **74**, 297 (2002) [arXiv:hep-ph/0107277].
[332] M. G. Albrow et al., "Physics at a Fermilab proton driver," arXiv:hep-ex/0509019.
[333] H. Nunokawa [Angra Neutrino Collaboration], *AIP Conf. Proc.* **981**, 208 (2008).
[334] V. D. Barger et al., arXiv:hep-ph/0103052; J. J. Gomez-Cadenas et al. [CERN working group on Super Beams Collaboration], arXiv:hep-ph/0105297; A. Blondel et al., *Nucl. Instrum. Meth. A* **503** (2001) 173.
[335] "The proton driver study," ed. by W. Chou, Fermilab-TM-2136.
[336] A. Rubbia and P. Sala, *JHEP* **0209**, 004 (2002) [arXiv:hep-ph/0207084].
[337] M. Ishitsuka, T. Kajita, H. Minakata, and H. Nunokawa, *Phys. Rev. D* **72**, 033003 (2005) [arXiv:hep-ph/0504026]; K. Hagiwara, N. Okamura, and K. I. Senda, *Phys. Lett. B* **637**, 266 (2006) [Erratum-ibid. B **641**, 491 (2006)] [arXiv:hep-ph/0504061]; T. Kajita, H. Minakata, S. Nakayama, and H. Nunokawa, *Phys. Rev. D* **75**, 013006 (2007) [arXiv:hep-ph/0609286].
[338] J. M. Conrad and M. H. Shaevitz, *Phys. Rev. Lett.* **104**, 141802 (2010) [arXiv:0912.4079 [hep-ex]]; J. Alonso et al., arXiv:1006.0260 [physics.ins-det].
[339] J. J. Gomez-Cadenas et al. [CERN working group on Super Beams], arXiv:hep-ph/0105297.
[340] M. Mezzetto, *J. Phys. G* **29**, 1781 (2003) [arXiv:hep-ex/0302005].
[341] J. E. Campagne and A. Cazes, *Eur. Phys. J. C* **45**, 643 (2006) [arXiv:hep-ex/0411062].
[342] V. Barger, D. Marfatia, and K. Whisnant, *Phys. Lett. B* **560**, 75 (2003) [arXiv:hep-ph/0210428].
[343] P. Huber, M. Lindner, and W. Winter, *Nucl. Phys. B* **654**, 3 (2003) [arXiv:hep-ph/0211300].
[344] H. Minakata, H. Nunokawa, and S. J. Parke, *Phys. Rev. D* **68**, 013010 (2003) [arXiv:hep-ph/0301210].
[345] For a recent discussion of iron scintillator detectors at Fermilab, see G. Barenboim et al., arXiv:hep-ex/0304017.
[346] For a discussion of iron scintillator detectors at the Indian Neutrino Observatory, see D. Indumathi [INO Collaboration], *AIP Conf. Proc.* **1382**, 29 (2011), and [174].
[347] D. Ayres et al., arXiv:hep-ex/0210005.
[348] P. Huber and J. Kopp, *JHEP* **1103**, 013 (2011) [arXiv:1010.3706 [hep-ph]].
[349] A. Blondel et al., *Nucl. Instrum. Meth. A* **451**, 102 (2000); A. Donini, M. B. Gavela, P. Hernandez, and S. Rigolin, *Nucl. Phys. B* **574**, 23 (2000) [arXiv:hep-ph/9909254]; M. Freund and T. Ohlsson, *Mod. Phys. Lett. A* **15**, 867 (2000) [arXiv:hep-ph/9909501]; M. Freund, M. Lindner, S. T. Petcov, and A. Romanino, *Nucl. Phys. B* **578**, 27 (2000) [arXiv:hep-ph/9912457]; C. Albright et al., arXiv:hep-ex/0008064.
[350] T. Adams et al., in *Proc. of the APS/DPF/DPB Summer Study on the Future of Particle Physics (Snowmass 2001)* ed. N. Graf, *eConf* **C010630**, E1001 (2001) [arXiv:hep-ph/0111030]; M. M. Alsharoa et al. [Muon Collider/Neutrino Factory Collaboration], *Phys. Rev. ST Accel. Beams* **6**, 081001 (2003) [arXiv:hep-ex/0207031]; M. Apollonio et al., arXiv:hep-ph/0210192.
[351] J. J. Gomez-Cadenas and D. A. Harris, *Ann. Rev. Nucl. Part. Sci.* **52**, 253 (2002); B. Autin, D. A. Harris, S. F. King, K. S. McFarland, and O. Yasuda, *J. Phys. G* **29**, 1743 (2003) [arXiv:hep-ph/0302218].

[352] C. H. Albright et al. [Neutrino Factory/Muon Collider Collaboration], arXiv:physics/0411123.
[353] A. Bandyopadhyay et al. "Physics at a future Neutrino Factory and super-beam facility" [ISS Physics Working Group], *Rept. Prog. Phys.* **72**, 106201 (2009) IOP Publishing, Ltd. [arXiv:0710.4947 [hep-ph]].
[354] A. Cervera, A. Donini, M. B. Gavela, J. J. Gomez Cadenas, P. Hernandez, O. Mena, and S. Rigolin, *Nucl. Phys. B* **579**, 17 (2000) [Erratum-ibid. B **593**, 731 (2001)] [arXiv:hep-ph/0002108]; M. Freund, P. Huber, and M. Lindner, *Nucl. Phys. B* **585**, 105 (2000) [arXiv:hep-ph/0004085]; J. Pinney and O. Yasuda, *Phys. Rev. D* **64**, 093008 (2001) [arXiv:hep-ph/0105087].
[355] V. D. Barger, S. Geer, R. Raja, and K. Whisnant, *Phys. Rev. D* **62**, 073002 (2000) [arXiv:hep-ph/0003184].
[356] M. Freund, P. Huber, and M. Lindner, *Nucl. Phys. B* **615**, 331 (2001) [arXiv:hep-ph/0105071].
[357] A. Bueno, M. Campanelli, S. Navas-Concha, and A. Rubbia, *Nucl. Phys. B* **631**, 239 (2002) [arXiv:hep-ph/0112297].
[358] A. Bueno, M. Campanelli, and A. Rubbia, *Nucl. Phys. B* **589**, 577 (2000) [arXiv:hep-ph/0005007].
[359] J. Burguet-Castell, M. B. Gavela, J. J. Gomez-Cadenas, P. Hernandez, and O. Mena, *Nucl. Phys. B* **646**, 301 (2002) [arXiv:hep-ph/0207080].
[360] S. Geer, O. Mena, and S. Pascoli, *Phys. Rev. D* **75**, 093001 (2007) [arXiv:hep-ph/0701258].
[361] A. D. Bross, M. Ellis, S. Geer, O. Mena, and S. Pascoli, *Phys. Rev. D* **77**, 093012 (2008) [arXiv:0709.3889 [hep-ph]].
[362] S. K. Agarwalla, P. Huber, J. Tang, and W. Winter, *JHEP* **1101**, 120 (2011) [arXiv:1012.1872 [hep-ph]].
[363] P. Huber, M. Lindner, and W. Winter, *Comput. Phys. Commun.* **167**, 195 (2005) [arXiv:hep-ph/0407333]; P. Huber, J. Kopp, M. Lindner, M. Rolinec, and W. Winter, *Comput. Phys. Commun.* **177**, 432 (2007) [arXiv:hep-ph/0701187]; see also http://www.mpi-hd.mpg.de/lin/globes/ for updates.
[364] A. Donini, D. Meloni, and P. Migliozzi, *Nucl. Phys. B* **646**, 321 (2002) [arXiv:hep-ph/0206034]; D. Autiero et al., *Eur. Phys. J. C* **33**, 243 (2004) [arXiv:hep-ph/0305185].
[365] B. Autin et al., *J. Phys. G* **29**, 1785 (2003) [arXiv:physics/0306106].
[366] M. Mezzetto, *J. Phys. G* **29**, 1771 (2003) [arXiv:hep-ex/0302007]; J. Bouchez, M. Lindroos, and M. Mezzetto, *AIP Conf. Proc.* **721**, 37 (2004) [arXiv:hep-ex/0310059].
[367] J. Burguet-Castell, D. Casper, E. Couce, J. J. Gomez-Cadenas, and P. Hernandez, *Nucl. Phys. B* **725**, 306 (2005) [arXiv:hep-ph/0503021].
[368] J. Burguet-Castell, D. Casper, J. J. Gomez-Cadenas, P. Hernandez, and F. Sanchez, *Nucl. Phys. B* **695**, 217 (2004) [arXiv:hep-ph/0312068].
[369] C. Rubbia, A. Ferrari, Y. Kadi, and V. Vlachoudis, *Nucl. Instrum. Meth. A* **568**, 475 (2006) [arXiv:hep-ph/0602032]; C. Rubbia, arXiv:hep-ph/0609235.
[370] S. K. Agarwalla, S. Choubey, and A. Raychaudhuri, *Nucl. Phys. B* **771**, 1 (2007) [arXiv:hep-ph/0610333].
[371] J. Bernabeu and C. Espinoza, *Phys. Lett. B* **664**, 285 (2008) [arXiv:0712.1034 [hep-ph]].
[372] A. Donini, E. Fernandez-Martinez, P. Migliozzi, S. Rigolin, and L. Scotto Lavina, *Nucl. Phys. B* **710**, 402 (2005) [arXiv:hep-ph/0406132].
[373] V. Barger, P. Huber, D. Marfatia, and W. Winter, *Phys. Rev. D* **76**, 053005 (2007) [arXiv:hep-ph/0703029]; S. Raby et al., arXiv:0810.4551 [hep-ph].
[374] J. E. Campagne, M. Maltoni, M. Mezzetto, and T. Schwetz, *JHEP* **0704**, 003 (2007) [arXiv:hep-ph/0603172].

[375] S. Choubey, P. Coloma, A. Donini, and E. Fernandez-Martinez, *JHEP* **0912**, 020 (2009) [arXiv:0907.2379 [hep-ph]].

[376] J. Bernabeu et al., arXiv:1005.3146 [hep-ph].

[377] O. W. Greenberg, *Phys. Rev. Lett.* **89**, 231602 (2002) [arXiv:hep-ph/0201258].

[378] M. Chaichian, A. D. Dolgov, V. A. Novikov, and A. Tureanu, arXiv:1103.0168 [hep-th].

[379] O. W. Greenberg, arXiv:1105.0927 [hep-ph].

[380] J. Arafune and J. Sato, *Phys. Rev. D* **55**, 1653 (1997) [arXiv:hep-ph/9607437]; V. D. Barger, Y. B. Dai, K. Whisnant, and B. L. Young, *Phys. Rev. D* **59**, 113010 (1999) [arXiv:hep-ph/9901388]; M. Koike and J. Sato, *Phys. Rev. D* **62**, 073006 (2000) [arXiv:hep-ph/9911258]; H. Yokomakura, K. Kimura, and A. Takamura, *Phys. Lett. B* **496**, 175 (2000) [arXiv:hep-ph/0009141]; S. J. Parke and T. J. Weiler, *Phys. Lett. B* **501**, 106 (2001) [arXiv:hep-ph/0011247]; T. Miura, E. Takasugi, Y. Kuno, and M. Yoshimura, *Phys. Rev. D* **64**, 013002 (2001) [arXiv:hep-ph/0102111]; E. K. Akhmedov, P. Huber, M. Lindner, and T. Ohlsson, *Nucl. Phys. B* **608**, 394 (2001) [arXiv:hep-ph/0105029]; T. Miura, T. Shindou, E. Takasugi, and M. Yoshimura, *Phys. Rev. D* **64**, 073017 (2001) [arXiv:hep-ph/0106086]; T. Ota, J. Sato, and Y. Kuno, *Phys. Lett. B* **520**, 289 (2001) [arXiv:hep-ph/0107007]; H. Minakata, H. Nunokawa, and S. Parke, *Phys. Lett. B* **537**, 249 (2002) [arXiv:hep-ph/0204171]; C. N. Leung and Y. Y. Wong, *Phys. Rev. D* **67**, 056005 (2003) [arXiv:hep-ph/0301211].

[381] M. Jacobson and T. Ohlsson, *Phys. Rev. D* **69**, 013003 (2004) [arXiv:hep-ph/0305064].

[382] V. D. Barger, S. Pakvasa, T. J. Weiler, and K. Whisnant, *Phys. Rev. Lett.* **85**, 5055 (2000) [arXiv:hep-ph/0005197]; S. Skadhauge, *Nucl. Phys. B* **639**, 281 (2002) [arXiv:hep-ph/0112189]; S. M. Bilenky, M. Freund, M. Lindner, T. Ohlsson, and W. Winter, *Phys. Rev. D* **65**, 073024 (2002) [arXiv:hep-ph/0112226]; J. N. Bahcall, V. Barger, and D. Marfatia, *Phys. Lett. B* **534**, 120 (2002) [arXiv:hep-ph/0201211]; I. Mocioiu and M. Pospelov, *Phys. Lett. B* **534**, 114 (2002) [arXiv:hep-ph/0202160].

[383] P. Langacker, *The Standard Model and Beyond*, Taylor & Francis, Boca Raton, 2009.

[384] S. Weinberg, *Phys. Rev. Lett.* **43** (1979) 1566.

[385] K. S. Babu and C. N. Leung, *Nucl. Phys. B* **619**, 667 (2001) [arXiv:hep-ph/0106054].

[386] E. Ma, *Phys. Rev. Lett.* **81**, 1171 (1998) [arXiv:hep-ph/9805219].

[387] P. Minkowski, *Phys. Lett. B* **67** (1977) 421; T. Yanagida, in *Proceedings of the Workshop on the Unified Theory and the Baryon Number in the Universe*, eds. O. Sawada et al., KEK Report 79-18, Tsukuba, 1979; M. Gell-Mann, P. Ramond, and R. Slansky, in *Supergravity*, eds. P. van Nieuwenhuizen et al., North-Holland, 1979 S. L. Glashow, in *Quarks and Leptons*, Cargèse, eds. M. Lévy et al., Plenum, 1980.

[388] R. N. Mohapatra and G. Senjanović, *Phys. Rev. Lett.* **44** (1980) 912.

[389] Z. Z. Xing, *Int. J. Mod. Phys. A* **23**, 4255 (2008) [arXiv:0810.1421 [hep-ph]].

[390] W. Konetschny and W. Kummer, *Phys. Lett. B* **70** (1977) 433; T. P. Cheng and L. F. Li, *Phys. Rev. D* **22** (1980) 2860; G. Lazarides, Q. Shafi, and C. Wetterich, *Nucl. Phys. B* **181** (1981) 287; J. Schechter and J.W.F. Valle, *Phys. Rev. D* **22** (1980) 2227; R. N. Mohapatra and G. Senjanović, *Phys. Rev. D* **23** (1981) 165.

[391] E. Ma and U. Sarkar, *Phys. Rev. Lett.* **80**, 5716 (1998) [arXiv:hep-ph/9802445].

[392] R. Foot, H. Lew, X. G. He, and G. C. Joshi, *Z. Phys. C* **44** (1989) 441.

[393] B. Bajc and G. Senjanović, *JHEP* **0708** (2007) 014 [arXiv:hep-ph/0612029]; P. Fileviez Pérez, *Phys. Lett. B* **654** (2007) 189 [arXiv:hep-ph/0702287]; *Phys. Rev. D* **76** (2007) 071701 [arXiv:0705.3589 [hep-ph]]. *JHEP* **0903**, 142 (2009) [arXiv:0809.1202 [hep-ph]].

[394] C. H. Albright and S. Nandi, *Mod. Phys. Lett. A* **11**, 737 (1996) [arXiv:hep-ph/9505383]; *Phys. Rev. D* **53**, 2699 (1996) [arXiv:hep-ph/9507376]; H. Nishiura,

K. Matsuda, and T. Fukuyama, *Phys. Rev. D* **60**, 013006 (1999) [arXiv:hep-ph/9902385]. K. S. Babu, B. Dutta, and R. N. Mohapatra, *Phys. Lett. B* **458**, 93 (1999) [arXiv:hep-ph/9904366]; G. Altarelli, F. Feruglio, and I. Masina, *Phys. Lett. B* **472**, 382 (2000) [arXiv:hep-ph/9907532].

[395] M. Jezabek and Y. Sumino, *Phys. Lett. B* **440**, 327 (1998) [arXiv:hep-ph/9807310]; G. Altarelli and F. Feruglio, *Phys. Lett. B* **439**, 112 (1998) [arXiv:hep-ph/9807353].

[396] G. Costa and E. Lunghi, *Nuovo Cim. A* **110**, 549 (1997) [arXiv:hep-ph/9709271]; E. K. Akhmedov, G. C. Branco, and M. N. Rebelo, *Phys. Lett. B* **478**, 215 (2000) [arXiv:hep-ph/9911364].

[397] S. M. Barr and I. Dorsner, *Nucl. Phys. B* **585**, 79 (2000) [arXiv:hep-ph/0003058].

[398] G. Altarelli and F. Feruglio, *Springer Tracts Mod. Phys.* **190**, 169 (2003) [arXiv:hep-ph/0206077]; arXiv:hep-ph/0306265.

[399] R. N. Mohapatra, arXiv:hep-ph/0211252.

[400] P. F. Harrison, D. H. Perkins, and W. G. Scott, *Phys. Lett. B* **349**, 137 (1995); *Phys. Lett. B* **396**, 186 (1997) [arXiv:hep-ph/9702243]; *Phys. Lett. B* **458**, 79 (1999) [arXiv:hep-ph/9904297].

[401] V. D. Barger, S. Pakvasa, T. J. Weiler, and K. Whisnant, *Phys. Lett. B* **437**, 107 (1998) [arXiv:hep-ph/9806387]; A. J. Baltz, A. S. Goldhaber, and M. Goldhaber, *Phys. Rev. Lett.* **81**, 5730 (1998) [arXiv:hep-ph/9806540]; F. Vissani, arXiv:hep-ph/9708483.

[402] G. Altarelli and F. Feruglio, *Phys. Lett. B* **439**, 112 (1998) [arXiv:hep-ph/9807353]; *Phys. Lett. B* **451**, 388 (1999) [arXiv:hep-ph/9812475].

[403] C. H. Albright and W. Rodejohann, *Phys. Lett. B* **665**, 378 (2008) [arXiv:0804.4581 [hep-ph]].

[404] L. Wolfenstein, *Phys. Rev. Lett.* **51**, 1945 (1983).

[405] A. Datta, L. Everett, and P. Ramond, *Phys. Lett. B* **620**, 42 (2005) [arXiv:hep-ph/0503222].

[406] M. Abbas and A. Y. Smirnov, *Phys. Rev. D* **82**, 013008 (2010) [arXiv:1004.0099 [hep-ph]].

[407] S. Goswami, S. T. Petcov, S. Ray, and W. Rodejohann, *Phys. Rev. D* **80**, 053013 (2009) [arXiv:0907.2869 [hep-ph]].

[408] A. J. Buras, J. R. Ellis, M. K. Gaillard, and D. V. Nanopoulos, *Nucl. Phys. B* **135**, 66 (1978).

[409] H. Georgi and C. Jarlskog, *Phys. Lett. B* **86**, 297 (1979).

[410] K. S. Babu and R. N. Mohapatra, *Phys. Rev. Lett.* **70**, 2845 (1993) [arXiv:hep-ph/9209215].

[411] K. S. Babu and S. M. Barr, *Phys. Lett. B* **381**, 202 (1996) [arXiv:hep-ph/9511446]; S. M. Barr, *Phys. Rev. D* **55**, 1659 (1997) [arXiv:hep-ph/9607419]; C. H. Albright, K. S. Babu, and S. M. Barr, *Phys. Rev. Lett.* **81**, 1167 (1998) [arXiv:hep-ph/9802314].

[412] C. D. Froggatt and H. B. Nielsen, *Nucl. Phys. B* **147**, 277 (1979).

[413] M. C. Chen and K. T. Mahanthappa, *Int. J. Mod. Phys.* A **18**, 5819 (2003) [arXiv:hep-ph/0305088].

[414] V. Antonelli, F. Caravaglios, R. Ferrari, and M. Picariello, *Phys. Lett. B* **549**, 325 (2002) [arXiv:hep-ph/0207347].

[415] C. H. Albright and S. M. Barr, *Phys. Lett. B* **461**, 218 (1999) [arXiv:hep-ph/9906297].

[416] I. Dorsner and S. M. Barr, *Nucl. Phys. B* **617**, 493 (2001) [arXiv:hep-ph/0108168].

[417] J. Sato and T. Yanagida, *Phys. Lett. B* **430**, 127 (1998) [arXiv:hep-ph/9710516]; C. H. Albright, K. S. Babu, and S. M. Barr, *Phys. Rev. Lett.* **81**, 1167 (1998) [arXiv:hep-ph/9802314]; Y. Nomura and T. Yanagida, *Phys. Rev. D* **59**, 017303 (1999) [arXiv:hep-ph/9807325]; G. Altarelli and F. Feruglio, *JHEP* **9811**, 021 (1998) [arXiv:hep-ph/9809596]; *Phys. Lett. B* **451**, 388 (1999) [arXiv:hep-ph/9812475]; K. S. Babu, J. C. Pati, and F. Wilczek, *Nucl. Phys. B* **566**, 33 (2000) [arXiv:hep-ph/9812538]; K. Hagiwara and N. Okamura, *Nucl. Phys. B* **548**, 60 (1999)

[arXiv:hep-ph/9811495]; Y. Nomura and T. Sugimoto, *Phys. Rev. D* **61**, 093003 (2000) [arXiv:hep-ph/9903334]; K. I. Izawa, K. Kurosawa, Y. Nomura, and T. Yanagida, *Phys. Rev. D* **60**, 115016 (1999) [arXiv:hep-ph/9904303]; Y. Nir and Y. Shadmi, *JHEP* **9905**, 023 (1999) [arXiv:hep-ph/9902293]; Q. Shafi and Z. Tavartkiladze, *Phys. Lett. B* **487**, 145 (2000) [arXiv:hep-ph/9910314]; P. H. Frampton and A. Rasin, *Phys. Lett. B* **478**, 424 (2000) [arXiv:hep-ph/9910522]; N. Maekawa, *Prog. Theor. Phys.* **106**, 401 (2001) [arXiv:hep-ph/0104200]; C. H. Albright and S. M. Barr, *Phys. Rev. D* **64**, 073010 (2001) [arXiv:hep-ph/0104294]; K. S. Babu and S. M. Barr, *Phys. Lett. B* **525**, 289 (2002) [arXiv:hep-ph/0111215]; C. S. Aulakh, B. Bajc, A. Melfo, G. Senjanovic, and F. Vissani, *Phys. Lett. B* **588**, 196 (2004) [arXiv:hep-ph/0306242]; C. H. Albright, *Phys. Rev. D* **72**, 013001 (2005) [Erratum-ibid. D **74**, 039903 (2006)] [arXiv:hep-ph/0502161]; X. Ji, Y. Li, and R. N. Mohapatra, *Phys. Lett. B* **633**, 755 (2006) [arXiv:hep-ph/0510353].

[418] J. K. Elwood, N. Irges, and P. Ramond, *Phys. Rev. Lett.* **81**, 5064 (1998) [arXiv:hep-ph/9807228]; N. Irges, S. Lavignac, and P. Ramond, *Phys. Rev. D* **58**, 035003 (1998) [arXiv:hep-ph/9802334].

[419] T. Blazek, S. Raby, and K. Tobe, *Phys. Rev. D* **62**, 055001 (2000) [arXiv:hep-ph/9912482]; W. Buchmuller and D. Wyler, *Phys. Lett.* B **521**, 291 (2001) [arXiv:hep-ph/0108216]; N. N. Singh and M. Patgiri, *Int. J. Mod. Phys. A* **17**, 3629 (2002) [arXiv:hep-ph/0111319]; M. C. Chen and K. T. Mahanthappa, *Phys. Rev. D* **68**, 017301 (2003) [arXiv:hep-ph/0212375]; S. Raby, *Phys. Lett. B* **561**, 119 (2003) [arXiv:hep-ph/0302027]; M. Bando and M. Obara, *Prog. Theor. Phys.* **109**, 995 (2003) [arXiv:hep-ph/0302034]; N. Oshimo, *Nucl. Phys. B* **668**, 258 (2003) [arXiv:hep-ph/0305166]; H. S. Goh, R. N. Mohapatra, and S. P. Ng, *Phys. Rev. D* **68**, 115008 (2003) [arXiv:hep-ph/0308197]; M. Bando, S. Kaneko, M. Obara, and M. Tanimoto, *Phys. Lett. B* **580**, 229 (2004) [arXiv:hep-ph/0309310]; W. M. Yang and Z. G. Wang, *Nucl. Phys. B* **707**, 87 (2005) [arXiv:hep-ph/0406221]; M. C. Chen and K. T. Mahanthappa, arXiv:hep-ph/0409165; S. Bertolini and M. Malinsky, *Phys. Rev. D* **72**, 055021 (2005) [arXiv:hep-ph/0504241]; K. S. Babu and C. Macesanu, *Phys. Rev. D* **72**, 115003 (2005) [arXiv:hep-ph/0505200]; R. Dermisek and S. Raby, *Phys. Lett. B* **622**, 327 (2005) [arXiv:hep-ph/0507045]; I. de Medeiros Varzielas and G. G. Ross, *Nucl. Phys. B* **733**, 31 (2006) [arXiv:hep-ph/0507176]; B. Dutta, Y. Mimura, and R. N. Mohapatra, *Phys. Rev. D* **72**, 075009 (2005) [arXiv:hep-ph/0507319]; Q. Shafi and Z. Tavartkiladze, *Phys. Lett. B* **633**, 595 (2006) [arXiv:hep-ph/0509237]; Z. Berezhiani and F. Nesti, *JHEP* **0603**, 041 (2006) [arXiv:hep-ph/0510011]; S. Bertolini, T. Schwetz, and M. Malinsky, *Phys. Rev. D* **73**, 115012 (2006) [arXiv:hep-ph/0605006]; R. Dermisek, M. Harada, and S. Raby, *Phys. Rev. D* **74**, 035011 (2006) [arXiv:hep-ph/0606055]; S. F. King and M. Malinsky, *JHEP* **0611**, 071 (2006) [arXiv:hep-ph/0608021]; Y. Cai and H. B. Yu, *Phys. Rev. D* **74**, 115005 (2006) [arXiv:hep-ph/0608022]; W. Grimus and H. Kuhbock, *Eur. Phys. J. C* **51**, 721 (2007) [arXiv:hep-ph/0612132]; T. Fukuyama, K. Matsuda, and H. Nishiura, *Int. J. Mod. Phys. A* **22**, 5325 (2007) [arXiv:hep-ph/0702284]; R. G. Moorhouse, *Phys. Rev. D* 77, 053006 (2008) [arXiv:0711.2626 [hep-ph]]; M. K. Parida, *Phys. Rev. D* **78**, 053004 (2008) [arXiv:0804.4571 [hep-ph]]; P. Di Bari and A. Riotto, *Phys. Lett. B* **671**, 462 (2009) [arXiv:0809.2285 [hep-ph]].

[420] C. H. Albright, arXiv:0905.0146 [hep-ph].

[421] W. Rodejohann, *Phys. Lett. B* **579**, 127 (2004) [arXiv:hep-ph/0308119].

[422] B. Bajc, A. Melfo, G. Senjanovic, and F. Vissani, *Phys. Lett. B* **634**, 272 (2006) [arXiv:hep-ph/0511352]; W. Grimus and H. Kuhbock, *Phys. Rev. D* 77, 055008 (2008) [arXiv:0710.1585 [hep-ph]]; B. Stech and Z. Tavartkiladze, *Phys. Rev. D* **77**, 076009 (2008) [arXiv:0802.0894 [hep-ph]]; M. Frigerio, P. Hosteins, S. Lavignac, and A. Romanino, *Nucl. Phys. B* **806**, 84 (2009) [arXiv:0804.0801 [hep-ph]].

[423] K. S. Babu, C. N. Leung, and J. Pantaleone, *Phys. Lett. B* **319**, 191 (1993) [arXiv:hep-ph/9309223].

[424] J. A. Casas, J. R. Espinosa, A. Ibarra, and I. Navarro, *Nucl. Phys. B* **573**, 652 (2000) [arXiv:hep-ph/9910420]; C. H. Albright and S. Geer, *Phys. Rev. D* **65**, 073004 (2002) [arXiv:hep-ph/0108070]; *Phys. Lett. B* **532**, 311 (2002) [arXiv:hep-ph/0112171]; S. Antusch, M. Drees, J. Kersten, M. Lindner, and M. Ratz, *Phys. Lett. B* **519**, 238 (2001) [arXiv:hep-ph/0108005]; *Phys. Lett. B* **525**, 130 (2002) [arXiv:hep-ph/0110366]. S. Antusch, J. Kersten, M. Lindner, and M. Ratz, *Nucl. Phys. B* **674**, 401 (2003) [arXiv:hep-ph/0305273]; T. Miura, T. Shindou, and E. Takasugi, *Phys. Rev. D* **66**, 093002 (2002) [arXiv:hep-ph/0206207].

[425] N. Haba, N. Okamura, and M. Sugiura, *Prog. Theor. Phys.* bf 103, 367 (2000) [arXiv:hep-ph/9810471]; J. R. Ellis and S. Lola, *Phys. Lett. B* **458**, 310 (1999) [arXiv:hep-ph/9904279]; J. A. Casas, J. R. Espinosa, A. Ibarra, and I. Navarro, *Nucl. Phys. B* **556**, 3 (1999) [arXiv:hep-ph/9904395]; *Nucl. Phys. B* **569**, 82 (2000) [arXiv:hep-ph/9905381]; N. Haba, Y. Matsui, N. Okamura, and M. Sugiura, *Prog. Theor. Phys.* **103**, 145 (2000) [arXiv:hep-ph/9908429]; K. R. Balaji, A. S. Dighe, R. N. Mohapatra, and M. K. Parida, *Phys. Rev. Lett.* **84**, 5034 (2000) [arXiv:hep-ph/0001310]; S. Antusch, J. Kersten, M. Lindner, and M. Ratz, *Phys. Lett. B* **544**, 1 (2002) [arXiv:hep-ph/0206078]; M. Frigerio and A. Y. Smirnov, *JHEP* **0302**, 004 (2003) [arXiv:hep-ph/0212263].

[426] F. S. Ling and P. Ramond, *Phys. Rev. D* **67**, 115010 (2003) [arXiv:hep-ph/0302264].

[427] S. F. King and G. G. Ross, *Phys. Lett. B* **520**, 243 (2001) [arXiv:hep-ph/0108112]; G. G. Ross and L. Velasco-Sevilla, *Nucl. Phys. B* **653**, 3 (2003) [arXiv:hep-ph/0208218]; A. Ibarra and G. G. Ross, *Phys. Lett. B* **575**, 279 (2003) [arXiv:hep-ph/0307051]; S. F. King and G. G. Ross, *Phys. Lett. B* **574**, 239 (2003) [arXiv:hep-ph/0307190].

[428] A. Datta, F. S. Ling, and P. Ramond, *Nucl. Phys. B* **671**, 383 (2003) [arXiv:hep-ph/0306002].

[429] Y. Koide and J. Sato, *Phys. Rev. D* **68**, 056004 (2003) [Erratum-ibid. D **69**, 019902 (2004)] [arXiv:hep-ph/0305291]; Y. Koide, *Phys. Lett. B* **574**, 82 (2003) [arXiv:hep-ph/0308097].

[430] A. Zee, *Phys. Lett. B* **93**, 389 (1980) [Erratum-ibid. B **95**, 461 (1980)]; *Phys. Lett. B* **161**, 141 (1985).

[431] P. H. Frampton and S. L. Glashow, *Phys. Lett. B* **461**, 95 (1999) [arXiv:hep-ph/9906375]; A. S. Joshipura and S. D. Rindani, *Phys. Lett. B* **464**, 239 (1999) [arXiv:hep-ph/9907390]; H. J. He, D. A. Dicus, and J. N. Ng, *Phys. Lett. B* **536**, 83 (2002) [arXiv:hep-ph/0203237]; H. S. Goh, R. N. Mohapatra, and S. P. Ng, *Phys. Lett. B* **542**, 116 (2002) [arXiv:hep-ph/0205131].

[432] C. Jarlskog, M. Matsuda, S. Skadhauge, and M. Tanimoto, *Phys. Lett. B* **449**, 240 (1999) [arXiv:hep-ph/9812282]; K. M. Cheung and O. C. Kong, *Phys. Rev. D* **61**, 113012 (2000) [arXiv:hep-ph/9912238]; X. G. He, *Eur. Phys. J. C* **34**, 371 (2004) [arXiv:hep-ph/0307172].

[433] H. S. Goh, R. N. Mohapatra, and S. P. Ng, *Phys. Lett. B* **542**, 116 (2002) [arXiv:hep-ph/0205131]; S. T. Petcov and W. Rodejohann, *Phys. Rev. D* **71**, 073002 (2005) [arXiv:hep-ph/0409135]; W. Grimus and L. Lavoura, *J. Phys. G***31**, 683 (2005) [arXiv:hep-ph/0410279].

[434] C. S. Lam, *Phys. Lett. B* **507**, 214 (2001) [arXiv:hep-ph/0104116]; W. Grimus and L. Lavoura, *JHEP* **0107**, 045 (2001) [arXiv:hep-ph/0105212]; *Phys. Lett. B* **572**, 189 (2003) [arXiv:hep-ph/0305046]; E. Ma, *Phys. Rev. D* **66**, 117301 (2002) [arXiv:hep-ph/0207352]; T. Kitabayashi and M. Yasue, *Phys. Rev. D* **67**, 015006 (2003) [arXiv: hep-ph/0209294]; P. F. Harrison and W. G. Scott, *Phys. Lett. B* **547**, 219 (2002) [arXiv:hep-ph/0210197]; Y. Koide, *Phys. Rev. D* **69**, 093001 (2004)

[arXiv:hep-ph/0312207]; W. Grimus, A. S. Joshipura, S. Kaneko, L. Lavoura, H. Sawanaka, and M. Tanimoto, *Nucl. Phys. B* **713**, 151 (2005) [arXiv:hep-ph/0408123]; R. N. Mohapatra, *JHEP* **0410**, 027 (2004) [arXiv:hep-ph/0408187]; R. N. Mohapatra, S. Nas, and H. B. Yu, *Phys. Lett. B* **615**, 231 (2005) [arXiv:hep-ph/0502026]; C. S. Lam, *Phys. Rev. D* **71**, 093001 (2005) [arXiv:hep-ph/0503159]; I. Aizawa and M. Yasue, *Phys. Rev. D* **73**, 015002 (2006) [arXiv:hep-ph/0510132]; K. Matsuda and H. Nishiura, *Phys. Rev. D* **73**, 013008 (2006) [arXiv:hep-ph/0511338]; Y. H. Ahn, S. K. Kang, C. S. Kim, and J. Lee, *Phys. Rev. D* **73**, 093005 (2006) [arXiv:hep-ph/0602160]; N. N. Singh, M. Rajkhowa, and A. Borah, *Pramana* **69**, 533 (2007) [arXiv:hep-ph/0603189]; T. Baba and M. Yasue, *Phys. Rev. D* **75**, 055001 (2007) [arXiv:hep-ph/0612034]; *Phys. Rev. D* **77**, 075008 (2008) [arXiv:0710.2713 [hep-ph]]; Y. Koide and E. Takasugi, *Phys. Rev. D* **77**, 016006 (2008) [arXiv:0706.4373 [hep-ph]].

[435] P. Binetruy, S. Lavignac, S. T. Petcov, and P. Ramond, *Nucl. Phys. B* **496**, 3 (1997) [arXiv:hep-ph/9610481]; N. F. Bell and R. R. Volkas, *Phys. Rev. D* **63**, 013006 (2001) [arXiv:hep-ph/0008177]; W. Rodejohann and M. A. Schmidt, *Phys. Atom. Nucl.* **69**, 1833 (2006) [arXiv:hep-ph/0507300]; S. Choubey and W. Rodejohann, *Eur. Phys. J. C* **40**, 259 (2005) [arXiv:hep-ph/0411190];

[436] C. D. Carone and M. Sher, *Phys. Lett. B* **420**, 83 (1998) [arXiv:hep-ph/9711259]; Y. L. Wu, *Phys. Rev. D* **60**, 073010 (1999) [arXiv:hep-ph/9810491]; E. Ma, *Phys. Lett. B* **456**, 48 (1999) [arXiv:hep-ph/9812344]; C. Wetterich, *Phys. Lett. B* **451**, 397 (1999) [arXiv:hep-ph/9812426]; R. Barbieri, L. J. Hall, G. L. Kane, and G. G. Ross, arXiv:hep-ph/9901228; I. Masina, *Phys. Lett. B* **633**, 134 (2006) [arXiv:hep-ph/0508031]; I. de Medeiros Varzielas, S. F. King, and G. G. Ross, *Phys. Lett. B* **644**, 153 (2007) [arXiv:hep-ph/0512313]; Y. L. Wu, *Int. J. Mod. Phys. A* **23**, 3376 (2008) [arXiv:0807.3847 [hep-ph]];

[437] M. J. Bowick and P. Ramond, *Phys. Lett. B* **103**, 338 (1981); D.R.T. Jones, G. L. Kane, and J. P. Leveille, *Nucl. Phys. B* **198**, 45 (1982); Z. G. Berezhiani, *Phys. Lett. B* **150**, 177 (1985); Z. G. Berezhiani and M. Y. Khlopov, *Sov. J. Nucl. Phys.* **51**, 739 (1990) [*Yad. Fiz.* **51**, 1157 (1990)]; Z. Berezhiani and A. Rossi, *Nucl. Phys. B* **594**, 113 (2001) [arXiv:hep-ph/0003084]; S. F. King and G. G. Ross, *Phys. Lett. B* **520**, 243 (2001) [arXiv:hep-ph/0108112]; G. G. Ross, L. Velasco-Sevilla, and O. Vives, *Nucl. Phys. B* **692**, 50 (2004) [arXiv:hep-ph/0401064]; T. Appelquist, Y. Bai, and M. Piai, *Phys. Rev. D* **74**, 076001 (2006) [arXiv:hep-ph/0607174]; Y. Koide, arXiv:0707.0899 [hep-ph]; Riazuddin, *Eur. Phys. J. C* **51**, 697 (2007) [arXiv:0707.0912 [hep-ph]].

[438] S. Pakvasa and H. Sugawara, *Phys. Lett. B* **73**, 61 (1978); H. Harari, H. Haut, and J. Weyers, *Phys. Lett. B* **78**, 459 (1978); Y. Yamanaka, H. Sugawara, and S. Pakvasa, *Phys. Rev. D* **25**, 1895 (1982) [Erratum-ibid. D **29**, 2135 (1984)]; N. Honzawa, I. Ito, S. Matsui, S. Y. Tsai, and H. Yoshida, *Prog. Theor. Phys.* **70**, 1624 (1983); Y. P. Yao, arXiv:hep-ph/9507207; L. J. Hall and H. Murayama, *Phys. Rev. Lett.* **75**, 3985 (1995) [arXiv:hep-ph/9508296]; K. Kang, J. E. Kim and P. Ko, *Z. Phys.* C **72**, 671 (1996) [arXiv:hep-ph/9503436]; C. D. Carone, *Nucl. Phys. Proc. Suppl.* **52A**, 177 (1997) [arXiv:hep-ph/9607323]; S. L. Adler, *Phys. Rev. D* **59**, 015012 (1999) [Erratum-ibid. D **59**, 099902 (1999)] [arXiv:hep-ph/9806518]; Y. Koide, *Phys. Rev. D* **60**, 077301 (1999) [arXiv:hep-ph/9905416]; M. Tanimoto, *Acta Phys. Polon. B* **30**, 3105 (1999) [arXiv:hep-ph/9910261]; E. Ma, *Phys. Rev. D* **61**, 033012 (2000) [arXiv:hep-ph/9909249]; J. I. Silva-Marcos, *JHEP* **0212**, 036 (2002) [arXiv:hep-ph/0204217]; K. Hamaguchi, M. Kakizaki, and M. Yamaguchi, *Phys. Rev. D* **68**, 056007 (2003) [arXiv:hep-ph/0212172]; P. F. Harrison and W. G. Scott, *Phys. Lett. B* **557**, 76 (2003) [arXiv:hep-ph/0302025]; T. Kobayashi, J. Kubo, and H. Terao, *Phys. Lett. B* **568**, 83 (2003) [arXiv:hep-ph/0303084]; J. Kubo, *Phys. Lett. B* **578**, 156 (2004) [Erratum-ibid. B **619**, 387 (2005)] [arXiv:hep-ph/0309167]; T. Araki, J. Kubo, and E. A. Paschos, *Eur. Phys. J. C* **45**, 465 (2006) [arXiv:hep-ph/0502164]; W. Grimus and L. Lavoura,

JHEP **0508**, 013 (2005) [arXiv:hep-ph/0504153]; *JHEP* **0601**, 018 (2006) [arXiv:hep-ph/0509239]; Y. Koide, *Phys. Rev. D* **73**, 057901 (2006) [arXiv:hep-ph/0509214]; *Eur. Phys. J. C* **50**, 809 (2007) [arXiv:hep-ph/0612058]; N. Haba and K. Yoshioka, *Nucl. Phys. B* **739**, 254 (2006) [arXiv:hep-ph/0511108]; J. E. Kim and J. C. Park, *JHEP* **0605**, 017 (2006) [arXiv:hep-ph/0512130]; N. Haba, A. Watanabe, and K. Yoshioka, *Phys. Rev. Lett.* **97**, 041601 (2006) [arXiv:hep-ph/0603116]; R. N. Mohapatra and H. B. Yu, *Phys. Lett. B* **644**, 346 (2007) [arXiv:hep-ph/0610023]; M. Picariello, *Int. J. Mod. Phys. A* **23**, 4435 (2008) [arXiv:hep-ph/0611189]; K. S. Babu, S. M. Barr, and I. Gogoladze, *Phys. Lett. B* **661**, 124 (2008) [arXiv:0709.3491 [hep-ph]]; C. Y. Chen and L. Wolfenstein, *Phys. Rev. D* **77**, 093009 (2008) [arXiv:0709.3767 [hep-ph]].

[439] J. Kubo, A. Mondragon, M. Mondragon, and E. Rodriguez-Jauregui, *Prog. Theor. Phys.* **109**, 795 (2003) [Erratum-ibid. **114**, 287 (2005)] [arXiv:hep-ph/0302196]; S. L. Chen, M. Frigerio, and E. Ma, *Phys. Rev. D* **70**, 073008 (2004) [Erratum-ibid. D **70**, 079905 (2004)] [arXiv:hep-ph/0404084]; T. Teshima, *Phys. Rev. D* **73**, 045019 (2006) [arXiv:hep-ph/0509094]; M. Tanimoto and T. Yanagida, *Phys. Lett. B* **633**, 567 (2006) [arXiv:hep-ph/0511336]; R. N. Mohapatra, S. Nasri, and H. B. Yu, *Phys. Lett. B* **639**, 318 (2006) [arXiv:hep-ph/0605020]; A. Mondragon, M. Mondragon, and E. Peinado, *Phys. Rev. D* **76**, 076003 (2007) [arXiv:0706.0354 [hep-ph]]; M. Mitra and S. Choubey, *Phys. Rev. D* **78**, 115014 (2008) [arXiv:0806.3254 [hep-ph]]; D. A. Dicus, S. F. Ge, and W. W. Repko, arXiv:1004.3266 [hep-ph]; R. Jora, J. Schechter, and M. N. Shahid, arXiv:1006.3307 [hep-ph].

[440] S. Pakvasa and H. Sugawara, *Phys. Lett. B* **82**, 105 (1979); E. Ma, *Phys. Lett. B* **632**, 352 (2006) [arXiv:hep-ph/0508231]; F. Caravaglios and S. Morisi, *Int. J. Mod. Phys. A* **22**, 2469 (2007) [arXiv:hep-ph/0611078]; Y. Koide, *JHEP* **0708**, 086 (2007) [arXiv:0705.2275 [hep-ph]]; S. Nandi and Z. Tavartkiladze, *Phys. Lett. B* **661**, 109 (2008) [arXiv:0708.4033 [hep-ph]].

[441] C. Hagedorn, M. Lindner, and R. N. Mohapatra, *JHEP* **0606**, 042 (2006) [arXiv:hep-ph/0602244]; H. Zhang, *Phys. Lett. B* **655**, 132 (2007) [arXiv:hep-ph/0612214].

[442] R. N. Mohapatra, M. K. Parida, and G. Rajasekaran, *Phys. Rev. D* **69**, 053007 (2004) [arXiv:hep-ph/0301234].

[443] C. S. Lam, *Phys. Rev. D* **78**, 073015 (2008) [arXiv:0809.1185 [hep-ph]].

[444] E. Ma and G. Rajasekaran, *Mod. Phys. Lett. A* **16**, 2207 (2001) [arXiv:hep-ph/0109236]; E. Ma, *Phys. Rev. D* **70**, 031901 (2004) [arXiv:hep-ph/0404199]; *New J. Phys.* **6**, 104 (2004) [arXiv:hep-ph/0405152]; *Mod. Phys. Lett. A* **20**, 2767 (2005) [arXiv:hep-ph/0506036]; *Phys. Rev. D* **73**, 057304 (2006) [arXiv:hep-ph/0511133]; *Mod. Phys. Lett. A* **21**, 2931 (2006) [arXiv:hep-ph/0607190]; *Mod. Phys. Lett. A* **22**, 101 (2007) [arXiv:hep-ph/0610342]; K. S. Babu, E. Ma, and J.W.F. Valle, *Phys. Lett. B* **552**, 207 (2003) [arXiv:hep-ph/0206292]; K. S. Babu, T. Kobayashi, and J. Kubo, *Phys. Rev. D* **67**, 075018 (2003) [arXiv:hep-ph/0212350]; M. Hirsch, J. C. Romao, S. Skadhauge, J.W.F. Valle, and A. Villanova del Moral, *Phys. Rev. D* **69**, 093006 (2004) [arXiv:hep-ph/0312265]; G. Altarelli and F. Feruglio, *Nucl. Phys. B* **720**, 64 (2005) [arXiv:hep-ph/0504165]; *Nucl. Phys. B* **741**, 215 (2006) [arXiv:hep-ph/0512103]; S. L. Chen, M. Frigerio, and E. Ma, *Nucl. Phys. B* **724**, 423 (2005) [arXiv:hep-ph/0504181]; M. Hirsch, A. Villanova del Moral, J.W.F. Valle, and E. Ma, *Phys. Rev. D* **72**, 091301 (2005) [Erratum-ibid. D **72**, 119904 (2005)] [arXiv:hep-ph/0507148]; X. G. He, Y. Y. Keum, and R. R. Volkas, *JHEP* **0604**, 039 (2006) [arXiv:hep-ph/0601001]; L. Lavoura and H. Kuhbock, *Mod. Phys. Lett. A* **22**, 181 (2007) [arXiv:hep-ph/0610050]; S. F. King and M. Malinsky, *Phys. Lett. B* **645**, 351 (2007) [arXiv:hep-ph/0610250]; X. G. He, *Nucl. Phys. Proc. Suppl.* **168**, 350 (2007) [arXiv:hep-ph/0612080]; Y. Koide, *Eur. Phys. J. C* **52**, 617 (2007) [arXiv:hep-ph/0701018]; S. Morisi, M. Picariello, and E. Torrente-Lujan, *Phys. Rev. D* **75**, 075015 (2007) [arXiv:hep-ph/0702034]; M. Hirsch, A. S. Joshipura, S. Kaneko,

and J.W.F. Valle, *Phys. Rev. Lett.* **99**, 151802 (2007) [arXiv:hep-ph/0703046]; F. Yin, *Phys. Rev. D* **75**, 073010 (2007) [arXiv:0704.3827 [hep-ph]]; F. Bazzocchi, S. Kaneko, and S. Morisi, *JHEP* **0803**, 063 (2008) [arXiv:0707.3032 [hep-ph]]; W. Chao, S. Luo, Z. Z. Xing, and S. Zhou, *Phys. Rev. D* **77**, 016001 (2008) [arXiv:0709.1069 [hep-ph]]; F. Bazzocchi, S. Morisi, and M. Picariello, *Phys. Lett. B* **659**, 628 (2008) [arXiv:0710.2928 [hep-ph]]; B. Brahmachari, S. Choubey, and M. Mitra, *Phys. Rev. D* **77**, 073008 (2008) [Erratum-ibid. D **77**, 119901 (2008)] [arXiv:0801.3554 [hep-ph]]; G. Altarelli, F. Feruglio, and C. Hagedorn, *JHEP* **0803**, 052 (2008) [arXiv:0802.0090 [hep-ph]]; C. Csaki, C. Delaunay, C. Grojean, and Y. Grossman, *JHEP* **0810**, 055 (2008) [arXiv:0806.0356 [hep-ph]]; P. H. Frampton and S. Matsuzaki, arXiv:0806.4592 [hep-ph]; F. Feruglio, C. Hagedorn, Y. Lin, and L. Merlo, *Nucl. Phys. B* **809**, 218 (2009) [arXiv:0807.3160 [hep-ph]]; H. Ishimori, T. Kobayashi, Y. Omura, and M. Tanimoto, *JHEP* **0812**, 082 (2008) [arXiv:0807.4625 [hep-ph]].

[445] E. Ma, *Mod. Phys. Lett. A* **20**, 2601 (2005) [arXiv:hep-ph/0508099]; *Phys. Lett. B* **671**, 366 (2009) [arXiv:0808.1729 [hep-ph]]; B. Adhikary, B. Brahmachari, A. Ghosal, E. Ma, and M. K. Parida, *Phys. Lett. B* **638**, 345 (2006) [arXiv:hep-ph/0603059]; B. Adhikary and A. Ghosal, *Phys. Rev. D* **78**, 073007 (2008) [arXiv:0803.3582 [hep-ph]]; M. Honda and M. Tanimoto, *Prog. Theor. Phys.* **119**, 583 (2008) [arXiv:0801.0181 [hep-ph]]; Y. Lin, *Nucl. Phys. B* **813**, 91 (2009) [arXiv:0804.2867 [hep-ph]].

[446] B. Adhikary and A. Ghosal, *Phys. Rev. D* **75**, 073020 (2007) [arXiv:hep-ph/0609193].

[447] D. B. Kaplan and M. Schmaltz, *Phys. Rev. D* **49**, 3741 (1994) [arXiv:hep-ph/9311281]; C. Luhn, S. Nasri, and P. Ramond, *J. Math. Phys.* **48**, 073501 (2007) [arXiv:hep-th/0701188].

[448] I. de Medeiros Varzielas, S. F. King, and G. G. Ross, *Phys. Lett. B* **648**, 201 (2007) [arXiv:hep-ph/0607045]; E. Ma, *Phys. Lett. B* **660**, 505 (2008) [arXiv:0709.0507 [hep-ph]].

[449] P. H. Frampton, T. W. Kephart, and S. Matsuzaki, *Phys. Rev. D* **78**, 073004 (2008) [arXiv:0807.4713 [hep-ph]]; D. A. Eby, P. H. Frampton, and S. Matsuzaki, *Phys. Lett. B* **671**, 386 (2009) [arXiv:0810.4899 [hep-ph]]; P. H. Frampton and S. Matsuzaki, *Phys. Lett. B* **679**, 347 (2009) [arXiv:0902.1140 [hep-ph]].

[450] C. Luhn, S. Nasri, and P. Ramond, *Phys. Lett. B* **652**, 27 (2007) [arXiv:0706.2341 [hep-ph]].

[451] E. Ma, *Phys. Lett. B* **649**, 287 (2007) [arXiv:hep-ph/0612022]; E. Ma, *Europhys. Lett.* **79**, 61001 (2007) [arXiv:hep-ph/0701016].

[452] Y. Kajiyama, M. Raidal, and A. Strumia, *Phys. Rev. D* **76**, 117301 (2007) [arXiv:0705.4559 [hep-ph]].

[453] L. L. Everett and A. J. Stuart, *Phys. Rev. D* **79**, 085005 (2009) [arXiv:0812.1057 [hep-ph]].

[454] P. H. Frampton, S. L. Glashow, and D. Marfatia, *Phys. Lett. B* **536**, 79 (2002) [arXiv:hep-ph/0201008].

[455] Z. Z. Xing, *Phys. Lett. B* **539**, 85 (2002) [arXiv:hep-ph/0205032].

[456] Z. Z. Xing, *Phys. Lett. B* **530**, 159 (2002) [arXiv:hep-ph/0201151]; B. R. Desai, D. P. Roy, and A. R. Vaucher, *Mod. Phys. Lett. A* **18**, 1355 (2003) [arXiv:hep-ph/0209035]; S. Kaneko and M. Tanimoto, *Phys. Lett. B* **551**, 127 (2003) [arXiv:hep-ph/0210155]; G. Bhattacharyya, A. Raychaudhuri, and A. Sil, *Phys. Rev. D* **67**, 073004 (2003) [arXiv:hep-ph/0211074]; R. Barbieri, T. Hambye, and A. Romanino, *JHEP* **0303**, 017 (2003) [arXiv:hep-ph/0302118]; M. Honda, S. Kaneko, and M. Tanimoto, *JHEP* **0309**, 028 (2003) [arXiv:hep-ph/0303227]; K. Hasegawa, C. S. Lim, and K. Ogure, *Phys. Rev. D* **68**, 053006 (2003) [arXiv:hep-ph/0303252]; S. Kaneko, M. Katsumata, and M. Tanimoto, *JHEP* **0307**, 025 (2003) [arXiv:hep-ph/0305014].

[457] W. Grimus and L. Lavoura, *J. Phys. G* **31**, 693 (2005) [arXiv:hep-ph/0412283]; A. Watanabe and K. Yoshioka, *JHEP* **0605**, 044 (2006) [arXiv:hep-ph/0601152].

[458] A. Kageyama, S. Kaneko, N. Shimoyama, and M. Tanimoto, *Phys. Lett. B* **538**, 96 (2002) [arXiv:hep-ph/0204291].

[459] P. H. Frampton, M. C. Oh, and T. Yoshikawa, *Phys. Rev. D* **66**, 033007 (2002) [arXiv:hep-ph/0204273].

[460] G. C. Branco, R. Gonzalez Felipe, F. R. Joaquim, and T. Yanagida, *Phys. Lett. B* **562**, 265 (2003) [arXiv:hep-ph/0212341]; G. C. Branco, M. N. Rebelo, and J. I. Silva-Marcos, *Nucl. Phys. B* **686**, 188 (2004) [arXiv:hep-ph/0311348]; I. Aizawa, T. Kitabayashi, and M. Yasue, *Phys. Rev. D* **71**, 075011 (2005) [arXiv:hep-ph/0502135]; B. C. Chauhan, J. Pulido, and M. Picariello, *Phys. Rev. D* **73**, 053003 (2006) [arXiv:hep-ph/0602084].

[461] E. Ma, *Phys. Rev. Lett.* **90**, 221802 (2003) [arXiv:hep-ph/0303126].

[462] E. Ma, arXiv:hep-ph/0208097.

[463] E. Ma and G. Rajasekaran, *Phys. Rev. D* **64**, 113012 (2001) [arXiv:hep-ph/0106291]; E. Ma, *Mod. Phys. Lett. A* **17**, 289 (2002) [arXiv:hep-ph/0201225]; K. S. Babu, E. Ma, and J. W. Valle, *Phys. Lett. B* **552**, 207 (2003) [arXiv:hep-ph/0206292]; E. Ma, *Mod. Phys. Lett. A* **17**, 2361 (2002) [arXiv:hep-ph/0211393].

[464] D. Black, A. H. Fariborz, S. Nasri, and J. Schechter, *Phys. Rev. D* **62**, 073015 (2000) [arXiv:hep-ph/0004105]; S. Nasri, J. Schechter, and S. Moussa, *Phys. Rev. D* **70**, 053005 (2004) [arXiv:hep-ph/0402176].

[465] X. G. He and A. Zee, *Phys. Lett. B* **560**, 87 (2003) [arXiv:hep-ph/0301092]; *Phys. Rev. D* **68**, 037302 (2003) [arXiv:hep-ph/0302201].

[466] H. Fritzsch, *Nucl. Phys. B* **155**, 189 (1979).

[467] H. Fritzsch and Z. Z. Xing, *Phys. Lett. B* **372**, 265 (1996) [arXiv:hep-ph/9509389]; *Phys. Lett. B* **440**, 313 (1998) [arXiv:hep-ph/9808272].

[468] C. S. Aulakh and R. N. Mohapatra, *Phys. Lett. B* **119**, 136 (1982); L. J. Hall and M. Suzuki, *Nucl. Phys. B* **231**, 419 (1984); J. R. Ellis, G. Gelmini, C. Jarlskog, G. G. Ross, and J. W. Valle, *Phys. Lett. B* **150**, 142 (1985); G. G. Ross and J. W. Valle, *Phys. Lett. B* **151**, 375 (1985); S. Dawson, *Nucl. Phys. B* **261**, 297 (1985).

[469] R. Hempfling, *Nucl. Phys. B* **478**, 3 (1996) [arXiv:hep-ph/9511288]; M. Drees, S. Pakvasa, X. Tata, and T. ter Veldhuis, *Phys. Rev. D* **57**, 5335 (1998) [arXiv:hep-ph/9712392]; E. J. Chun, S. K. Kang, C. W. Kim, and U. W. Lee, *Nucl. Phys. B* **544**, 89 (1999) [arXiv:hep-ph/9807327]; A. S. Joshipura and S. K. Vempati, *Phys. Rev. D* **60**, 111303 (1999) [arXiv:hep-ph/9903435]; A. S. Joshipura, R. D. Vaidya, and S. K. Vempati, *Phys. Rev. D* **65**, 053018 (2002) [arXiv:hep-ph/0107204]; O. C. Kong, *Mod. Phys. Lett. A* **14**, 903 (1999) [arXiv:hep-ph/9808304]; D. E. Kaplan and A. E. Nelson, *JHEP* **0001**, 033 (2000) [arXiv:hep-ph/9901254]; J. C. Romao, M. A. Diaz, M. Hirsch, W. Porod, and J. W. Valle, *Phys. Rev. D* **61**, 071703 (2000) [arXiv:hep-ph/9907499]; O. Haug, J. D. Vergados, A. Faessler, and S. Kovalenko, *Nucl. Phys. B* **565**, 38 (2000) [arXiv:hep-ph/9909318]; M. Hirsch, M. A. Diaz, W. Porod, J. C. Romao, and J. W. Valle, *Phys. Rev. D* **62**, 113008 (2000) [Erratum-ibid. D **65**, 119901 (2002)] [arXiv:hep-ph/0004115]; M. A. Diaz, M. Hirsch, W. Porod, J. C. Romao, and J.W.F. Valle, *Phys. Rev. D* **68**, 013009 (2003) [Erratum-ibid. D **71**, 059904 (2005)] [arXiv:hep-ph/0302021].

[470] M. Hirsch, H. V. Klapdor-Kleingrothaus, and S. G. Kovalenko, *Phys. Lett. B* **398**, 311 (1997) [arXiv:hep-ph/9701253]; arXiv:hep-ph/9701273; *Phys. Rev. D* **57**, 1947 (1998) [arXiv:hep-ph/9707207]; M. Hirsch, H. V. Klapdor-Kleingrothaus, S. Kolb, and S. G. Kovalenko, *Phys. Rev. D* **57**, 2020 (1998).

[471] B. Mukhopadhyaya, S. Roy, and F. Vissani, *Phys. Lett. B* **443**, 191 (1998) [arXiv:hep-ph/9808265]; W. Porod, M. Hirsch, J. Romao, and J. W. Valle, *Phys. Rev. D* **63**, 115004 (2001) [arXiv:hep-ph/0011248].

[472] E. J. Chun and J. S. Lee, *Phys. Rev. D* **60**, 075006 (1999) [arXiv:hep-ph/9811201]; S. Y. Choi, E. J. Chun, S. K. Kang, and J. S. Lee, *Phys. Rev. D* **60**, 075002 (1999)

[arXiv:hep-ph/9903465]; E. J. Chun, D. W. Jung, S. K. Kang, and J. D. Park, *Phys. Rev. D* **66**, 073003 (2002) [arXiv:hep-ph/0206030].

[473] D. Aristizabal Sierra, M. Hirsch, J.W.F. Valle, and A. Villanova del Moral, *Phys. Rev. D* **68**, 033006 (2003) [arXiv:hep-ph/0304141].

[474] V. D. Barger, T. Han, S. Hesselbach, and D. Marfatia, *Phys. Lett. B* **538**, 346 (2002) [arXiv:hep-ph/0108261].

[475] M. B. Magro, F. de Campos, O.J.P. Eboli, W. Porod, D. Restrepo, and J.W.F. Valle, *JHEP* **0309**, 071 (2003) [arXiv:hep-ph/0304232].

[476] P. Fileviez Perez and M. B. Wise, *Phys. Rev. D* **80**, 053006 (2009) [arXiv:0906.2950 [hep-ph]].

[477] A. Zee, *Nucl. Phys. B* **264**, 99 (1986); K. S. Babu, *Phys. Lett. B* **203**, 132 (1988).

[478] K. S. Babu and C. Macesanu, *Phys. Rev. D* **67**, 073010 (2003) [arXiv:hep-ph/0212058].

[479] G. Cleaver, M. Cvetic, J. R. Espinosa, L. L. Everett, and P. Langacker, *Phys. Rev. D* **57**, 2701 (1998) [arXiv:hep-ph/9705391]; P. Langacker, *Phys. Rev. D* **58**, 093017 (1998) [arXiv:hep-ph/9805281]; N. Arkani-Hamed, L. J. Hall, H. Murayama, D. Tucker-Smith, and N. Weiner, *Phys. Rev. D* **64**, 115011 (2001) [arXiv:hep-ph/0006312]; F. Borzumati and Y. Nomura, *Phys. Rev. D* **64**, 053005 (2001) [arXiv:hep-ph/0007018]; I. Gogoladze and A. Perez-Lorenzana, *Phys. Rev. D* **65**, 095011 (2002) [arXiv:hep-ph/0112034]; M. C. Chen, A. de Gouvea, and B. A. Dobrescu, *Phys. Rev. D* **75**, 055009 (2007) [arXiv:hep-ph/0612017].

[480] P. Langacker, G. Paz, L. T. Wang, and I. Yavin, *Phys. Rev. Lett.* **100**, 041802 (2008) [arXiv:0710.1632 [hep-ph]]; D. A. Demir, L. L. Everett, and P. Langäcker, *Phys. Rev. Lett.* **100**, 091804 (2008) [arXiv:0712.1341 [hep-ph]].

[481] V. Barger, P. Langacker, and H.-S. Lee, *Phys. Rev. D* **67**, 075009 (2003) [arXiv:hep-ph/0302066].

[482] E. Ma, *Phys. Lett. B* **380**, 286–290 (1996); [arXiv:hep-ph/9507348]; S. F. King, S. Moretti, R. Nevzorov, *Phys. Rev. D* **73**, 035009 (2006) [arXiv:hep-ph/0510419].

[483] E. Keith and E. Ma, *Phys. Rev. D* **54**, 3587–3593 (1996) [arXiv:hep-ph/9603353].

[484] P. Langacker, *Rev. Mod. Phys.* **81**, 1199 (2009) [arXiv:0801.1345 [hep-ph]].

[485] V. Barger, P. Fileviez Perez, and S. Spinner, *Phys. Rev. Lett.* **102**, 181802 (2009) [arXiv: 0812.3661 [hep-ph]]; *Phys. Lett. B* **696**, 509–512 (2011) [arXiv:1010.4023 [hep-ph]]; M. Ambroso and B. A. Ovrut, [arXiv:1005.5392 [hep-th]]; D. K. Ghosh, G. Senjanovic, and Y. Zhang, *Phys. Lett. B* **698**, 420–424 (2011) [arXiv:1010.3968 [hep-ph]].

[486] R. N. Mohapatra and S. Nussinov, *Phys. Rev. D* **60**, 013002 (1999) [arXiv:hep-ph/9809415].

[487] T. Appelquist and R. Shrock, *Phys. Lett. B* **548**, 204 (2002) [arXiv:hep-ph/0204141]; T. Appelquist, M. Piai, and R. Shrock, *Phys. Rev. D* **69**, 015002 (2004) [arXiv:hep-ph/0308061].

[488] S. Antusch, J. Kersten, M. Lindner, and M. Ratz, *Nucl. Phys. B* **658**, 203 (2003) [arXiv:hep-ph/0211385].

[489] R. Blumenhagen, M. Cvetic, and T. Weigand, *Nucl. Phys. B* **771**, 113 (2007) [arXiv:hep-th/0609191]; M. Cvetic, R. Richter, and T. Weigand, *Phys. Rev. D* **76**, 086002 (2007) [arXiv:hep-th/0703028]; R. Blumenhagen, M. Cvetic, D. Lust, R. Richter, and T. Weigand, *Phys. Rev. Lett.* **100**, 061602 (2008) [arXiv:0707.1871 [hep-th]].

[490] M. Cvetic and P. Langacker, *Phys. Rev. D* **78**, 066012 (2008) [arXiv:0803.2876 [hep-th]].

[491] N. Arkani-Hamed, S. Dimopoulos, and G. R. Dvali, *Phys. Lett. B* **429**, 263 (1998) [arXiv:hep-ph/9803315]; *Phys. Rev. D* **59**, 086004 (1999) [arXiv:hep-ph/9807344]; I. Antoniadis, N. Arkani-Hamed, S. Dimopoulos, and G. R. Dvali, *Phys. Lett. B* **436**, 257 (1998) [arXiv:hep-ph/9804398].

[492] L. Randall and R. Sundrum, *Phys. Rev. Lett.* **83**, 3370 (1999) [arXiv:hep-ph/9905221]; *Phys. Rev. Lett.* **83**, 4690 (1999) [arXiv:hep-th/9906064].

[493] K. R. Dienes, E. Dudas, and T. Gherghetta, *Nucl. Phys. B* **557**, 25 (1999) [arXiv:hep-ph/9811428]; N. Arkani-Hamed, S. Dimopoulos, G. R. Dvali, and J. March-Russell, *Phys. Rev. D* **65**, 024032 (2002) [arXiv:hep-ph/9811448]; A. E. Faraggi, and M. Pospelov, *Phys. Lett. B* **458**, 237 (1999) [arXiv:hep-ph/9901299]; R. N. Mohapatra, S. Nandi, and A. Perez-Lorenzana, *Phys. Lett. B* **466**, 115 (1999) [arXiv:hep-ph/9907520]; R. N. Mohapatra and A. Perez-Lorenzana, *Nucl. Phys. B* **576**, 466 (2000) [arXiv:hep-ph/9910474]; R. N. Mohapatra and A. Perez-Lorenzana, *Nucl. Phys. B* **593**, 451 (2001) [arXiv:hep-ph/0006278]; *Phys. Rev. D* **66**, 035005 (2002) [arXiv:hep-ph/0205347]; *Phys. Rev. D* **67**, 075015 (2003) [arXiv:hep-ph/0212254]; R. N. Mohapatra, A. Perez-Lorenzana, and C. A. de S. Pires, *Phys. Lett. B* **491**, 143 (2000) [arXiv:hep-ph/0008158]; Y. Grossman and M. Neubert, *Phys. Lett. B* **474**, 361 (2000) [arXiv:hep-ph/9912408]; A. Lukas, P. Ramond, A. Romanino, and G. G. Ross, *Phys. Lett. B* **495**, 136 (2000) [arXiv:hep-ph/0008049]; *JHEP* **0104**, 010 (2001) [arXiv:hep-ph/0011295]; K. R. Dienes and I. Sarcevic, *Phys. Lett. B* **500**, 133 (2001) [arXiv:hep-ph/0008144]; D. O. Caldwell, R. N. Mohapatra, and S. J. Yellin, *Phys. Rev. Lett.* **87**, 041601 (2001) [arXiv:hep-ph/0010353]; *Phys. Rev. D* **64**, 073001 (2001) [arXiv:hep-ph/0102279]; H. Davoudiasl, P. Langacker, and M. Perelstein, *Phys. Rev. D* **65**, 105015 (2002) [arXiv:hep-ph/0201128]; T. Appelquist, B. A. Dobrescu, E. Ponton, and H. U. Yee, *Phys. Rev. D* **65**, 105019 (2002) [arXiv:hep-ph/0201131]; P. Q. Hung, *Phys. Rev. D* **67**, 095011 (2003) [arXiv:hep-ph/0210131].

[494] N. Arkani-Hamed, A. G. Cohen, and H. Georgi, *Phys. Rev. Lett.* **86**, 4757 (2001) [arXiv:hep-th/0104005]; *Phys. Lett. B* **513**, 232 (2001) [arXiv:hep-ph/0105239].

[495] K. R. S. Balaji, M. Lindner, and G. Seidl, *Phys. Rev. Lett.* **91**, 161803 (2003) [arXiv:hep-ph/0303245]; T. Hallgren, T. Ohlsson, and G. Seidl, *JHEP* **0502**, 049 (2005) [arXiv:hep-ph/0411312]; T. Enkhbat and G. Seidl, *Nucl. Phys.* B **730**, 223 (2005) [arXiv:hep-ph/0504104]; J. Kubo, *Phys. Lett. B* **622**, 303 (2005) [arXiv:hep-ph/0506043].

[496] R. Dermisek, *Phys. Rev. D* **70**, 033007 (2004) [arXiv:hep-ph/0312206]; *Phys. Rev. D* **70**, 073016 (2004) [arXiv:hep-ph/0406017].

[497] S. F. King, *JHEP* **0508**, 105 (2005) [arXiv:hep-ph/0506297].

[498] N. Haba, *JHEP* **0605**, 030 (2006) [arXiv:hep-ph/0603119]; M. T. Eisele and N. Haba, *Phys. Rev. D* **74**, 073007 (2006) [arXiv:hep-ph/0603158].

[499] L. J. Hall, H. Murayama, and N. Weiner, *Phys. Rev. Lett.* **84**, 2572 (2000) [arXiv:hep-ph/9911341]; N. Haba and H. Murayama, *Phys. Rev. D* **63**, 053010 (2001) [arXiv:hep-ph/0009174]; A. de Gouvea and H. Murayama, *Phys. Lett. B* **573**, 94 (2003) [arXiv:hep-ph/0301050]; L. J. Hall, M. P. Salem, and T. Watari, *Phys. Rev. D* **76**, 093001 (2007) [arXiv:0707.3446 [hep-ph]]; J. Jenkins, *Phys. Rev. D* **79**, 113003 (2009) [arXiv:0808.1702 [hep-ph]].

[500] J. R. Espinosa, arXiv:hep-ph/0306019.

[501] G. Altarelli, F. Feruglio, and I. Masina, *JHEP* **0301**, 035 (2003) [arXiv:hep-ph/0210342]; M. Hirsch and S. F. King, *Phys. Lett. B* **516**, 103 (2001) [arXiv:hep-ph/0102103].

[502] For recent reviews see, M. Dine and A. Kusenko, *Rev. Mod. Phys.* **76**, 1 (2004) [arXiv:hep-ph/0303065]; J. M. Cline, arXiv:hep-ph/0609145; S. Davidson, E. Nardi, and Y. Nir, *Phys. Rept.* **466**, 105 (2008) [arXiv:0802.2962 [hep-ph]].

[503] M. Fukugita and T. Yanagida, *Phys. Lett. B* **174**, 45 (1986).

[504] M. Flanz, E. A. Paschos, and U. Sarkar, *Phys. Lett. B* **345**, 248 (1995) [Erratum-ibid. B **382**, 447 (1996)] [arXiv:hep-ph/9411366]; M. Flanz, E. A. Paschos, U. Sarkar, and J. Weiss, *Phys. Lett. B* **389**, 693 (1996) [arXiv:hep-ph/9607310].

[505] F. R. Klinkhamer and N. S. Manton, *Phys. Rev. D* **30**, 2212 (1984); V. A. Kuzmin, V. A. Rubakov, and M. E. Shaposhnikov, *Phys. Lett. B* **155**, 36 (1985).

[506] G. 't Hooft, *Phys. Rev. Lett.* **37**, 8 (1976); *Phys. Rev. D* **14**, 3432 (1976) [Erratum-ibid. D **18**, 2199 (1978)].

[507] J. A. Harvey and M. S. Turner, *Phys. Rev. D* **42**, 3344 (1990).

[508] E. K. Akhmedov, M. Frigerio, and A. Y. Smirnov, *JHEP* **0309**, 021 (2003) [arXiv:hep-ph/0305322].

[509] R. Barbieri, P. Creminelli, A. Strumia, and N. Tetradis, *Nucl. Phys. B* **575**, 61 (2000) [arXiv:hep-ph/9911315].

[510] A. Abada, S. Davidson, F. X. Josse-Michaux, M. Losada, and A. Riotto, *JCAP* **0604**, 004 (2006) [arXiv:hep-ph/0601083]; E. Nardi, Y. Nir, E. Roulet, and J. Racker, *JHEP* **0601**, 164 (2006) [arXiv:hep-ph/0601084].

[511] P. Di Bari, *Nucl. Phys. B* **727**, 318 (2005) [arXiv:hep-ph/0502082]; O. Vives, *Phys. Rev. D* **73**, 073006 (2006) [arXiv:hep-ph/0512160].

[512] A. Abada, P. Hosteins, F. X. Josse-Michaux, and S. Lavignac, *Nucl. Phys. B* **809**, 183 (2009) [arXiv:0808.2058 [hep-ph]].

[513] P. Di Bari and A. Riotto, *Phys. Lett. B* **671** (2009) 462 [arXiv:0809.2285 [hep-ph]].

[514] P. Hosteins, S. Lavignac, and C. A. Savoy, *Nucl. Phys. B* **755**, 137 (2006) [arXiv:hep-ph/0606078].

[515] S. Blanchet, D. Marfatia, and A. Mustafayev, *JHEP* **1011**, 038 (2010) [arXiv:1006.2857 [hep-ph]]; P. Di Bari and A. Riotto, arXiv:1012.2343 [hep-ph].

[516] G. C. Branco, R. Gonzalez Felipe, F. R. Joaquim, I. Masina, M. N. Rebelo, and C. A. Savoy, *Phys. Rev. D* **67**, 073025 (2003) [arXiv:hep-ph/0211001].

[517] G. C. Branco, R. Gonzalez Felipe, F. R. Joaquim, and M. N. Rebelo, *Nucl. Phys. B* **640**, 202 (2002) [arXiv:hep-ph/0202030]; J. R. Ellis, M. Raidal, and T. Yanagida, *Phys. Lett. B* **546**, 228 (2002) [arXiv:hep-ph/0206300]; S. Davidson and A. Ibarra, *Nucl. Phys. B* **648**, 345 (2003) [arXiv:hep-ph/0206304]; W. Rodejohann, *Phys. Lett. B* **542**, 100 (2002) [arXiv:hep-ph/0207053]; M. N. Rebelo, *Phys. Rev. D* **67**, 013008 (2003) [arXiv:hep-ph/0207236]; P. H. Frampton, S. L. Glashow, and T. Yanagida, *Phys. Lett. B* **548**, 119 (2002) [arXiv:hep-ph/0208157]; T. Endoh, S. Kaneko, S. K. Kang, T. Morozumi, and M. Tanimoto, *Phys. Rev. Lett.* **89**, 231601 (2002) [arXiv:hep-ph/0209020]; S. F. King, *Phys. Rev. D* **67**, 113010 (2003) [arXiv:hep-ph/0211228]; S. Pascoli, S. T. Petcov, and W. Rodejohann, *Phys. Rev. D* **68**, 093007 (2003) [arXiv:hep-ph/0302054].

[518] M. L. Brooks et al. [MEGA Collaboration], *Phys. Rev. Lett.* **83**, 1521 (1999) [arXiv:hep-ex/9905013].

[519] K. Hayasaka et al. [Belle Collaboration], *Phys. Lett. B* **666**, 16 (2008) [arXiv:0705.0650 [hep-ex]].

[520] B. Aubert et al. [BABAR Collaboration], arXiv:0908.2381 [hep-ex].

[521] U. Bellgardt et al. [SINDRUM Collaboration], *Nucl. Phys. B* **299**, 1 (1988).

[522] Y. Miyazaki et al. [Belle Collaboration], *Phys. Lett. B* **660**, 154 (2008) [arXiv:0711.2189 [hep-ex]].

[523] C. Dohmen et al. [SINDRUM II Collaboration.], *Phys. Lett. B* **317**, 631 (1993).

[524] S. Ritt [MEG Collaboration], *Nucl. Phys. Proc. Suppl.* **162**, 279 (2006); T. Mori, *Nucl. Phys. Proc. Suppl.* **169**, 166 (2007).

[525] M. Bona et al., arXiv:0709.0451 [hep-ex]; A. G. Akeroyd et al. [SuperKEKB Physics Working Group], arXiv:hep-ex/0406071.

[526] With no experiment planned, we use theoretical estimates based on the PSI beam intensity from W. J. Marciano, T. Mori, and J. M. Roney, *Ann. Rev. Nucl. Part. Sci.* **58**, 315 (2008).

[527] Y. Mori et al. [The PRIME Working Group], "An Experimental Search for the $\mu^- - e^-$ Conversion Process at an Ultimate Sensitivity of the Order of 10^{-18} with PRISM", LOI-25, http://www-ps.kek.jp/jhf-np/LOIlist/LOIlist.html.

[528] E. C. Dukes et al. [Mu2e Collaboration], "Proposal to Search for $\mu^- N \rightarrow e^- N$ with a Single Event Sensitivity Below 10^{-16}", http://mu2e.fnal.gov/public/hep/index.shtml.

[529] R. Buras, H. T. Janka, M. T. Keil, G. G. Raffelt, and M. Rampp, *Astrophys. J.* **587**, 320 (2003) [arXiv:astro-ph/0205006].

[530] T. A. Thompson, A. Burrows, and P. A. Pinto, *Astrophys. J.* **592**, 434 (2003) [arXiv:astro-ph/0211194].

[531] See, e.g., A. Dighe, *J. Phys. Conf. Ser.* **136**, 022041 (2008) [arXiv:0809.2977 [hep-ph]], and references therein.

[532] H. Duan, G. M. Fuller, and Y. Z. Qian, *Phys. Rev. D* 77, 085016 (2008) [arXiv:0801.1363 [hep-ph]].

[533] B. Dasgupta, A. Dighe, A. Mirizzi, and G. G. Raffelt, *Phys. Rev. D* 77, 113007 (2008) [arXiv:0801.1660 [hep-ph]].

[534] H. Duan, G. M. Fuller, and Y. Z. Qian, arXiv:1001.2799 [hep-ph].

[535] T. Totani, K. Sato, H. E. Dalhed, and J. R. Wilson, *Astrophys. J.* **496**, 216 (1998) [arXiv:astro-ph/9710203].

[536] G. G. Raffelt, M. T. Keil, R. Buras, H. T. Janka, and M. Rampp, arXiv:astro-ph/0303226.

[537] A. Dighe, *J. Phys. Conf. Ser.* **203**, 012015 (2010) [arXiv:0912.4167 [hep-ph]].

[538] H. Duan, G. M. Fuller, J. Carlson, and Y. Z. Qian, *Phys. Rev. D* **74**, 105014 (2006) [arXiv:astro-ph/0606616]; *Phys. Rev. Lett.* **97**, 241101 (2006) [arXiv:astro-ph/0608050]; G. L. Fogli, E. Lisi, A. Marrone, and A. Mirizzi, *JCAP* **0712**, 010 (2007) [arXiv:0707.1998 [hep-ph]]; A. Esteban-Pretel, S. Pastor, R. Tomas, G. G. Raffelt, and G. Sigl, *Phys. Rev. D* **76**, 125018 (2007) [arXiv:0706.2498 [astro-ph]].

[539] H. Duan and A. Friedland, arXiv:1006.2359 [hep-ph].

[540] J. F. Cherry, J. Carlson, A. Friedland, G. M. Fuller, and A. Vlasenko, arXiv:1203.1607 [hep-ph].

[541] B. Dasgupta, A. Dighe, G. G. Raffelt, and A. Y. Smirnov, *Phys. Rev. Lett.* **103**, 051105 (2009) [arXiv:0904.3542 [hep-ph]].

[542] A. Friedland, *Phys. Rev. Lett.* **104**, 191102 (2010) [arXiv:1001.0996 [hep-ph]].

[543] G. Brown, H. Bethe, and G. Baym, *Nucl. Phys.A* **375**, 481 (1982).

[544] A. S. Dighe and A. Y. Smirnov, *Phys. Rev. D* **62**, 033007 (2000) [arXiv:hep-ph/9907423].

[545] G. L. Fogli, E. Lisi, D. Montanino, and A. Palazzo, *Phys. Rev. D* **65**, 073008 (2002) [Erratum-ibid. D **66**, 039901 (2002)] [arXiv:hep-ph/0111199].

[546] T. K. Kuo and J. Pantaleone, *Phys. Rev. D* **37** 298 (1988).

[547] R. C. Schirato and G. M. Fuller, arXiv:astro-ph/0205390.

[548] K. Hirata et al. [KAMIOKANDE-II Collaboration], *Phys. Rev. Lett.* **58**, 1490 (1987); K. S. Hirata et al., *Phys. Rev. D* **38**, 448 (1988).

[549] R. M. Bionta et al., *Phys. Rev. Lett.* **58**, 1494 (1987); C. B. Bratton et al. [IMB Collaboration], *Phys. Rev. D* **37**, 3361 (1988).

[550] V. Barger, D. Marfatia, and B. P. Wood, *Phys. Lett. B* **532**, 19 (2002) [arXiv:hep-ph/0202158].

[551] T. J. Loredo and D. Q. Lamb, *Phys. Rev. D* **65**, 063002 (2002) [arXiv:astro-ph/0107260].

[552] M. L. Cherry et al., *J. Phys. G* **8**, 879 (1982); G. T. Zatsepin and O. G. Ryazhskaya, *Usp. Fiz. Nauk* **146**, 713 (1985) [*Sov. Phys. Usp.* **28**, 726 (1985)]; E. N. Alexeyev et al., *Zh. Eksp. Teor. Fiz.* **104**, 2897 (1993) [*JETP* **77**, 339 (1993)]; R. S. Miller, R. Becker-Szendy, C. B. Bratton, J. Breault, D. Casper, S. T. Dye, W. Gajewski, M. Goldhaber et al., *Astrophys. J.* **428**, 629–632 (1994); M. Ambrosio et al. [MACRO Collaboration], *Astropart. Phys.* **8**, 123–133 (1998); R. V. Novoseltseva, M. M. Boliev, I. M. Dzaparova, M. M. Kochkarov, S. P. Mikheyev, Y. F. Novoseltsev, V. B. Petkov, P. S. Striganov et al., arXiv:0910.0738 [astro-ph.HE].

[553] C. Vigorito et al. [LVD Collaboration], *J. Phys. Conf. Ser.* **136**, 042074 (2008); M. Selvi [LVD Collaboration], [arXiv:hep-ex/0608061].
[554] M. Ikeda et al. [Super-Kamiokande Collaboration], *Astrophys. J.* **669**, 519–524 (2007) [arXiv:0706.2283 [astro-ph]].
[555] A. A. Aguilar-Arevalo et al. [MiniBooNE Collaboration], *Phys. Rev. D* **81**, 032001 (2010) [arXiv:0910.3182 [astro-ph.HE]].
[556] P. Antonioli et al., *New J. Phys.* **6**, 114 (2004) [arXiv:astro-ph/0406214].
[557] F. Halzen and G. G. Raffelt, *Phys. Rev. D* **80**, 087301 (2009) [arXiv:0908.2317 [astro-ph.HE]].
[558] E. K. Akhmedov, C. Lunardini, and A. Y. Smirnov, *Nucl. Phys. B* **643**, 339 (2002) [arXiv:hep-ph/0204091].
[559] B. Jegerlehner, F. Neubig, and G. Raffelt, *Phys. Rev. D* **54**, 1194 (1996) [arXiv:astro-ph/9601111]; S. Choubey, D. Majumdar, and K. Kar, *J. Phys. G* **25**, 1001 (1999) [arXiv:hep-ph/9809424]; G. Dutta, D. Indumathi, M. V. Murthy, and G. Rajasekaran, *Phys. Rev. D* **61**, 013009 (2000) [arXiv:hep-ph/9907372]; *Phys. Rev. D* **64**, 073011 (2001) [arXiv:hep-ph/0101093]; C. Lunardini and A. Y. Smirnov, *Phys. Rev. D* **63**, 073009 (2001) [arXiv:hep-ph/0009356]; *Nucl. Phys. B* **616**, 307 (2001) [arXiv:hep-ph/0106149]; *JCAP* **0306**, 009 (2003) [arXiv:hep-ph/0302033]; H. Minakata and H. Nunokawa, *Phys. Lett. B* **504**, 301 (2001) [arXiv:hep-ph/0010240]; M. Kachelriess, R. Tomas, and J. W. Valle, *JHEP* **0101**, 030 (2001) [arXiv:hep-ph/0012134]; K. Takahashi, M. Watanabe, K. Sato, and T. Totani, *Phys. Rev. D* **64**, 093004 (2001) [arXiv:hep-ph/0105204]; K. Takahashi and K. Sato, *Prog. Theor. Phys.* **109**, 919 (2003) [arXiv:hep-ph/0205070].
[560] V. Barger, D. Marfatia, and B. P. Wood, *Phys. Lett. B* **547**, 37 (2002) [arXiv:hep-ph/0112125].
[561] M. Kachelriess, R. Tomas, R. Buras, H. T. Janka, A. Marek, and M. Rampp, *Phys. Rev. D* **71**, 063003 (2005) [arXiv:astro-ph/0412082].
[562] V. Barger, P. Huber, and D. Marfatia, *Phys. Lett. B* **617**, 167 (2005) [arXiv:hep-ph/0501184].
[563] J. M. Cordes and D. F. Chernoff, *Astrophys. J.* **505**, 315 (1998) [arXiv:astro-ph/9707308]; Z. Arzoumanian, D. F. Chernoffs, and J. M. Cordes, *Astrophys. J.* **568**, 289 (2002) [arXiv:astro-ph/0106159].
[564] J. M. Blondin, A. Mezzacappa, and C. DeMarino, *Astrophys. J.* **584**, 971 (2003) [arXiv:astro-ph/0210634].
[565] P. Goldreich, D. Lai, and M. Sahrling, *Unsolved Problems in Astrophysics*, pp. 269–280, Princeton University Press, 1997.
[566] For a recent review, see J. F. Beacom, *Ann. Rev. Nucl. Part. Sci.* **60**, 439 (2010) [arXiv:1004.3311 [astro-ph.HE]].
[567] S. Ando, K. Sato, and T. Totani, *Astropart. Phys.* **18**, 307 (2003) [arXiv:astro-ph/0202450]; M. Fukugita and M. Kawasaki, *Mon. Not. Roy. Astron. Soc.* **340**, L7 (2003) [arXiv:astro-ph/0204376]; J. F. Beacom and M. R. Vagins, *Phys. Rev. Lett.* **93**, 171101 (2004) [arXiv:hep-ph/0309300].
[568] H. Yuksel, S. Ando, and J. F. Beacom, *Phys. Rev. C* **74**, 015803 (2006) [arXiv:astro-ph/0509297].
[569] J. F. Beacom and L. E. Strigari, *Phys. Rev. C* **73**, 035807 (2006) [arXiv:hep-ph/0508202].
[570] L. E. Strigari, M. Kaplinghat, G. Steigman, and T. P. Walker, *JCAP* **0403**, 007 (2004) [arXiv:astro-ph/0312346].
[571] M. Malek et al. [Super-Kamiokande Collaboration], *Phys. Rev. Lett.* **90**, 061101 (2003) [arXiv:hep-ex/0209028].
[572] H. Watanabe et al. [Super-Kamiokande Collaboration], arXiv:0811.0735 [hep-ex].

[573] J. W. Cronin, S. P. Swordy, and T. K. Gaisser, *Sci. Am.* **276**, 32 (1997); S. P. Swordy, *Space Science Reviews* **99**, 85 (2001).
[574] F. Halzen, *Astrophys. Space Sci.* **309**, 407 (2007) [arXiv:astro-ph/0611915].
[575] K. Greisen, *Phys. Rev. Lett.* **16**, 748 (1966); G. T. Zatsepin and V. A. Kuzmin, *JETP Lett.* **4**, 78 (1966) [*Pisma Zh. Eksp. Teor. Fiz.* **4**, 114 (1966)].
[576] F. W. Stecker, *Phys. Rev. Lett.* **21**, 1016 (1968); F. A. Aharonian and J. W. Cronin, *Phys. Rev. D* **50**, 1892 (1994); J. W. Elbert and P. Sommers, *Astrophys. J.* **441**, 151 (1995) [arXiv:astro-ph/9410069].
[577] S. Yoshida and M. Teshima, *Prog. Theor. Phys.* **89**, 833 (1993).
[578] R. Abbasi et al. [HiRes Collaboration], *Phys. Rev. Lett.* **100**, 101101 (2008) [arXiv:astro-ph/0703099].
[579] J. Abraham et al. [Pierre Auger Collaboration], *Nucl. Instrum. Meth.* A **523**, 50 (2004); *Phys. Rev. Lett.* **101**, 061101 (2008) [arXiv:0806.4302 [astro-ph]].
[580] J. Abraham et al. [The Pierre Auger Collaboration], *Phys. Lett. B* **685**, 239 (2010) [arXiv:1002.1975 [astro-ph.HE]].
[581] V. S. Berezinsky and G. T. Zatsepin, *Yad. Fiz.* **11**, 200 (1970).
[582] J. Abraham et al. [Pierre Auger Observatory Collaboration], *Phys. Rev. Lett.* **104**, 091101 (2010) [arXiv:1002.0699 [astro-ph.HE]]; however, a contrary conclusion is suggested by R. U. Abbasi et al. [HiRes Collaboration], *Phys. Rev. Lett.* **104**, 161101 (2010) [arXiv:0910.4184 [astro-ph.HE]].
[583] F. Halzen and S. R. Klein, *Rev. Sci. Instrum.* **81**, 081101 (2010) [arXiv:1007.1247 [astro-ph.HE]].
[584] F. Halzen, arXiv:0910.0436 [astro-ph.HE].
[585] R. Abbasi et al. [IceCube Collaboration], *Phys. Rev. D* **79**, 102005 (2009) [arXiv:0902.0675 [astro-ph.HE]].
[586] M. C. Gonzalez-Garcia, M. Maltoni, and J. Rojo, *JHEP* **0610**, 075 (2006) [arXiv:hep-ph/0607324].
[587] K. Daum et al. [Frejus Collaboration], *Z. Phys. C* **66**, 417 (1995).
[588] W. Rhode et al. [Frejus Collaboration], *Astropart. Phys.* **4**, 217 (1996).
[589] A. Achterberg et al. [IceCube Collaboration], *Phys. Rev. D* **76**, 042008 (2007) [Erratum-ibid. D 77, 089904 (2008)] [arXiv:0705.1315 [astro-ph]].
[590] R. Abbasi et al [IceCube Collaboration], *Astropart. Phys.* **34**, 48 (2010) [arXiv:1004.2357 [astro-ph.HE]].
[591] For a summary of current limits, see R. Abbasi et al. [ICECUBE Collaboration], *Phys. Rev. Lett.* **102**, 201302 (2009) [arXiv:0902.2460 [astro-ph.CO]].
[592] D. F. Cowen [IceCube Collaboration], Prepared for *13th International Workshop on Neutrino Telescopes: Un altro modo di guardare il cielo: Tribute to Galileo*, Venice, Italy, 10–13 Mar 2009.
[593] E. Waxman and J. N. Bahcall, *Phys. Rev. D* **59**, 023002 (1999) [arXiv:hep-ph/9807282]; J. N. Bahcall and E. Waxman, *Phys. Rev. D* **64**, 023002 (2001) [arXiv:hep-ph/9902383].
[594] I. Kravchenko et al., *Phys. Rev. D* **73**, 082002 (2006) [arXiv:astro-ph/0601148].
[595] P. Miocinovic et al. [ANITA Collaboration], In the *Proceedings of 22nd Texas Symposium on Relativistic Astrophysics at Stanford University*, Stanford, California, 13–17 Dec 2004, [arXiv:astro-ph/0503304].
[596] P. W. Gorham et al. [The ANITA Collaboration], *Phys. Rev. D* **82**, 022004 (2010) [arXiv:1003.2961 [astro-ph.HE]]; *Phys. Rev. Lett.* **103**, 051103 (2009) [arXiv:0812.2715 [astro-ph]].
[597] V. Barger, P. Huber, and D. Marfatia, *Phys. Lett. B* **642**, 333 (2006) [arXiv:hep-ph/0606311].
[598] D. V. Semikoz and G. Sigl, *JCAP* **0404**, 003 (2004) [arXiv:hep-ph/0309328].
[599] L. A. Anchordoqui and T. Montaruli, *Ann. Rev. Nucl. Part. Sci.* **60**, 129 (2010).

[600] S. Pakvasa, W. Rodejohann, and T. J. Weiler, *JHEP* **0802**, 005 (2008) [arXiv:0711.4517 [hep-ph]].

[601] J. F. Beacom, N. F. Bell, D. Hooper, S. Pakvasa, and T. J. Weiler, *Phys. Rev. D* **68**, 093005 (2003) [Erratum-ibid. D **72**, 019901 (2005)] [arXiv:hep-ph/0307025]; *Phys. Rev. D* **69**, 017303 (2004) [arXiv:hep-ph/0309267].

[602] G. B. Gelmini, A. Kusenko, and T. J. Weiler, *Sci. Am.* **302N5**, 20 (2010).

[603] M. Drees, R. Godbole, and P. Roy, *Theory and phenomenology of sparticles: An account of four-dimensional N=1 supersymmetry in high energy physics*, Hackensack, USA: World Scientific (2004).

[604] H. Baer and X. Tata, *Weak scale supersymmetry: From superfields to scattering events*, Cambridge, UK: Univ. Pr. (2006); P. Binetruy, *Supersymmetry: Theory, experiment and cosmology*, Oxford, UK: Oxford Univ. Pr. (2006).

[605] H. E. Haber and G. L. Kane, *Phys. Rept.* **117**, 75 (1985).

[606] Dark Matter Scientific Assessment Group (DMSAG), "Report on the Direct Detection and Study of Dark Matter," http://www.science.energy.gov/~/media/hep/pdf/files/pdfs/dmsagreportjuly18_2007.pdf

[607] G. Jungman, M. Kamionkowski, and K. Griest, *Phys. Rept.* **267**, 195 (1996) [arXiv:hep-ph/9506380].

[608] O. Mena, S. Palomares-Ruiz, and S. Pascoli, *Phys. Lett. B* **664**, 92 (2008) [arXiv:0706.3909 [hep-ph]].

[609] E. A. Baltz, M. Battaglia, M. E. Peskin, and T. Wizansky, *Phys. Rev. D* **74**, 103521 (2006) [arXiv:hep-ph/0602187].

[610] D. Hooper and L. T. Wang, *Phys. Rev. D* **69**, 035001 (2004) [arXiv:hep-ph/0309036].

[611] R. L. Arnowitt, B. Dutta, T. Kamon, N. Kolev, and D. A. Toback, *Phys. Lett. B* **639**, 46 (2006) [arXiv:hep-ph/0603128].

[612] L. Roszkowski, R. R. de Austri, J. Silk, and R. Trotta, *Phys. Lett. B* **671**, 10 (2009) [arXiv:0707.0622 [astro-ph]].

[613] K. Griest and D. Seckel, *Nucl. Phys.* B **283**, 681 (1987) [Erratum-ibid. B **296**, 1034 (1988)].

[614] J. Ellis, K. A. Olive, C. Savage, and V. C. Spanos, *Phys. Rev. D* **81**, 085004 (2010) [arXiv:0912.3137 [hep-ph]].

[615] A. Gould, *Astrophys. J.* **388**, 338 (1992).

[616] M. Kamionkowski, *Phys. Rev. D* **44**, 3021 (1991).

[617] M. Cirelli, N. Fornengo, T. Montaruli, I. Sokalski, A. Strumia, and F. Vissani, *Nucl. Phys. B* **727**, 99 (2005) [Erratum-ibid. B **790**, 338 (2008)] [arXiv:hep-ph/0506298].

[618] P. Lipari and T. Stanev, *Phys. Rev. D* **44**, 3543 (1991).

[619] D. Chirkin and W. Rhode, arXiv:hep-ph/0407075.

[620] V. Barger, Y. Gao, and D. Marfatia, *Phys. Rev. D* Vol. **83**, 055012 (2011), © 2011 American Physical Society [arXiv:1101.4410 [hep-ph]].

[621] M. C. Gonzalez-Garcia, F. Halzen, and S. Mohapatra, *Astropart. Phys.* **31**, 437 (2009) [arXiv:0902.1176 [astro-ph.HE]].

[622] V. Barger, J. Kumar, D. Marfatia, and E. M. Sessolo, *Phys. Rev. D* **81**, 115010 (2010) [arXiv:1004.4573 [hep-ph]].

[623] E. Resconi [IceCube Collaboration], *Nucl. Instrum. Meth. A* **602**, 7 (2009) [arXiv:0807.3891 [astro-ph]].

[624] T. DeYoung, private communication.

[625] M. Honda, T. Kajita, K. Kasahara, S. Midorikawa, and T. Sanuki, *Phys. Rev. D* **75**, 043006 (2007) [arXiv:astro-ph/0611418].

[626] A. E. Erkoca, M. H. Reno, and I. Sarcevic, *Phys. Rev. D* **80**, 043514 (2009) [arXiv:0906.4364 [hep-ph]].

[627] C. Athanassopoulos et al. [LSND Collaboration], *Phys. Rev. C* **54**, 2685 (1996) [arXiv:nucl-ex/9605001]; *Phys. Rev. Lett.* **77**, 3082 (1996) [arXiv:nucl-ex/9605003]; A. Aguilar et al., *Phys. Rev. D* **64**, 112007 (2001) [arXiv:hep-ex/0104049].

[628] C. Athanassopoulos et al. [LSND Collaboration], *Phys. Rev. C* **58**, 2489 (1998) [arXiv:nucl-ex/9706006]; *Phys. Rev. Lett.* **81**, 1774 (1998) [arXiv:nucl-ex/9709006].

[629] K. Eitel [KARMEN Collaboration], *Nucl. Phys. Proc. Suppl.* **91**, 191 (2000) [arXiv:hep-ex/0008002]; B. Armbruster et al. [KARMEN Collaboration], *Phys. Rev. D* **65**, 112001 (2002) [arXiv:hep-ex/0203021].

[630] Y. Declais et al., *Nucl. Phys. B* **434**, 503 (1995).

[631] I. E. Stockdale et al., *Phys. Rev. Lett.* **52**, 1384 (1984); A. Romosan et al. [CCFR/NuTeV Collaboration], *Phys. Rev. Lett.* **78**, 2912 (1997) [arXiv:hep-ex/9611013].

[632] P. Astier et al. [NOMAD Collaboration], *Nucl. Phys. B* **611**, 3 (2001) [arXiv:hep-ex/0106102]; P. Astier et al. [NOMAD Collaboration], *Phys. Lett. B* **570**, 19 (2003) [arXiv:hep-ex/0306037].

[633] E. D. Church, K. Eitel, G. B. Mills, and M. Steidl, *Phys. Rev. D* **66**, 013001 (2002) [arXiv:hep-ex/0203023].

[634] D. O. Caldwell and R. N. Mohapatra, *Phys. Rev. D* **48**, 3259 (1993); J. T. Peltoniemi and J. W. Valle, *Nucl. Phys. B* **406**, 409 (1993) [arXiv:hep-ph/9302316]; E. J. Chun, A. S. Joshipura, and A. Y. Smirnov, *Phys. Lett. B* **357**, 608 (1995) [arXiv:hep-ph/9505275]; *Phys. Rev. D* **54**, 4654 (1996) [arXiv:hep-ph/9507371]; R. Foot and R. R. Volkas, *Phys. Rev. D* **52**, 6595 (1995) [arXiv:hep-ph/9505359]; Z. G. Berezhiani and R. N. Mohapatra, *Phys. Rev. D* **52**, 6607 (1995) [arXiv:hep-ph/9505385]; E. Ma and P. Roy, *Phys. Rev. D* **52**, 4780 (1995) [arXiv:hep-ph/9504342]; K. Benakli and A. Y. Smirnov, *Phys. Rev. Lett.* **79**, 4314 (1997) [arXiv:hep-ph/9703465].

[635] V. D. Barger, P. Langacker, J. P. Leveille, and S. Pakvasa, *Phys. Rev. Lett.* **45**, 692 (1980).

[636] N. Okada and O. Yasuda, *Int. J. Mod. Phys. A* **12**, 3669 (1997) [arXiv:hep-ph/9606411]; S. M. Bilenky, C. Giunti, and W. Grimus, *Eur. Phys. J. C* **1**, 247 (1998) [arXiv:hep-ph/9607372]; S. M. Bilenky, C. Giunti, W. Grimus, and T. Schwetz, *Phys. Rev. D* **60**, 073007 (1999) [arXiv:hep-ph/9903454].

[637] V. D. Barger, S. Pakvasa, T. J. Weiler, and K. Whisnant, *Phys. Rev. D* **58**, 093016 (1998) [arXiv:hep-ph/9806328].

[638] V. D. Barger, T. J. Weiler, and K. Whisnant, *Phys. Lett. B* **427**, 97 (1998) [arXiv:hep-ph/9712495]; S. C. Gibbons, R. N. Mohapatra, S. Nandi, and A. Raychaudhuri, *Phys. Lett. B* **430**, 296 (1998) [arXiv:hep-ph/9803299]; N. Gaur, A. Ghosal, E. Ma, and P. Roy, *Phys. Rev. D* **58**, 071301 (1998) [arXiv:hep-ph/9806272].

[639] M. Maltoni, T. Schwetz, M. A. Tortola, and J.W.F. Valle, *New J. Phys.* **6**, 122 (2004) [arXiv:hep-ph/0405172].

[640] S. Fukuda et al. [Super-Kamiokande Collaboration], *Phys. Rev. Lett.* **85**, 3999 (2000) [arXiv:hep-ex/0009001].

[641] T. Nakaya [Super-Kamiokande Collaboration], *eConf* **C020620**, SAAT01 (2002) [arXiv:hep-ex/0209036].

[642] F. Dydak et al., *Phys. Lett. B* **134**, 281 (1984).

[643] P. Adamson et al. [MINOS Collaboration], *Phys. Rev. Lett.* **107**, 011802 (2011) [arXiv:1104.3992 [hep-ex]]; *Phys. Rev. D* **81**, 052004 (2010) [arXiv:1001.0336 [hep-ex]]; *Phys. Rev. Lett.* **101**, 221804 (2008) [arXiv:0807.2424 [hep-ex]].

[644] K.B.M. Mahn et al. [SciBooNE and MiniBooNE Collaboration], arXiv:1106.5685 [hep-ex].

[645] V. D. Barger, B. Kayser, J. Learned, T. J. Weiler, and K. Whisnant, *Phys. Lett. B* **489**, 345 (2000) [arXiv:hep-ph/0008019]; C. Giunti and M. Laveder, *JHEP* **0102**, 001 (2001) [arXiv:hep-ph/0010009].

[646] O. L. Peres and A. Y. Smirnov, *Nucl. Phys. B* **599**, 3 (2001) [arXiv:hep-ph/0011054].

[647] W. Grimus and T. Schwetz, *Eur. Phys. J. C* **20**, 1 (2001) [arXiv:hep-ph/0102252]; M. Maltoni, T. Schwetz, and J. W. Valle, *Phys. Lett. B* **518**, 252 (2001) [arXiv:hep-ph/0107150].

[648] M. Maltoni, T. Schwetz, M. A. Tortola, and J. W. Valle, *Nucl. Phys. B* **643**, 321 (2002) [arXiv:hep-ph/0207157].

[649] H. Pas, L. G. Song, and T. J. Weiler, *Phys. Rev. D* **67**, 073019 (2003) [arXiv:hep-ph/0209373].

[650] A. A. Aguilar-Arevalo et al. [The MiniBooNE Collaboration], *Phys. Rev. Lett.* **98**, 231801 (2007) [arXiv:0704.1500 [hep-ex]].

[651] A. A. Aguilar-Arevalo et al. [MiniBooNE Collaboration], *Phys. Rev. D* **78**, 012007 (2008) [arXiv:0805.1764 [hep-ex]].

[652] A. A. Aguilar-Arevalo et al. [MiniBooNE Collaboration], *Phys. Rev. Lett.* **102**, 101802 (2009) [arXiv:0812.2243 [hep-ex]].

[653] A. A. Aguilar-Arevalo et al. [MiniBooNE Collaboration], *Phys. Rev. Lett.* **103**, 111801 (2009) [arXiv:0904.1958 [hep-ex]].

[654] A. A. Aguilar-Arevalo et al. [MiniBooNE Collaboration], arXiv:1007.1150 [hep-ex].

[655] M. Sorel, J. M. Conrad, and M. Shaevitz, *Phys. Rev. D* **70**, 073004 (2004) [arXiv:hep-ph/0305255].

[656] G. Karagiorgi, A. Aguilar-Arevalo, J. M. Conrad, M. H. Shaevitz, K. Whisnant, M. Sorel, and V. Barger, *Phys. Rev. D* 75, 013011 (2007) [Erratum-ibid. D **80**, 099902 (2009)] [arXiv:hep-ph/0609177].

[657] M. Maltoni and T. Schwetz, *Phys. Rev. D* **76**, 093005 (2007) [arXiv:0705.0107 [hep-ph]].

[658] A. A. Aguilar-Arevalo et al. [MiniBooNE Collaboration], *Phys. Rev. Lett.* **103**, 061802 (2009) [arXiv:0903.2465 [hep-ex]].

[659] G. Karagiorgi, Z. Djurcic, J. M. Conrad, M. H. Shaevitz, and M. Sorel, *Phys. Rev. D* **80**, 073001 (2009) [Erratum-ibid. D **81**, 039902 (2010)] [arXiv:0906.1997 [hep-ph]].

[660] E. Akhmedov and T. Schwetz, arXiv:1007.4171 [hep-ph].

[661] T. Schwetz, *JHEP* **0802**, 011 (2008) [arXiv:0710.2985 [hep-ph]].

[662] W. S. Hou and A. Soddu, *Phys. Lett. B* **638**, 229 (2006) [arXiv:hep-ph/0512278]; W. S. Hou and F. F. Lee, arXiv:1004.2359 [hep-ph].

[663] A. S. Riis, S. Hannestad, *JCAP* **1102**, 011 (2011) [arXiv:1008.1495 [astro-ph.CO]].

[664] H. Murayama and T. Yanagida, *Phys. Lett. B* **520**, 263 (2001) [arXiv:hep-ph/0010178]; G. Barenboim, L. Borissov, J. Lykken, and A. Y. Smirnov, *JHEP* **0210**, 001 (2002) [arXiv:hep-ph/0108199]; G. Barenboim, L. Borissov, and J. Lykken, *Phys. Lett. B* **534**, 106 (2002) [arXiv:hep-ph/0201080].

[665] G. Barenboim and J. Lykken, *Phys. Lett. B* **554**, 73 (2003) [arXiv:hep-ph/0210411]; O. W. Greenberg, *Phys. Lett. B* **567**, 179 (2003) [arXiv:hep-ph/0305276].

[666] G. Barenboim, L. Borissov, and J. Lykken, arXiv:hep-ph/0212116.

[667] A. Strumia, *Phys. Lett. B* **539**, 91 (2002) [arXiv:hep-ph/0201134].

[668] M. C. Gonzalez-Garcia, M. Maltoni, and T. Schwetz, *Phys. Rev. D* **68**, 053007 (2003) [arXiv:hep-ph/0306226].

[669] G. Barenboim and J. D. Lykken, *Phys. Rev. D* **80**, 113008 (2009) [arXiv:0908.2993 [hep-ph]].

[670] V. Barger, D. Marfatia, and K. Whisnant, *Phys. Lett. B* **576**, 303 (2003) [arXiv:hep-ph/0308299].

[671] S. Hollenberg, O. Micu, H. Pas, and T. J. Weiler, *Phys. Rev. D* **80**, 093005 (2009) [arXiv:0906.0150 [hep-ph]]; *AIP Conf. Proc.* **1200**, 952 (2010) [arXiv:0908.3986 [hep-ph]].

[672] S. Hollenberg and H. Pas, arXiv:0904.2167 [hep-ph]; S. Esposito and G. Salesi, *Mod. Phys. Lett. A* **25**, 597 (2010) [arXiv:0906.5542 [hep-ph]].

[673] D. Colladay and V. A. Kostelecky, *Phys. Rev. D* **58**, 116002 (1998) [arXiv:hep-ph/9809521]; V. A. Kostelecky, *Phys. Rev. D* **69**, 105009 (2004) [arXiv:hep-th/0312310]; R. Bluhm, *Lect. Notes Phys.* **702**, 191 (2006) [arXiv:hep-ph/0506054].

[674] V. A. Kostelecky and M. Mewes, *Phys. Rev. D* **69**, 016005 (2004) [arXiv:hep-ph/0309025].

[675] S. Hollenberg, O. Micu, and H. Pas, *Phys. Rev. D* **80**, 053010 (2009) [arXiv:0906.5072 [hep-ph]].

[676] S. Hollenberg, O. Micu, and H. Pas, *Prog. Part. Nucl. Phys.* **64**, 193 (2010) [arXiv:0911.1018 [hep-ph]].

[677] J. N. Abdurashitov et al. [SAGE Collaboration], *Phys. Rev. C* **80**, 015807 (2009) [arXiv:0901.2200 [nucl-ex]]; F. Kaether, W. Hampel, G. Heusser, J. Kiko, and T. Kirsten, *Phys. Lett. B* **685**, 47 (2010) [arXiv:1001.2731 [hep-ex]].

[678] C. Giunti and M. Laveder, arXiv:1006.3244 [hep-ph].

[679] C. Giunti and M. Laveder, *Phys. Rev. D* **77**, 093002 (2008) [arXiv:0707.4593 [hep-ph]]; M. A. Acero, C. Giunti, and M. Laveder, *Phys. Rev. D* **78**, 073009 (2008) [arXiv:0711.4222 [hep-ph]]; C. Giunti and M. Laveder, arXiv:1005.4599 [hep-ph]; *Phys. Rev. D* **80**, 013005 (2009) [arXiv:0902.1992 [hep-ph]].

[680] S. N. Gninenko, *Phys. Rev. Lett.* **103**, 241802 (2009) [arXiv:0902.3802 [hep-ph]].

[681] S. N. Gninenko and D. S. Gorbunov, *Phys. Rev. D* **81**, 075013 (2010) [arXiv:0907.4666 [hep-ph]].

[682] S. N. Gninenko, *Phys. Rev. D* **83**, 015015 (2011) [arXiv:1009.5536 [hep-ph]].

[683] J. A. Harvey, C. T. Hill, and R. J. Hill, *Phys. Rev. Lett.* **99**, 261601 (2007) [arXiv:0708.1281 [hep-ph]].

[684] I. Stancu et al., arXiv:0910.2698 [hep-ex].

[685] S. K. Agarwalla and P. Huber, arXiv:1007.3228 [hep-ph].

[686] A. Kalliomaki, J. Maalampi, and M. Tanimoto, *Phys. Lett. B* **469**, 179 (1999) [arXiv:hep-ph/9909301]; A. Donini and D. Meloni, *Eur. Phys. J. C* **22**, 179 (2001) [arXiv:hep-ph/0105089]; A. Donini, M. Lusignoli, and D. Meloni, *Nucl. Phys. B* **624**, 405 (2002) [arXiv:hep-ph/0107231]; A. Dighe and S. Ray, *Phys. Rev. D* **76**, 113001 (2007) [arXiv:0709.0383 [hep-ph]]; S. Goswami and T. Ota, *Phys. Rev. D* **78**, 033012 (2008) [arXiv:0802.1434 [hep-ph]]; A. Donini, K. i. Fuki, J. Lopez-Pavon, D. Meloni, and O. Yasuda, *JHEP* **0908**, 041 (2009) [arXiv:0812.3703 [hep-ph]]; C. Giunti, M. Laveder, and W. Winter, *Phys. Rev. D* **80**, 073005 (2009) [arXiv:0907.5487 [hep-ph]]; D. Meloni, J. Tang, and W. Winter, arXiv:1007.2419 [hep-ph].

[687] R. Fardon, A. E. Nelson, and N. Weiner, *JCAP* **0410**, 005 (2004) [arXiv:astro-ph/0309800].

[688] R. Takahashi and M. Tanimoto, *Phys. Lett. B* **633**, 675 (2006) [arXiv:hep-ph/0507142]; R. Fardon, A. E. Nelson, and N. Weiner, *JHEP* **0603**, 042 (2006) [arXiv:hep-ph/0507235].

[689] R. D. Peccei, *Phys. Rev. D* **71**, 023527 (2005) [arXiv:hep-ph/0411137].

[690] D. B. Kaplan, A. E. Nelson, and N. Weiner, *Phys. Rev. Lett.* **93**, 091801 (2004) [arXiv:hep-ph/0401099].

[691] K. M. Zurek, *JHEP* **0410**, 058 (2004) [arXiv:hep-ph/0405141].

[692] V. Barger, P. Huber, and D. Marfatia, *Phys. Rev. Lett.* **95**, 211802 (2005) [arXiv:hep-ph/0502196].

[693] N. Weiner and K. M. Zurek, *Phys. Rev. D* **74**, 023517 (2006) [arXiv:hep-ph/0509201].

[694] A. Ringwald and Y.Y.Y. Wong, *JCAP* **0412**, 005 (2004) [arXiv:hep-ph/0408241].

[695] N. Afshordi, M. Zaldarriaga, and K. Kohri, *Phys. Rev. D* **72**, 065024 (2005) [arXiv:astro-ph/0506663]; R. Takahashi and M. Tanimoto, *JHEP* **0605**, 021 (2006) [arXiv:astro-ph/0601119]; O. E. Bjaelde, A. W. Brookfield, C. van de Bruck, S. Hannestad, D. F. Mota, L. Schrempp, and D. Tocchini-Valentini, *JCAP* **0801**, 026 (2008) [arXiv:0705.2018 [astro-ph]].

[696] O. E. Bjaelde, A. W. Brookfield, C. van de Bruck, S. Hannestad, D. F. Mota, L. Schrempp, and D. Tocchini-Valentini, *JCAP* **0801**, 026 (2008) [arXiv:0705.2018 [astro-ph]].

[697] M. C. Gonzalez-Garcia, P. C. de Holanda, and R. Z. Funchal, *Phys. Rev. D* **73**, 033008 (2006) [arXiv:hep-ph/0511093].

[698] V. Barger, D. Marfatia, and K. Whisnant, *Phys. Rev. D* Vol. **73**, 013005 (2006), © 2006 The American Physical Society, [arXiv:hep-ph/0509163v3].

[699] T. Blazek, *Phys. Rev. D* **62**, 055001 (2000) [arXiv:hep-ph/9912482]. T. Schwetz and W. Winter, *Phys. Lett. B* **633**, 557 (2006) [arXiv:hep-ph/0511177].

[700] P. H. Gu, X. J. Bi, B. Feng, B. L. Young, and X. Zhang, *Chin. Phys. C* **32**, 530 (2008) [arXiv:hep-ph/0512076].

[701] P. B. Pal and L. Wolfenstein, *Phys. Rev. D* **25**, 766 (1982).

[702] Y. Hosotani, *Nucl. Phys. B* **191**, 411 (1981) [Erratum-ibid. B **197**, 546 (1982)].

[703] V. D. Barger, W. Y. Keung, and S. Pakvasa, *Phys. Rev. D* **25**, 907 (1982).

[704] G. B. Gelmini and M. Roncadelli, *Phys. Lett. B* **99**, 411 (1981); H. M. Georgi, S. L. Glashow, and S. Nussinov, *Nucl. Phys. B* **193**, 297 (1981); J. Schechter and J.W.F. Valle, *Phys. Rev. D* **25**, 774 (1982).

[705] F. Wilczek, *Phys. Rev. Lett.* **49**, 1549 (1982); G. B. Gelmini and J.W.F. Valle, *Phys. Lett. B* **142**, 181 (1984); G. Gelmini, D. N. Schramm, and J.W.F. Valle, *Phys. Lett. B* **146**, 311 (1984); A. Kumar and R. N. Mohapatra, *Phys. Lett. B* **150**, 191 (1985); B. Grinstein, J. Preskill, and M. B. Wise, *Phys. Lett. B* **159**, 57 (1985); H. Harari and Y. Nir, *Nucl. Phys. B* **292**, 251 (1987); Z. G. Berezhiani and M. Y. Khlopov, *Z. Phys. C* **49**, 73 (1991).

[706] M. Lindner, T. Ohlsson, and W. Winter, *Nucl. Phys. B* **607**, 326 (2001) [arXiv:hep-ph/0103170].

[707] P. Lipari and M. Lusignoli, *Phys. Rev. D* **60**, 013003 (1999) [arXiv:hep-ph/9901350].

[708] G. L. Fogli, E. Lisi, A. Marrone, and G. Scioscia, *Phys. Rev. D* **59**, 117303 (1999) [arXiv:hep-ph/9902267].

[709] S. Choubey and S. Goswami, *Astropart. Phys.* **14**, 67 (2000) [arXiv:hep-ph/9904257].

[710] M. C. Gonzalez-Garcia and M. Maltoni, *Phys. Lett. B* **663**, 405 (2008) [arXiv:0802.3699 [hep-ph]].

[711] S. Choubey, S. Goswami, and D. Majumdar, *Phys. Lett. B* **484**, 73 (2000) [arXiv:hep-ph/0004193].

[712] A. S. Joshipura, E. Masso, and S. Mohanty, *Phys. Rev. D* **66**, 113008 (2002) [arXiv:hep-ph/0203181].

[713] J. F. Beacom and N. F. Bell, *Phys. Rev. D* **65**, 113009 (2002) [arXiv:hep-ph/0204111].

[714] S. Ando, *Phys. Lett. B* **570**, 11 (2003) [arXiv:hep-ph/0307169]; G. L. Fogli, E. Lisi, A. Mirizzi, and D. Montanino, *Phys. Rev. D* **70**, 013001 (2004) [arXiv:hep-ph/0401227].

[715] M. Lindner, T. Ohlsson, and W. Winter, *Nucl. Phys. B* **622**, 429 (2002) [arXiv:astro-ph/0105309].

[716] A. B. Balantekin and G. M. Fuller, *J. Phys. G* **29**, 2513 (2003) [arXiv:astro-ph/0309519].

[717] S. Dodelson and L. M. Widrow, *Phys. Rev. Lett.* **72**, 17 (1994) [arXiv:hep-ph/9303287].

[718] S. Palomares-Ruiz, S. Pascoli, and T. Schwetz, *JHEP* **0509**, 048 (2005) [arXiv:hep-ph/0505216].

[719] V. D. Barger, R.J.N. Phillips, and S. Sarkar, *Phys. Lett. B* **352**, 365 (1995) [Erratum-ibid. B **356**, 617 (1995)] [arXiv:hep-ph/9503295].

[720] P. B. Pal and L. Wolfenstein, *Phys. Rev. D* **25**, 766 (1982).

[721] K. Abazajian, G. M. Fuller, and W. H. Tucker, *Astrophys. J.* **562**, 593 (2001) [arXiv:astro-ph/0106002].

[722] L. L. Cowie, G. P. Garmire, M. W. Bautz, A. J. Barger, W. N. Brandt, and A. E. Hornschemeier, arXiv:astro-ph/0201186.

[723] M. Loewenstein and A. Kusenko, *Astrophys. J.* **714**, 652 (2010) [arXiv:0912.0552 [astro-ph.HE]]; M. Loewenstein, A. Kusenko, and P. L. Biermann, *Astrophys. J.* **700**, 426 (2009) [arXiv:0812.2710 [astro-ph]].

[724] H. Yuksel, J. F. Beacom, and C. R. Watson, *Phys. Rev. Lett.* **101**, 121301 (2008) [arXiv:0706.4084 [astro-ph]].

[725] A. Boyarsky, A. Neronov, O. Ruchayskiy, and M. Shaposhnikov, *Mon. Not. Roy. Astron. Soc.* **370**, 213 (2006) [arXiv:astro-ph/0512509].

[726] C. R. Watson, J. F. Beacom, H. Yuksel, and T. P. Walker, *Phys. Rev. D* **74**, 033009 (2006) [arXiv:astro-ph/0605424].

[727] U. Seljak, A. Makarov, P. McDonald, and H. Trac, *Phys. Rev. Lett.* **97**, 191303 (2006) [arXiv:astro-ph/0602430]; M. Viel, J. Lesgourgues, M. G. Haehnelt, S. Matarrese, and A. Riotto, *Phys. Rev. Lett.* **97**, 071301 (2006) [arXiv:astro-ph/0605706].

[728] A. Kusenko, *Phys. Rev. Lett.* **97**, 241301 (2006) [arXiv:hep-ph/0609081]; K. Petraki and A. Kusenko, *Phys. Rev. D* **77**, 065014 (2008) [arXiv:0711.4646 [hep-ph]].

[729] K. N. Abazajian and G. M. Fuller, *Phys. Rev. D* **66**, 023526 (2002) [arXiv:astro-ph/0204293].

[730] A. Kusenko and M. Loewenstein, arXiv:1001.4055 [astro-ph.CO].

[731] S. W. Hawking, *Commun. Math. Phys.* **43**, 199 (1975) [Erratum-ibid. **46**, 206 (1976)]; *Phys. Rev. D* **14**, 2460 (1976); *Commun. Math. Phys.* **87**, 395 (1982).

[732] S. B. Giddings and A. Strominger, *Nucl. Phys. B* **307**, 854 (1988); W. H. Zurek, *Phys. Today* **44N10**, 36 (1991); G. Amelino-Camelia, J. R. Ellis, N. E. Mavromatos, and D. V. Nanopoulos, *Int. J. Mod. Phys. A* **12**, 607 (1997) [arXiv:hep-th/9605211]; L. J. Garay, *Int. J. Mod. Phys. A* **14**, 4079 (1999) [arXiv:gr-qc/9911002].

[733] J. R. Ellis, J. S. Hagelin, D. V. Nanopoulos, and M. Srednicki, *Nucl. Phys. B* **241**, 381 (1984).

[734] T. Banks, L. Susskind, and M. E. Peskin, *Nucl. Phys. B* **244**, 125 (1984); P. Huet and M. E. Peskin, *Nucl. Phys. B* **434**, 3 (1995) [arXiv:hep-ph/9403257]; J. R. Ellis, J. L. Lopez, N. E. Mavromatos, and D. V. Nanopoulos, *Phys. Rev. D* **53**, 3846 (1996) [arXiv:hep-ph/9505340]; F. Benatti and R. Floreanini, *Phys. Lett. B* **389**, 100 (1996) [arXiv:hep-th/9607059]; *Phys. Lett. B* **401**, 337 (1997) [arXiv:hep-ph/9704283]; J. Bernabeu, N. E. Mavromatos, and S. Sarkar, *Phys. Rev. D* **74**, 045014 (2006) [arXiv:hep-th/0606137].

[735] F. Benatti and R. Floreanini, *Phys. Lett. B* **451**, 422 (1999) [arXiv:quant-ph/9902026].

[736] Y. Liu, L. Z Hu, and M. L. Ge, *Phys. Rev. D* **56**, 6648 (1997); Y. Liu, J. L. Chen, and M. L. Ge, *J. Phys. G* **24**, 2289 (1998) [arXiv:hep-ph/9711381]; C. P. Sun and D. L. Zhou, arXiv:hep-ph/9808334.

[737] C. H. Chang, W. S. Dai, X. Q. Li, Y. Liu, F. C. Ma, and Z. J. Tao, *Phys. Rev. D* **60**, 033006 (1999) [arXiv:hep-ph/9809371].

[738] Y. Grossman and M. P. Worah, arXiv:hep-ph/9807511.

[739] A. M. Gago, E. M. Santos, W.J.C. Teves, and R. Z. Funchal, *Phys. Rev. D* **63**, 113013 (2001) [arXiv:hep-ph/0010092]; arXiv:hep-ph/0208166; M. Blennow, T. Ohlsson, and W. Winter, *JHEP* **0506**, 049 (2005) [arXiv:hep-ph/0502147]; *Eur. Phys. J. C* **49**, 1023 (2007) [arXiv:hep-ph/0508175]; N. E. Mavromatos, A. Meregaglia, A. Rubbia, A. Sakharov, and S. Sarkar, *Phys. Rev. D* **77**, 053014 (2008) [arXiv:0801.0872 [hep-ph]].

[740] F. Benatti and R. Floreanini, *JHEP* **0002**, 032 (2000) [arXiv:hep-ph/0002221]; *Phys. Rev. D* **64**, 085015 (2001) [arXiv:hep-ph/0105303]; G. Barenboim and N. E. Mavromatos, *Phys. Rev. D* **70**, 093015 (2004) [arXiv:hep-ph/0406035]; N. Mavromatos and S. Sarkar, *Phys. Rev. D* **72**, 065016 (2005) [arXiv:hep-th/0506242]; S. Sarkar, arXiv:hep-ph/0610010; J. Alexandre, K. Farakos, N. E. Mavromatos,

and P. Pasipoularides, *Phys. Rev. D* **79**, 107701 (2009) [arXiv:0902.3386 [hep-ph]].

[741] G. Lindblad, *Commun. Math. Phys.* **48**, 119 (1976).

[742] G. L. Fogli, E. Lisi, A. Marrone, D. Montanino, and A. Palazzo, *Phys. Rev. D* **76**, 033006 (2007) [arXiv:0704.2568 [hep-ph]].

[743] J. R. Ellis, N. E. Mavromatos, D. V. Nanopoulos, and E. Winstanley, *Mod. Phys. Lett. A* **12**, 243 (1997) [arXiv:gr-qc/9602011]; J. R. Ellis, N. E. Mavromatos, and D. V. Nanopoulos, *Mod. Phys. Lett. A* **12**, 1759 (1997) [arXiv:hep-th/9704169]; R. Gambini, R. A. Porto, and J. Pullin, *Class. Quant. Grav.* **21**, L51 (2004) [arXiv:gr-qc/0305098].

[744] F. N. Loreti and A. B. Balantekin, *Phys. Rev. D* **50**, 4762 (1994) [arXiv:nucl-th/9406003]; H. Nunokawa, A. Rossi, V. B. Semikoz, and J.W.F. Valle, *Nucl. Phys. B* **472**, 495 (1996) [arXiv:hep-ph/9602307]; A. B. Balantekin, J. M. Fetter, and F. N. Loreti, *Phys. Rev. D* **54**, 3941 (1996) [arXiv:astro-ph/9604061]; C. P. Burgess and D. Michaud, *Annals Phys.* **256**, 1 (1997) [arXiv:hep-ph/9606295]; A. A. Bykov, M. C. Gonzalez-Garcia, C. Pena-Garay, V. Y. Popov, and V. B. Semikoz, arXiv:hep-ph/0005244; C. Burgess, N. S. Dzhalilov, M. Maltoni, T. I. Rashba, V. B. Semikoz, M. A. Tortola, and J.W.F. Valle, *Astrophys. J.* **588**, L65 (2003) [arXiv:hep-ph/0209094]; M. M. Guzzo, P. C. de Holanda, and N. Reggiani, *Phys. Lett. B* **569**, 45 (2003) [arXiv:hep-ph/0303203]; N. Reggiani, M. M. Guzzo, and P. C. de Holanda, *Braz. J. Phys.* **34**, 1729 (2004); A. B. Balantekin and H. Yuksel, *Phys. Rev. D* **68**, 013006 (2003) [arXiv:hep-ph/0303169]; C. P. Burgess, N. S. Dzhalilov, M. Maltoni, T. I. Rashba, V. B. Semikoz, M. A. Tortola, and J.W.F. Valle, *JCAP* **0401**, 007 (2004) [arXiv:hep-ph/0310366]; F. Benatti and R. Floreanini, *Phys. Rev. D* **71**, 013003 (2005) [arXiv:hep-ph/0412311].

[745] T. Ohlsson, *Phys. Lett. B* **502**, 159 (2001) [arXiv:hep-ph/0012272].

[746] T. Schwetz, *Phys. Lett. B* **577**, 120 (2003) [arXiv:hep-ph/0308003].

[747] G. Barenboim and N. E. Mavromatos, *JHEP* **0501**, 034 (2005) [arXiv:hep-ph/0404014]; G. Barenboim, N. E. Mavromatos, S. Sarkar, and A. Waldron-Lauda, *Nucl. Phys. B* **758**, 90 (2006) [arXiv:hep-ph/0603028]; Y. Farzan, T. Schwetz, and A. Y. Smirnov, *JHEP* **0807**, 067 (2008) [arXiv:0805.2098 [hep-ph]].

[748] D. V. Ahluwalia, *Mod. Phys. Lett. A* **16**, 917 (2001) [arXiv:hep-ph/0104316]; D. Hooper, D. Morgan, and E. Winstanley, *Phys. Lett. B* **609**, 206 (2005) [arXiv:hep-ph/0410094]; *Phys. Rev. D* **72**, 065009 (2005) [arXiv:hep-ph/0506091]; D. Morgan, E. Winstanley, J. Brunner, and L. F. Thompson, *Astropart. Phys.* **25**, 311 (2006) [arXiv:astro-ph/0412618]; L. A. Anchordoqui, H. Goldberg, M. C. Gonzalez-Garcia, F. Halzen, D. Hooper, S. Sarkar, and T. J. Weiler, *Phys. Rev. D* **72**, 065019 (2005) [arXiv:hep-ph/0506168]; Y. Farzan and A. Y. Smirnov, *Nucl. Phys. B* **805**, 356 (2008) [arXiv:0803.0495 [hep-ph]].

[749] F. N. Loreti, Y. Z. Qian, G. M. Fuller, and A. B. Balantekin, *Phys. Rev. D* **52**, 6664 (1995) [arXiv:astro-ph/9508106].

[750] G. G. Raffelt and I. Tamborra, arXiv:1006.0002 [hep-ph].

[751] V. A. Kostelecky and S. Samuel, *Phys. Rev. D* **39**, 683 (1989); *Phys. Rev. Lett.* **66**, 1811 (1991); V. A. Kostelecky and R. Potting, *Nucl. Phys. B* **359**, 545 (1991); *Phys. Lett. B* **381**, 89 (1996) [arXiv:hep-th/9605088]; V. A. Kostelecky and R. Potting, *Phys. Rev. D* **63**, 046007 (2001) [arXiv:hep-th/0008252]; V. A. Kostelecky, M. Perry, and R. Potting, *Phys. Rev. Lett.* **84**, 4541 (2000) [arXiv:hep-th/9912243].

[752] See, e.g., R. Gambini and J. Pullin, *Phys. Rev. D* **59**, 124021 (1999) [arXiv:gr-qc/9809038]; J. Alfaro, H. A. Morales-Tecotl, and L. F. Urrutia, *Phys. Rev. D* **66**, 124006 (2002) [arXiv:hep-th/0208192]; D. Sudarsky, L. Urrutia, and H. Vucetich, *Phys. Rev. Lett.* **89**, 231301 (2002) [arXiv:gr-qc/0204027]; G. Amelino-Camelia, *Mod. Phys. Lett. A* **17**, 899 (2002) [arXiv:gr-qc/0204051]; Y. J. Ng, *Mod. Phys. Lett. A* **18**,

1073 (2003) [arXiv:gr-qc/0305019]; R. C. Myers and M. Pospelov, *Phys. Rev. Lett.* **90**, 211601 (2003) [arXiv:hep-ph/0301124]; N. E. Mavromatos, *Nucl. Instrum. Meth.* B **214**, 1 (2004) [arXiv:hep-ph/0305215].

[753] B. R. Heckel, E. G. Adelberger, C. E. Cramer, T. S. Cook, S. Schlamminger, and U. Schmidt, *Phys. Rev. D* **78**, 092006 (2008) [arXiv:0808.2673 [hep-ex]].

[754] S. Choubey and S. F. King, *Phys. Lett. B* **586**, 353 (2004) [arXiv:hep-ph/0311326].

[755] G. L. Fogli, E. Lisi, A. Marrone, and G. Scioscia, *Phys. Rev. D* **60**, 053006 (1999) [arXiv:hep-ph/9904248]; M. C. Gonzalez-Garcia and M. Maltoni, *Phys. Rev. D* **70**, 033010 (2004) [arXiv:hep-ph/0404085].

[756] A. Datta, R. Gandhi, P. Mehta, and S. Uma Sankar, *Phys. Lett. B* **597**, 356 (2004) [arXiv:hep-ph/0312027]; G. Battistoni et al., *Phys. Lett. B* **615**, 14 (2005) [arXiv:hep-ex/0503015].

[757] V. A. Kostelecky and M. Mewes, *Phys. Rev. D* **70**, 031902 (2004) [arXiv:hep-ph/0308300].

[758] V. Barger, D. Marfatia, and K. Whisnant, *Phys. Lett. B* **653**, 267 (2007) [arXiv:0706.1085 [hep-ph]].

[759] T. Katori, V. A. Kostelecky, and R. Tayloe, *Phys. Rev. D* **74**, 105009 (2006) [arXiv:hep-ph/0606154]; J. S. Diaz, V. A. Kostelecky, and M. Mewes, *Phys. Rev. D* **80**, 076007 (2009) [arXiv:0908.1401 [hep-ph]].

[760] S. Ando, M. Kamionkowski, and I. Mocioiu, *Phys. Rev. D* **80**, 123522 (2009) [arXiv:0910.4391 [hep-ph]].

[761] A. de Gouvea and Y. Grossman, *Phys. Rev. D* **74**, 093008 (2006) [arXiv:hep-ph/0602237].

[762] J. S. Diaz and A. Kostelecky, arXiv:1012.5985 [hep-ph].

[763] D. Hooper, D. Morgan, and E. Winstanley, *Phys. Rev. D* **72**, 065009 (2005) [arXiv:hep-ph/0506091]; D. Morgan, E. Winstanley, J. Brunner, and L. F. Thompson, arXiv:0705.1897 [astro-ph].

[764] V. D. Barger, S. Pakvasa, T. J. Weiler, and K. Whisnant, *Phys. Rev. Lett.* **85**, 5055 (2000) [arXiv:hep-ph/0005197].

[765] See, e.g., L. B. Auerbach et al. [LSND Collaboration], *Phys. Rev. D* **72**, 076004 (2005) [arXiv:hep-ex/0506067]; P. Adamson et al. [MINOS Collaboration], *Phys. Rev. Lett.* **101**, 151601 (2008) [arXiv:0806.4945 [hep-ex]]; *Phys. Rev. Lett.* **105**, 151601 (2010) [arXiv:1007.2791 [hep-ex]].

[766] T. Katori [MiniBooNE Collaboration], arXiv:1008.0906 [hep-ph].

[767] F. J. Botella, C. S. Lim, and W. J. Marciano, *Phys. Rev. D* **35**, 896 (1987).

[768] J. W. Valle, *Phys. Lett. B* **199**, 432 (1987); Y. Grossman, *Phys. Lett. B* **359**, 141 (1995) [arXiv:hep-ph/9507344].

[769] For a recent, detailed review of NSI of neutrinos, see H. Minakata, arXiv:0905.1387 [hep-ph].

[770] T. Ota and J. Sato, *Phys. Rev. D* **71**, 096004 (2005) [arXiv:hep-ph/0502124].

[771] R. Adhikari, S. K. Agarwalla, and A. Raychaudhuri, *Phys. Lett. B* **642**, 111 (2006) [arXiv:hep-ph/0608034].

[772] For a comprehensive summary, see S. Davidson, C. Pena-Garay, N. Rius, and A. Santamaria, *JHEP* **0303**, 011 (2003) [arXiv:hep-ph/0302093].

[773] L. B. Auerbach et al. [LSND Collaboration], *Phys. Rev. D* **63**, 112001 (2001) [arXiv:hep-ex/0101039].

[774] Z. Daraktchieva et al. [MUNU Collaboration], *Phys. Lett. B* **564**, 190 (2003) [arXiv:hep-ex/0304011]; *Phys. Lett. B* **615**, 153 (2005) [arXiv:hep-ex/0502037].

[775] P. Vilain et al. [CHARM-II Collaboration], *Phys. Lett. B* **335**, 246 (1994).

[776] J. Dorenbosch et al. [CHARM Collaboration], *Phys. Lett. B* **180**, 303 (1986).

[777] G. P. Zeller et al. [NuTeV Collaboration], *Phys. Rev. Lett.* **88**, 091802 (2002) [Erratum-ibid. **90**, 239902 (2003)] [arXiv:hep-ex/0110059].

[778] Z. Berezhiani and A. Rossi, *Phys. Lett. B* **535**, 207 (2002) [arXiv:hep-ph/0111137].
[779] J. Barranco, O. G. Miranda, C. A. Moura, and J.W.F. Valle, *Phys. Rev. D* **77**, 093014 (2008) [arXiv:0711.0698 [hep-ph]]; *Phys. Rev. D* **73**, 113001 (2006) [arXiv:hep-ph/0512195].
[780] M. Deniz et al. [TEXONO Collaboration], *Phys. Rev. D* **82**, 033004 (2010) [arXiv:1006.1947 [hep-ph]].
[781] A. de Gouvea and J. Jenkins, *Phys. Rev. D* **74**, 033004 (2006) [arXiv:hep-ph/0603036].
[782] J. Barranco, O. G. Miranda, and T. I. Rashba, *JHEP* **0512**, 021 (2005) [arXiv:hep-ph/0508299].
[783] C. H. Chen, C. Q. Geng, and T. C. Yuan, *Phys. Rev. D* **75**, 077301 (2007) [arXiv:hep-ph/0703196].
[784] Z. Berezhiani, R. S. Raghavan, and A. Rossi, *Nucl. Phys. B* **638**, 62 (2002) [arXiv:hep-ph/0111138].
[785] M. B. Gavela, D. Hernandez, T. Ota, and W. Winter, *Phys. Rev. D* **79**, 013007 (2009) [arXiv:0809.3451 [hep-ph]].
[786] A. de Gouvea, S. Lola, and K. Tobe, *Phys. Rev. D* **63**, 035004 (2001) [arXiv:hep-ph/0008085].
[787] C. Biggio, M. Blennow, and E. Fernandez-Martinez, *JHEP* **0903**, 139 (2009) [arXiv:0902.0607 [hep-ph]]; *JHEP* **0908**, 090 (2009) [arXiv:0907.0097 [hep-ph]].
[788] E. Roulet, *Phys. Rev. D* **44**, 935 (1991); M. M. Guzzo, A. Masiero, and S. T. Petcov, *Phys. Lett. B* **260**, 154 (1991); M. M. Guzzo and S. T. Petcov, *Phys. Lett. B* **271**, 172 (1991); V. D. Barger, R. J. Phillips, and K. Whisnant, *Phys. Rev. D* **44**, 1629 (1991).
[789] S. Bergmann, M. M. Guzzo, P. C. de Holanda, P. I. Krastev, and H. Nunokawa, *Phys. Rev. D* **62**, 073001 (2000) [arXiv:hep-ph/0004049]; A. M. Gago, M. M. Guzzo, P. C. de Holanda, H. Nunokawa, O.L.G. Peres, V. Pleitez, and R. Z. Funchal, *Phys. Rev. D* **65**, 073012 (2002) [arXiv:hep-ph/0112060]; A. Friedland, C. Lunardini, and C. Pena-Garay, *Phys. Lett. B* **594**, 347 (2004) [arXiv:hep-ph/0402266]; M. M. Guzzo, P. C. de Holanda, and O.L.G. Peres, *Phys. Lett. B* **591**, 1 (2004) [arXiv:hep-ph/0403134].
[790] O. G. Miranda, M. A. Tortola, and J.W.F. Valle, *JHEP* **0610**, 008 (2006) [arXiv:hep-ph/0406280].
[791] A. Bolanos, O. G. Miranda, A. Palazzo, M. A. Tortola, and J.W.F. Valle, *Phys. Rev. D* **79**, 113012 (2009) [arXiv:0812.4417 [hep-ph]].
[792] F. J. Escrihuela, O. G. Miranda, M. A. Tortola, and J.W.F. Valle, *Phys. Rev. D* **80**, 105009 (2009) [Erratum-ibid. D **80**, 129908 (2009)] [arXiv:0907.2630 [hep-ph]].
[793] M. C. Gonzalez-Garcia et al., *Phys. Rev. Lett.* **82**, 3202 (1999) [arXiv:hep-ph/9809531]; N. Fornengo, M. C. Gonzalez-Garcia, and J.W.F. Valle, *JHEP* **0007**, 006 (2000) [arXiv:hep-ph/9906539].
[794] N. Fornengo, M. Maltoni, R. Tomas, and J.W.F. Valle, *Phys. Rev. D* **65**, 013010 (2002) [arXiv:hep-ph/0108043].
[795] M. C. Gonzalez-Garcia and M. Maltoni, *Phys. Rev. D* **70**, 033010 (2004) [arXiv:hep-ph/0404085].
[796] M. Maltoni, *Nucl. Phys. Proc. Suppl.* **114**, 191 (2003) [arXiv:hep-ph/0210111].
[797] A. Friedland, C. Lunardini, and M. Maltoni, *Phys. Rev. D* **70**, 111301 (2004) [arXiv:hep-ph/0408264].
[798] J. Kopp, M. Lindner, T. Ota, and J. Sato, *Phys. Rev. D* **77**, 013007 (2008) [arXiv:0708.0152 [hep-ph]].
[799] N. C. Ribeiro, H. Nunokawa, T. Kajita, S. Nakayama, P. Ko, and H. Minakata, *Phys. Rev. D* **77**, 073007 (2008) [arXiv:0712.4314 [hep-ph]].
[800] A. Esteban-Pretel, J.W.F. Valle, and P. Huber, *Phys. Lett. B* **668**, 197 (2008) [arXiv:0803.1790 [hep-ph]].

[801] M. C. Gonzalez-Garcia, Y. Grossman, A. Gusso, and Y. Nir, *Phys. Rev. D* **64**, 096006 (2001) [arXiv:hep-ph/0105159].
[802] A. M. Gago, M. M. Guzzo, H. Nunokawa, W.J.C. Teves, and R. Z. Funchal, *Phys. Rev. D* **64**, 073003 (2001) [arXiv:hep-ph/0105196]; P. Huber and J.W.F. Valle, *Phys. Lett. B* **523**, 151 (2001) [arXiv:hep-ph/0108193]; T. Ota, J. Sato, and N. Yamashita, *Phys. Rev. D* **65**, 093015 (2002) [arXiv:hep-ph/0112329]; M. Campanelli and A. Romanino, *Phys. Rev. D* **66**, 113001 (2002) [arXiv:hep-ph/0207350]; M. Blennow, T. Ohlsson, and W. Winter, *Eur. Phys. J. C* **49**, 1023 (2007) [arXiv:hep-ph/0508175].
[803] P. Huber, T. Schwetz, and J.W.F. Valle, *Phys. Rev. Lett.* **88**, 101804 (2002) [arXiv:hep-ph/0111224].
[804] P. Huber, T. Schwetz, and J.W.F. Valle, *Phys. Rev. D* **66**, 013006 (2002) [arXiv:hep-ph/0202048].
[805] J. Kopp, M. Lindner, and T. Ota, *Phys. Rev. D* **76**, 013001 (2007) [arXiv:hep-ph/0702269].
[806] N. C. Ribeiro, H. Minakata, H. Nunokawa, S. Uchinami, and R. Z. Funchal, *JHEP* **0712**, 002 (2007) [arXiv:0709.1980 [hep-ph]].
[807] J. Kopp, T. Ota, and W. Winter, *Phys. Rev. D* **78**, 053007 (2008) [arXiv:0804.2261 [hep-ph]].
[808] W. Winter, *Phys. Lett. B* **671**, 77 (2009) [arXiv:0808.3583 [hep-ph]].
[809] G. Altarelli and D. Meloni, *Nucl. Phys. B* **809**, 158 (2009) [arXiv:0809.1041 [hep-ph]].
[810] A. Esteban-Pretel, R. Tomas, J.W.F. Valle, J. Tang, and W. Winter, *Phys. Rev. D* **80**, 053001 (2009) [arXiv:0903.3039 [hep-ph]].
[811] A. M. Gago, H. Minakata, H. Nunokawa, S. Uchinami, and R. Z. Funchal, *JHEP* **1001**, 049 (2010) [arXiv:0904.3360 [hep-ph]].
[812] D. Meloni, T. Ohlsson, W. Winter, and H. Zhang, *JHEP* **1004**, 041 (2010) [arXiv:0912.2735 [hep-ph]].
[813] A. Palazzo and J.W.F. Valle, *Phys. Rev. D* **80**, 091301 (2009) [arXiv:0909.1535 [hep-ph]].
[814] H. Nunokawa, Y. Z. Qian, A. Rossi, and J.W.F. Valle, *Phys. Rev. D* **54**, 4356 (1996) [arXiv:hep-ph/9605301]; G. L. Fogli, E. Lisi, A. Mirizz, and D. Montanino, *Phys. Rev. D* **66**, 013009 (2002) [arXiv:hep-ph/0202269]; A. Esteban-Pretel, R. Tomas, and J.W.F. Valle, *Phys. Rev. D* **76**, 053001 (2007) [arXiv:0704.0032 [hep-ph]]; *Phys. Rev. D* **81**, 063003 (2010) [arXiv:0909.2196 [hep-ph]].
[815] D. Meloni, T. Ohlsson, and H. Zhang, *JHEP* **0904**, 033 (2009) [arXiv:0901.1784 [hep-ph]].
[816] T. Kikuchi, H. Minakata, and S. Uchinami, *JHEP* **0903**, 114 (2009) [arXiv:0809.3312 [hep-ph]].
[817] W. Y. Keung and G. Senjanovic, *Phys. Rev. Lett.* **50**, 1427 (1983).
[818] T. Han and B. Zhang, *Phys. Rev. Lett.* **97**, 171804 (2006) [arXiv:hep-ph/0604064].
[819] A. Atre, T. Han, S. Pascoli, and B. Zhang, *JHEP* **0905**, 030 (2009) [arXiv:0901.3589 [hep-ph]].
[820] P. Fileviez Perez, T. Han, G. Y. Huang, T. Li, and K. Wang, *Phys. Rev. D* **78**, 071301 (2008) [arXiv:0803.3450 [hep-ph]]; *Phys. Rev. D* **78**, 015018 (2008) [arXiv:0805.3536 [hep-ph]]; R. Franceschini, T. Hambye, and A. Strumia, *Phys. Rev. D* **78**, 033002 (2008) [arXiv:0805.1613 [hep-ph]]; P. Fileviez Perez, T. Han, T. Li, and M. J. Ramsey-Musolf, *Nucl. Phys. B* **819**, 139 (2009) [arXiv:0810.4138 [hep-ph]]; S. Blanchet, Z. Chacko, and R. N. Mohapatra, *Phys. Rev. D* **80**, 085002 (2009) [arXiv:0812.3837 [hep-ph]]; W. Chao, Z. Si, Y. Zheng, and S. Zhou, *Phys. Lett. B* **683**, 26 (2010) [arXiv:0907.0935 [hep-ph]]; P. Fileviez Perez, T. Han, and T. Li, *Phys. Rev. D* **80**, 073015 (2009) [arXiv:0907.4186 [hep-ph]]; T. Li and X. G. He, *Phys. Rev. D* **80**, 093003 (2009) [arXiv:0907.4193 [hep-ph]]; P. Nath et al., *Nucl. Phys. Proc. Suppl.* **200–202**, 185 (2010) [arXiv:1001.2693 [hep-ph]].

[821] F. del Aguila and J. A. Aguilar-Saavedra, *Nucl. Phys. B* **813**, 22 (2009) [arXiv:0808.2468 [hep-ph]]; J. A. Aguilar-Saavedra, *Nucl. Phys. B* **828**, 289 (2010) [arXiv:0905.2221 [hep-ph]].
[822] W. J. Marciano and A. I. Sanda, *Phys. Lett. B* **67**, 303 (1977); B. W. Lee and R. E. Shrock, *Phys. Rev. D* **16**, 1444 (1977); K. Fujikawa and R. Shrock, *Phys. Rev. Lett.* **45**, 963 (1980).
[823] J. F. Nieves, *Phys. Rev. D* **26**, 3152 (1982).
[824] S. T. Petcov, *Sov. J. Nucl. Phys.* **25**, 340 (1977) [*Yad. Fiz.* **25**, 641 (1977)] [Erratum-ibid. **25**, 698 (1977)] [Erratum-ibid. **25**, 1336 (1977)].
[825] N. F. Bell, V. Cirigliano, M. J. Ramsey-Musolf, P. Vogel, and M. B. Wise, *Phys. Rev. Lett.* **95**, 151802 (2005) [arXiv:hep-ph/0504134]; N. F. Bell, M. Gorchtein, M. J. Ramsey-Musolf, P. Vogel, and P. Wang, *Phys. Lett. B* **642**, 377 (2006) [arXiv:hep-ph/0606248]; N. F. Bell, *Int. J. Mod. Phys.* A **22**, 4891 (2007) [arXiv:0707.1556 [hep-ph]].
[826] C. Giunti and A. Studenikin, *Phys. Atom. Nucl.* **72**, 2089 (2009) [arXiv:0812.3646 [hep-ph]].
[827] P. Vogel and J. Engel, *Phys. Rev. D* **39**, 3378 (1989).
[828] J. F. Beacom and P. Vogel, *Phys. Rev. Lett.* **83**, 5222 (1999) [arXiv:hep-ph/9907383].
[829] G. G. Raffelt, *Phys. Rept.* **320**, 319 (1999).
[830] D. W. Liu et al. [Super-Kamiokande Collaboration], *Phys. Rev. Lett.* **93**, 021802 (2004) [arXiv:hep-ex/0402015].
[831] H. T. Wong et al. [TEXONO Collaboration], *Phys. Rev. D* **75**, 012001 (2007) [arXiv:hep-ex/0605006]; M. Deniz et al. [TEXONO Collaboration], *Phys. Rev. D* **81**, 072001 (2010) [arXiv:0911.1597 [hep-ex]].
[832] A. G. Beda et al., *Phys. Atom. Nucl.* **70**, 1873 (2007) [arXiv:0705.4576 [hep-ex]]; A. G. Beda, V. B. Brudanin, V. G. Egorov, D. V. Medvedev, V. S. Pogosov, M. V. Shirchenko, and A. S. Starostin, arXiv:1005.2736 [hep-ex].
[833] D. Montanino, M. Picariello, and J. Pulido, *Phys. Rev. D* **77**, 093011 (2008) [arXiv:0801.2643 [hep-ph]].
[834] G. G. Raffelt, *Phys. Rev. Lett.* **64**, 2856 (1990).
[835] C. S. Lim and W. J. Marciano, *Phys. Rev. D* **37**, 1368 (1988); E. K. Akhmedov, *Phys. Lett. B* **213**, 64 (1988); R. Barbieri and G. Fiorentini, *Nucl. Phys. B* **304**, 909 (1988); A. B. Balantekin, P. J. Hatchell, and F. Loreti, *Phys. Rev. D* **41**, 3583 (1990).
[836] R. S. Raghavan, A. B. Balantekin, F. Loreti, A. J. Baltz, S. Pakvasa and J. T. Pantaleone, *Phys. Rev. D* **44**, 3786 (1991); E. K. Akhmedov, *Phys. Lett. B* **255**, 84 (1991); A. B. Balantekin and F. Loreti, *Phys. Rev. D* **45**, 1059 (1992).
[837] Y. Gando et al. [Super-Kamiokande Collaboration], *Phys. Rev. Lett.* **90**, 171302 (2003) [arXiv:hep-ex/0212067].
[838] B. Aharmim et al. [SNO Collaboration], *Phys. Rev. D* **70**, 093014 (2004) [arXiv:hep-ex/0407029].
[839] K. Eguchi et al. [KamLAND Collaboration], *Phys. Rev. Lett.* **92**, 071301 (2004) [arXiv:hep-ex/0310047].
[840] These constraints depend on assumptions about the solar magnetic field; see O. G. Miranda, T. I. Rashba, A. I. Rez, and J.W.F. Valle, *Phys. Rev. Lett.* **93**, 051304 (2004) [arXiv:hep-ph/0311014]; O. G. Miranda, T. I. Rashba, A. I. Rez, and J.W.F. Valle, *Phys. Rev. D* **70**, 113002 (2004) [arXiv:hep-ph/0406066]; M. M. Guzzo, P. C. de Holanda, and O.L.G. Peres, *Phys. Rev. D* **72**, 073004 (2005) [arXiv:hep-ph/0504185]; A. Friedland, arXiv:hep-ph/0505165.
[841] A. B. Balantekin, *AIP Conf. Proc.* **847**, 128 (2006) [arXiv:hep-ph/0601113].
[842] J. Pulido, *Phys. Rept.* **211**, 167 (1992); X. Shi, D. N. Schramm, R. Rosner, and D. S. Dearborn, *Comments Nucl. Part. Phys.* **21**, 151 (1993).

[843] See, e.g., A. Cisneros, *Atrophys. Space Sci.* **10**, 87 (1971); E. K. Akhmedov and M. Y. Khlopov, *Mod. Phys. Lett. A* **3**, 451 (1988); *Sov. J. Nucl. Phys.* **47**, 689 (1988) [*Yad. Fiz.* **47**, 1079 (1988)]; J. Vidal and J. Wudka, *Phys. Lett. B* **249**, 473 (1990); A. Y. Smirnov, *Phys. Lett. B* **260**, 161 (1991); E. K. Akhmedov, S. T. Petcov, and A. Y. Smirnov, *Phys. Rev. D* **48**, 2167 (1993) [arXiv:hep-ph/9301211]; G. G. Likhachev and A. I. Studenikin, *J. Exp. Theor. Phys.* **81**, 419 (1995) [Zh. Eksp. Teor. Fiz. **108**, 769 (1995)]; A. M. Egorov, A. E. Lobanov, and A. I. Studenikin, *Phys. Lett. B* **491**, 137 (2000) [arXiv:hep-ph/9910476]; A. E. Lobanov and A. I. Studenikin, *Phys. Lett. B* **515**, 94 (2001) [arXiv:hep-ph/0106101]; M. S. Dvornikov and A. I. Studenikin, *Phys. Atom. Nucl.* **64**, 1624 (2001) [*Yad. Fiz.* **64**, 1705 (2001)]; *Phys. Atom. Nucl.* **67**, 719 (2004) [*Yad. Fiz.* **67**, 741 (2004)]; E. K. Akhmedov and J. Pulido, *Phys. Lett. B* **553**, 7 (2003) [arXiv:hep-ph/0209192].

[844] C. R. Das, J. Pulido, and M. Picariello, *Phys. Rev. D* **79**, 073010 (2009) [arXiv:0902.1310 [hep-ph]]; for a contrary viewpoint, see [218].

[845] See, e.g., M. B. Voloshin and M. I. Vysotsky, *Sov. J. Nucl. Phys.* **44**, 544 (1986) [*Yad. Fiz.* **44**, 845 (1986)]; L. B. Okun, M. B. Voloshin, and M. I. Vysotsky, *Sov. Phys. JETP* **64**, 446 (1986); [*Zh. Eksp. Teor. Fiz.* **91**, 754 (1986)]; *Sov. J. Nucl. Phys.* **44**, 440 (1986) [*Yad. Fiz.* **44**, 677 (1986)]; J. Pulido, B. C. Chauhan, and R. S. Raghavan, arXiv:hep-ph/0511341.

[846] See, e.g., T. Ibrahim and P. Nath, *Phys. Rev. D* **78**, 075013 (2008) [arXiv:0806.3880 [hep-ph]].

[847] See, e.g., V. D. Barger, H. Baer, K. Hagiwara, and R.J.N. Phillips, *Phys. Rev. D* **30**, 947 (1984); V. D. Barger, J. L. Hewett, and T. G. Rizzo, *Mod. Phys. Lett. A* **5**, 743 (1990).

[848] P. H. Frampton, P. Q. Hung, and M. Sher, *Phys. Rept.* **330**, 263 (2000) [arXiv:hep-ph/9903387].

[849] J. E. Dubicki and C. D. Froggatt, *Phys. Lett. B* **567**, 46 (2003) [arXiv:hep-ph/0305007]; P. Q. Hung and M. Sher, *Phys. Rev. D* **77**, 037302 (2008) [arXiv:0711.4353 [hep-ph]]; R. Fok and G. D. Kribs, *Phys. Rev. D* **78**, 075023 (2008) [arXiv:0803.4207 [hep-ph]]; P. Q. Hung and C. Xiong, arXiv:0911.3890 [hep-ph]; arXiv:0911.3892 [hep-ph]; G. Burdman, L. Da Rold, and R. D. Matheus, arXiv:0912.5219 [hep-ph]; M. Hashimoto, *Phys. Rev. D* **81**, 075023 (2010) [arXiv:1001.4335 [hep-ph]]; A. J. Buras, B. Duling, T. Feldmann, T. Heidsieck, C. Promberger, and S. Recksiegel, arXiv:1002.2126 [hep-ph]; *JHEP* **1007**, 094 (2010) [arXiv:1004.4565 [hep-ph]]; A. J. Buras, B. Duling, T. Feldmann, T. Heidsieck, and C. Promberger, arXiv:1006.5356 [hep-ph]; G.W.S. Hou, arXiv:1007.2288 [hep-ph].

[850] Z. Murdock, S. Nandi, and Z. Tavartkiladze, *Phys. Lett. B* **668**, 303 (2008) [arXiv:0806.2064 [hep-ph]]; K. Kong, S. C. Park, and T. G. Rizzo, *JHEP* **1007**, 059 (2010) [arXiv:1004.4635 [hep-ph]];

[851] For early discussions of models with mirror symmetry, see J. Maalampi and M. Roos, *Phys. Rept.* **186**, 53 (1990); R. Foot, H. Lew, and R. R. Volkas, *Mod. Phys. Lett. A* **7**, 2567 (1992); R. Foot, *Mod. Phys. Lett. A* **9**, 169 (1994) [arXiv:hep-ph/9402241]; R. Foot and R. R. Volkas, *Phys. Rev. D* **52**, 6595 (1995) [arXiv:hep-ph/9505359].

[852] A. Rajaraman and D. Whiteson, arXiv:1005.4407 [hep-ph].

[853] See, e.g., G. D. Kribs, T. Plehn, M. Spannowsky, and T.M.P. Tait, *Phys. Rev. D* **76**, 075016 (2007) [arXiv:0706.3718 [hep-ph]]; V. A. Novikov, L. B. Okun, A. N. Rozanov, and M. I. Vysotsky, *JETP Lett.* **76**, 127 (2002) [*Pisma Zh. Eksp. Teor. Fiz.* **76**, 158 (2002)] [arXiv:hep-ph/0203132]; D. Choudhury, T.M.P. Tait and C.E.M. Wagner, *Phys. Rev. D* **65**, 053002 (2002) [arXiv:hep-ph/0109097]; H. J. He, N. Polonsky, and S. Su, *Phys. Rev. D* **64**, 053004 (2001) [arXiv:hep-ph/0102144]; C. Csaki and F. Csikor, *Phys. Lett. B* **309**, 103 (1993) [arXiv:hep-ph/9303219]; J. Erler and P. Langacker, *Phys. Rev. Lett.* **105**, 031801 (2010) [arXiv:1003.3211 [hep-ph]].

[854] L. M. Carpenter and A. Rajaraman, arXiv:1005.0628 [hep-ph].

[855] For a summary of the current status of a sequential fourth generation, see B. Holdom, W. S. Hou, T. Hurth, M. L. Mangano, S. Sultansoy, and G. Unel, *PMC Phys. A* **3**, 4 (2009) [arXiv:0904.4698 [hep-ph]].

[856] M. S. Chanowitz, M. A. Furman, and I. Hinchliffe, *Nucl. Phys. B* **153**, 402 (1979).

[857] For discussions of fourth-generation searches at the LHC, see B. Holdom, *JHEP* **0608**, 076 (2006) [arXiv:hep-ph/0606146]; *JHEP* **0703**, 063 (2007) [arXiv:hep-ph/0702037]; *JHEP* **0708**, 069 (2007) [arXiv:0705.1736 [hep-ph]]; *Nuovo Cim.* **123B**, 1205 (2008); O. Cakir, H. Duran Yildiz, R. Mehdiyev, and I. Turk Cakir, *Eur. Phys. J. C* **56**, 537 (2008) [arXiv:0801.0236 [hep-ph]]; G. Burdman, L. Da Rold, O. Eboli, and R. D. Matheus, *Phys. Rev. D* **79**, 075026 (2009) [arXiv:0812.0368 [hep-ph]]; I. T. Cakir, H. Duran Yildiz, O. Cakir, and G. Unel, *Phys. Rev. D* **80**, 095009 (2009) [arXiv:0908.0123 [hep-ph]]; B. Holdom and Q. S. Yan, arXiv:1004.3031 [hep-ph].

- Index -

A_4 symmetry, 109–110
accelerator neutrinos, *see* long-baseline neutrinos
adiabatic propagation, 5, 39, 40
AGASA, 134
AMANDA, 128, 130, 131, 133, 135
ambiguities, *see* parameter degeneracies
anarchy, *see* neutrino anarchy
Angra, 82, 85, 160
ANITA, 133–135
ANTARES, 9, 135
ArgoNeut, 23
ATLAS, 64
atmospheric neutrinos, 26, 27, 59–63, 129, 150, 165; constraint on θ_{13}, 63; flux, 59, 128–130, 145; matter effects, *see* matter effects; oscillations of, *see* neutrino oscillations; production, 59; with decoherence, 163–164; with non-standard interactions, 168; zenith angle dependence, 60–61
Auger experiment, 127, 129, 133–135

BBN (Big Bang Nucleosynthesis), 8, 12–13, 159
beta beam, *see* neutrino beam
beta decay, 1–2, 31, 116, 154; inverse, 2, 26, 58, 80, 157; neutrinoless $\beta\beta$, *see* neutrinoless $\beta\beta$ decay; neutron, 15; tritium, 3, 8, 71, 74, 103, 160
bimaximal mixing, *see* neutrino mixing
Borexino, 6, 52–58, 123, 166, 170
Bugey, 148–152

Cabibbo angle, 18, 71, 105, 109, 111
CCFR, 21, 23, 148, 151
CDHS, 21, 150–152
CHARM, 166, 168
Chlorine experiment (Homestake), 4, 24, 47, 49, 52, 54, 150
CHOOZ, 7, 61, 63, 66, 68, 69, 77, 150, 160
CKM mixing matrix, 4, 102, 105–106
CLEAN, 57, 123
cloud chamber, emulsion, 27
CMB (Cosmic Microwave Background), 8, 12–13, 72
CNGS, 28, 77
cosmogenic neutrinos, 126–127, 133
CP phase, 33; Dirac, 7, 33, 66, 70, 77, 93–97, 103, 115, 123, 138; in 3 + 2 model, 154; Majorana, 8, 33, 74, 103, 105
CP symmetry; conservation of, 35, 76, 153; violation of, 3, 8–9, 33, 35, 74, 77–79, 82, 85–97, 154, 168
CPT symmetry, 97; conservation of, 6, 35, 49, 153; violation of, 9, 66, 155–157, 165
critical density, 37–40, 43
cross section, 23; deep inelastic, 19–23; Δ resonance, 18; neutrino capture, 23; on electrons, 16; on nucleons, elastic and quasielastic, 17–18; ultra high-energy, 133
CUORE, 74

dark energy, 9, 73, 158, 160
dark matter, 13, 135, 158, 160, 162; annihilation in the sun, 9, 13, 131, 138–145; neutrino mass constraint, 72
Daya Bay, 8, 69, 82–84, 160
decoherence, 7, 43–44; due to quantum gravity, 62, 64, 138, 163–164
DeepCore, 9, 129–132, 139–140, 144–146
degeneracies, *see* parameter degeneracies
δm^2_{21}, 8, 34, 64; combined fit from solar and reactor data, 49–51; from KamLAND, 49; global fit, 69–70; in four-neutrino models, 149; in long-baseline neutrino oscillations, 41; in neutrino mass models, 105; in supernova neutrino oscillations, 121; with non-standard interactions, 168
δm^2_{31}, 34; global fit, 69–70; in $0\nu\beta\beta$ decay, 74; in atmospheric neutrino oscillations, 60; in CPT violating models, 155; in four-neutrino models, 149, 151; in long-baseline neutrino oscillations, 41, 66, 76–79, 85–96; in neutrino mass models, 105–108; in supernova neutrino oscillations, 121

Dirac neutrino, 11, 102, 104, 112, 113, 170–171
DONUT, 4, 28
"double bang" events, 135
Double Chooz, 8, 82–83, 160
DUSEL, 86, 93

EUSO, 135
EXO, 74
extra dimensions, 113, 156

family symmetry, *see* horizontal symmetry
FGST (Fermi Gamma-ray Space Telescope), 126
FINeSSE, 23
flavor democracy, 110
flavor symmetry, *see* horizontal symmetry
flavor-changing neutral currents, 111, 161
fourth generation neutrino, 171
Frejus, 86, 91, 92, 131
Fritzsch ansatz, 110
Froggatt-Nielsen diagrams, 106

Gadolimium loading, 81, 82, 125, 157
GALLEX, 5, 24, 47, 54, 157
GENIUS, 74
geoneutrinos, 27, 57–58
GLoBES, 90
GNO, 5, 24, 47, 54
golden channel, 87–90
Gran Sasso, 6, 28, 89, 91
Grand Unified Theory (GUT), 9, 76, 105–107, 113–115; SO(10), 101, 105, 114; SU(5), 105
GZK energy, 126–127, 135

HALO, 123
Hanohano, 26, 58
Higgs boson, 99–101, 106–115, 171; triplet, 100–101, 107, 110–112
HiRes, 127, 129, 133–134
horizontal symmetry, 107–109, 112, 113
Hubble Space Telescope, 13
Hyper-Kamiokande, 26

ICARUS, 27, 70, 77
IceCube, 9, 123, 128–133, 135, 142–145
IceTop, 129
IMB (Irvine-Michigan-Brookhaven) experiment, 6, 26, 59, 122
INO, 27
inverted hierarchy, *see* mass spectrum
iron calorimeter, 27, 87, 91

J-PARC, 79
Jarlskog invariant, 35

K2K, 7, 23, 64, 77, 150, 152, 160
Kaluza-Klein modes, 113
Kamiokande, 5, 26, 49, 59, 122
KamLAND, 6, 26; as a solar neutrino detector, 57; as a supernova neutrino detector, 123; as a supernova relic neutrino detector, 124–125; geoneutrino data, 58; reactor data, 44, 49–52, 57, 68–69, 104, 106, 150, 155, 160, 162, 164, 165, 168–169; solar antineutrino data, 171
KARMEN, 148–150, 153, 156–157
KATRIN, 72

LBNE, 86
$L_\mu - L_\tau$ symmetry, 109
$L_e - L_\mu - L_\tau$ symmetry, 108
LENA, 26, 58
LENS, 57
leptogenesis, 76, 114–115
lepton flavor violation, 114–115, 166
lepton number violation, 73, 76, 99, 111, 114, 169
level crossing, 6, 39, 121
liquid argon TPC, 26, 63, 87, 92, 95, 124, 157, 173
liquid scintillator detector, 6, 26–27, 44, 52, 57, 58, 79, 80, 147
LMA (Large Mixing Angle solution), 5, 40, 49–53, 55–56, 107, 108, 121, 160, 168
long-baseline neutrinos, 41, 64–67, 150, 160–163; oscillations of, *see* neutrino oscillations
lopsided models, 106–107
Lorentz invariance violation, 97, 164–166
LOW (low δm^2 solution), 5, 49, 51
LSND, 147–157, 160, 162, 164–166, 171
LVD, 26, 122
Lyman α forest, 73, 163

MACRO, 7, 61
magnetic moment, 170–171
Majorana experiment, 74
Majorana neutrino, 8, 11, 73, 99–104, 106–108, 110–113, 169–171
mass matrix, 12, 100–110, 112–114; textures, 107–110, 112, 115
mass spectrum; inverted hierarchy, 67, 70, 74, 76, 103–105, 109–111, 120–122, 138, 149; normal hierarchy, 67, 70, 74, 76, 93, 103–107, 109–111, 120–121, 138, 149, 151; quasidegenerate, 70, 71, 103–105, 109–110
mass-varying neutrinos, 158–161
matter effects, 5, 8, 36; for atmospheric neutrinos, 63, 150; for mass-varying neutrinos, 159–161; for sterile neutrinos, 42, 150; in decoherence, 164; in supernovae, 121–122, 164; in the earth, 41–43, 49, 63, 66, 77, 78, 87, 90, 91, 98, 150; in the sun, 38–41, 140, 171; MSW,

see MSW effect; with Lorentz invariance violation, 165; with neutrino magnetic moment, 171; with non-standard interactions, 167–169
MEMPHYS, 26, 86
MicroBooNE, 23, 157
MINERνA, 23
MiniBooNE, 18, 23, 123, 138, 151–157; low-energy anomaly, 155–157, 165–166
MINOS, 7, 21, 23, 27, 61, 64–68, 77, 150, 155, 160
mixing matrix, *see* neutrino mixing
MNS mixing matrix, *see* neutrino mixing
model building, 99
MOON, 57, 74
MSW effect, 5, 40, 44, 49, 117, 120–122, 171
MSW resonance, *see* resonance
μ-τ symmetry, 109

neutrino anarchy, 108, 113
neutrino beam, 28–32; beta beam, 8, 31–32, 91–95; conventional, 28–30, 77–79; neutrino factory, 8, 30–31, 87–90, 92–97, 158, 168; off-axis, 8, 29, 79, 86; superbeam, 8, 86–89, 92–97, 161, 168; wide-band, 8, 78, 86, 92–97
neutrino capture, 23, 24
neutrino decay, 62, 64, 138; active, 161–162, 171; sterile, 157, 162
neutrino factory, *see* neutrino beam
neutrino mass bound, 71–74; $0\nu\beta\beta$ decay, 73; cosmology, 72–73; tritium beta decay, 71
neutrino mixing, 33–34, 76, 161; and neutrino magnetic moment, 170; bimaximal, 104, 109; in matter, 37; in vacuum, 33–34; MNS (PMNS), 3–4, 33; sterile, 150–151, 153, 162; tribimaximal, 9, 104, 108–109, 137–138
neutrino oscillations, 3–9, 33–42; atmospheric, 33, 35–36, 43, 149–151, 155, 162, 165–168; in matter, *see* matter effects; in vacuum, 5, 34–36; long-baseline, 35, 41–42, 66, 76–79, 86–98; reactor, 33, 36; solar, 33, 36, 38–41, 43, 44, 68, 150, 151, 155, 168; supernova, 120–122; ultra high-energy, 136
neutrino spectra; from muon decay, 14, 30; from neutron beta decay, 15; from pion decay, 14; from tau decay, 15
neutrinoless $\beta\beta$ decay, 8, 9, 73–75, 103, 105, 160
NOMAD, 19, 21, 148, 151
non-standard interactions, 50, 154, 160, 166–169
nonadiabatic propagation, 5, 39, 40
normal hierarchy, *see* mass spectrum
NOνA, 8, 23, 26, 30, 79, 82, 85
number of neutrinos; from cosmology, 12; from LEP, 4, 12
NuMI, 29, 64, 79, 86
NuTeV, 19, 21, 166

off-axis beam, *see* neutrino beam
OMNIS, 123
OPERA, 7, 28, 70, 77
oscillation probability, 3–4, 33–36, 39–44, 53–55, 77, 137, 151, 154; appearance, 4; disappearance, 3; survival, 3
oscillations, *see* neutrino oscillations

Palo Verde, 7, 61, 150, 160
parameter degeneracies, 64, 77–80, 86–90, 95, 123; (δ, θ_{13}), 77, 85; eight-fold, 77, 88; $\mathrm{sgn}(\delta m^2_{31})$, 78, 85; θ_{23} octant, 78–79; with non-standard interactions, 169
PMNS mixing matrix, *see* neutrino mixing
probability, *see* oscillation probability

quantum gravity, 138, 159, 163
quark mixing, *see* CKM mixing matrix
quasidegenerate mass spectrum, *see* mass spectrum
QVO (quasi-vacuum oscillations), 5

R-parity violation, 73, 107, 111–112, 166
radiochemical detector, 5, 24, 57
reactor neutrinos, 2, 6, 7, 26, 27, 44, 49, 57, 80–85, 160; oscillations of, *see* neutrino oscillations; spectrum, 81
relic neutrinos; cosmological, 13; supernova, 27, 124, 162
RENO, 8, 69, 82, 85
resonance, 37–40, 43, 92, 120, 121, 130, 150, 156, 165, 171; parametric, 63
resonant spin-flavor precession, *see* spin-flavor precession
RICE, 133–135
right-handed neutrino, 9, 99–102, 105–108, 112, 154, 166; at colliders, 169–170; fourth generation, 171; hierarchical, 113; in leptogenesis, 114–115

SAGE, 5, 24, 47, 54, 157
Sakharov conditions, 114
SciBooNE, 23, 151
seesaw mechanism, 99–102, 105, 110, 112–115, 154, 161, 165, 170, 171; Type I, 100, 101, 114; Type II, 100, 101; Type III, 101
silver channel, 90
single pion production, 18
SIREN, 57
Sloan Digital Sky Survey, 13, 73
SMA (Small Mixing Angle solution), 49, 51, 162
SNO, 5, 47–56, 58, 106, 124, 171
SNO+, 57
solar neutrinos, 4, 25–27, 45–49, 53–57, 165, 168, 170; CNO cycle, 45; flux, 24, 46–47; oscillations of, *see* neutrino oscillations; *pp* chain, 45; Standard Solar Model, 4, 5, 46–48, 52–57
Soudan-2, 7, 61
spacetime foam, *see* quantum gravity
spin-flavor precession, 50, 52, 171
SPL, 86, 92–94
Standard Model, 9, 16, 73, 76–77, 99, 107, 134, 164, 166, 169–171
Standard Model Extension (SME), 164, 165
Standard Solar Model, *see* solar neutrinos
sterile neutrinos, 3, 9, 12, 42, 113, 124, 149–158, 162–163, 171, 174
Super-Kamiokande, 5, 7, 26; accelerator neutrinos, 64, 79, 86, 157; atmospheric neutrinos, 59–66, 68, 131, 150, 162; solar neutrinos, 48–50, 150, 171; supernova neutrinos, 123; supernova relic neutrinos, 124–125
superbeam, *see* neutrino beam
SuperNEMO, 74
supernova, 116; Chandrasekar limit, 116; collective effects, 119; constraint on neutrino mass, 73; detection, 122; Early Warning System (SNEWS), 123; MSW conversions in, 120–122; neutrino flux, 118; neutrinos, 27, 116–125; oscillations of, *see* neutrino oscillations; relic neutrinos, 27, 124; SN1987A, 6, 122

T symmetry, 97
T2K, 8, 23, 30, 66–69, 79, 80, 82, 85–86
T2KK, 86, 92–97
θ_{12}; combined fit from solar and reactor data, 49–52; from future reactor experiment, 57; from KamLAND, 49; global fit, 69; in $0\nu\beta\beta$ decay, 74; in astrophysical neutrino oscillations, 137; in long-baseline neutrino oscillations, 41; in MNS matrix, 33–34; in neutrino models, 103–105, 109; in reactor neutrino oscillations, 80; in solar neutrino oscillations, 36, 57; in sterile neutrino oscillations, 150; with non-standard interactions, 168
θ_{13}, 67, 77; ambiguity with δ, *see* parameter degeneracies; and *CP* violation, 35, 77; and matter effects, 64; and non-standard interactions, 168; bound on, 61–63; for mass-varying neutrinos, 160; global fit, 68–70; in GUT models, 107; in long-baseline neutrino oscillations, 41, 86–97; in MNS matrix, 33–34; in neutrino anarchy, 113; in reactor neutrino oscillations, 80, 82; in solar neutrino oscillations, 36; in supernova neutrino oscillations, 121, 123
θ_{23}; global fit, 69–70; in atmospheric neutrino oscillations, 35, 60; in long-baseline neutrino oscillations, 35, 41, 64–66; in MNS matrix, 33–34; in neutrino models, 103–105; octant ambiguity, *see* parameter degeneracies; with non-standard interactions, 168
3 + 1 model, 149, 151–152, 154, 157; *CPT* violating, 155
3 + 2 model, 153–154
3 + 3 model, 154
tribimaximal mixing, *see* neutrino mixing
triplet Higgs boson, *see* Higgs boson
2 + 2 model, 149–151, 153; *CPT* violating, 155

UNO, 26, 123

V − A interaction, 2, 14, 15, 20, 171

water Cherenkov detector, 5, 6, 25–27, 46, 86–87, 91–93, 95, 124, 135
Waxman-Bahcall flux, 132
wide-band beam, *see* neutrino beam
WMAP, 13, 73

XAX, 57
XMASS, 57, 74

Z-burst mechanism, 13, 134
Zee model, 108